AF519787

ESSENTIALS
of
VIROLOGY

OTHER RELATED TEXT BOOKS

A Text Book of Environmental Science	*Vidya Thakur*
Agricultural Microbiology	*R.P. Pareek*
An Introduction to Botanical Nomenclature	*A.R. Naqshi*
An Introduction to Environmental Biotechnology	*A. Bharani*
An Introduction to Fungi, 4th Edition	*H.C. Dube*
Applied Microbiology (Agriculture, Environmental, Food and Industrial Microbiology)	*S.M. Reddy*
Biodiversity Traditional Knowledge and Intellectual Property Rights	*S. Ram Reddy*
Colour Atlas of Medicinal Plants	*R.K. Bhutya*
Comprehensive Laboratory Manual of Life Science	*Jyoti Saxena*
Crop Improvement And Mutation Breeding	*A.K. Sharma*
Cytochalasin: Incidence and Biological Activities	*S. Kiran*
Ecology Workbook	*R. Misra*
Gene Editing Principles and Applications	*B.D. Singh*
Growth and Development in Plants	*K.V. Krishnamurthy*
Indira's Objective Agricultural Biotechnology 2nd. Ed	*R.L. Arya*
Instrumentation in Biosciences	*S.M. Reddy*
Introduction to Agricultural Botany	*M. Pandiyan*
Laboratory Manual of Microbiology, Biochemistry and Molecular Biology	*Jyoti Saxena*
Manual of Ethnobotany 2nd Revised Edition	*S.K. Jain*
Manual of Phycology	*G.M. Smith*
MCQs in Microbiology: Advanced	*Balaram Mohapatra*
Medical Microbiology and Immunology	*S.M. Reddy*
Microbial Diversity : Exploration & Bioprospecting	*S. Ram Reddy*
Microbial Physiology 2nd Ed	*S. Ram Reddy*
Mycology and Microbiology (A Text Book For UG & PG Student)	*C. Manoharachary*
Mycotoxins Problem And Its Management	*S.M. Reddy*
Objective Botany: MCQ's (For Civil Service Examinations, NET, SET, Ph.D. and Allied Examinations)	*D.A. Patil*
Objective Life Science 4ed	*Kailash Choudhary*
Plant Biotechnology and Molecular Biology: A Laboratory Manual	*M.S. Punia*
Plant Biotechnology: A Textbook	*S.M. Paul Khurana*
Plant Breeding In 21st Century	*B.D. Singh*
Plant Reproduction 2nd Ed	*T. Pullaiah*
Plant Tissue Culture: Theory and Practicals 2nd Edition	*T. Pullaiah*
Practicals in Plant Physiology and Biochemistry	*Manju Bala*
Research Methodology in Biotechnology	*P.S. Narayana*
Research Methodology in Plant Science 2nd Ed	*P.S. Narayana*

ESSENTIALS
of
VIROLOGY

[Second Edition]

S. Ram Reddy

Department of Microbiology,
Kakatiya University, Warangal (India)

S. M. Reddy

Department of Botany,
Kakatiya University, Warangal (India)

Published by:

Jodhpur
5 A, New Pali Road, P.O. Box 91
Jodhpur - 342 001 Rajasthan (India)
E-mail: info@scientificpub.com
Website: www.scientificpubonline.com

New Delhi
4806/24, Ansari Road, Daryaganj,
New Delhi - 110 002 (India)

Print 2026

ISBN: 9789394645547 (H/B)

First Edition 2007
Second Impression 2012

Printed in India

PREFACE TO THE SECOND EDITION

The first edition of this book was bought out in 2006 and it was well received by students and teachers as evident from the fact that few reprints were also made to meet the demand. However, during the last sixteen years, the subject of virology has undergone phenomenal developments, witnessed new discoveries and concepts that lead to look virology as one of the important disciplines in modern biology. Taxonomy of viruses is one of the important areas that is undergoing revolutionary changes for every ICTV report. The recent ICTV classification of viruses (Virus Taxonomy, Release 2020) is a radical deviation from the earlier classifications.

Discovery of giant viruses and virophages in the beginning of this century surprised the virologists. Understanding these two entities is forcing the virologists to investigate them in a new perspective. Further, it became necessary to alter the age old concepts and definitions of viruses. Another, welcome approach to discover the diversity of viruses in a range of habits is the robust metagenomic surveillance. This approach is yielding an amazing diversity and abundance of viruses in biosphere. Emerging virus and viral diseases are reported since several countries. However, the recent COVID-19 pandemic, that caused millions of death and unprecedented global social and economic regressions, has alerted the global community about the impending dangers of future emerging and reemerging viruses and viral diseases. Until the fourth quarter of last century, viruses were looked upon as pathogens of plants, animals and humans. Nevertheless, the latest research findings, several beneficial potentials of these organisms in many fields especially biotechnology are unfolding. All three major groups of viruses, phages, plant viruses and animal viruses are explored in this direction.

In light of these new developments, it became necessary to update the book incorporating latest developments in the subject and bring out a new and revised second edition. We are pleased to bring out this new edition adding new chapters like giant viruses, virophages, viral metagenomics and emerging viruses. The latest ICTV classification (Release 2020) and nomenclature rules are also provided in detail. A new chapter on applications of plant viruses is also added. Additional latest information is added to some old chapters.

With these additions, we hope the new edition meets the basic requirements of all sections interested in virology and provides a strong foundation. We wish to thank all students and teacher community for encouraging and patronizing the first edition. We wish to extend our appreciation to our families for their patience and encouragement during the preparation of the book.

Finally, but most important, we are grateful to **Sri Pawan Kumar Sharma** and **Sri Tanay Sharma M/s Scientific Publishers** (India), Jodhpur for bringing out this book timely and in an excellent format.

1st March, 2023 **Authors**

Warangal (T.S)

PREFACE TO FIRST EDITION

Virology, in less than hundred years of organized scientific investigations, has grown leaps and bounds and ramified into all the disciplines of biology. New information is pouring in at an accelerated pace. Modern disciplines like molecular biology, biotechnology have enriched themselves borrowing much of information from virology. It has wide applications both in basic and applied sciences. Fortunately, excellent advanced treatises are available on the subject providing the latest information from time-to-time. However, a book that can provide basic and holistic information of all viruses to a beginner is a long-felt need. Through this book we made a humble effort to build a strong conceptual foundation of virology so that interested students can navigate through the advanced treatises.

The subject matter of this book is presented in four parts. Part-I deals with history, principles, and taxonomy of viruses. Part-II deals with the biology of bacteriophages. Different aspects of plant viruses are presented in part-III. Part-IV is devoted to animal and human viruses. Mycoviruses, phycoviruses, subviral agents, viroids and prions are introduced in the last part. Though, a wealth of information is available on the subject, keeping in view the scope of the book, we have restricted subject matter to certain limits. The subject matter has been presented in a lucid style well supported by figures, illustrations, tables and graphs so that it could be used for both teaching and learning.

We hope that the contents of the book fulfill the needs of undergraduate and post-graduate students of all biological disciplines including agriculture, veterinary, pharmacy and medicine. We have taken every possible care to present the subject matter in factual form. Nevertheless, human errors are likely to creep in inadvertently. We look forward to our teacher fraternity to point out such errors and also to give constructive suggestions for the improvement of this book in future.

24th December, 2006 **Authors**
Warangal.

CONTENTS

PART ONE

PRINCIPLES
of
VIROLOGY

1

HISTORY OF VIROLOGY

The term `virus' has originated from a latin word which means poison. The term virus, now-a-days, is used in different contexts: in computers, social sciences and biology. In all these contexts it conveys the same meaning: an agent or entity that disrupts the normal functioning of a system. In a biological sense, the word has been in use for about hundred years. Originally the word was used to describe such agents which were thought to be responsible for causing certain diseases in plants, animals and humans whose causal agent was found to be mysterious, dangerous and at the same time did not fit into the paradigm of known living organisms. The term gradually came to signify a special type of entity entirely different from those of typical living microorganisms. The information gathered over the years substantiated the unique status of these entities and paved the way for creation of a new branch of biology referred to as **virology**. Today, virology is one of the most important disciplines of biological sciences.

Owing to ultramicroscopic nature, the development of our knowledge of viruses has depended on the techniques available. Initially viruses, like other microbes, were studied by pathologists interested in the causes of the infectious diseases of humans and their domesticated animals and plants. Even today, these causes remain as the main driving force for advancing the subject. Later, when bacterial viruses were discovered during the period 1915-1917, it was the hope that they control bacterial diseases that stimulated their investigation. The idea that viruses might be used to probe fundamental problems of biology generated in the early 1940s, with the developement of phage group. These studies heralded the new field of molecular biology during the period 1950-1970, and molecular biology has in turn revolutionized the study of viruses that led to an explosion of knowledge about them.

1. DISEASES AND MYTHS

Mythology, history and ancient religious scripts record several incidences involving what we now know to be viruses. Some diseases like smallpox, influenza and common cold are particularly mentioned in ancient literature. Though some are mild, diseases like smallpox created a terror in the minds of people. Unable to trace out the causal agent, unaware of preventive and control measures, people attributed such disease to be caused by supernatural powers or displeasure of a God or Goddess. Pujas and offerings were performed to propitiate these Gods and Goddesses. Goddess `Sitala' was considered to be responsible for smallpox and worshipped as such by millions of our countrymen. The idea of worshipping was to propitiate the Goddess so that her worshippers were not harmed. In remote villages of our land the practice still continues. According to one estimate, sixty million people died of smallpox alone during the eighteenth century. Influenza, the second major viral disease in the past, used to claim lives in millions. It still occurs in epidemic form all over the world. Yellow fever also used to cause deaths in thousands in tropical countries of Central America and Africa. Ancient sculptures and paintings depicting the polio inflicted people are also available indicating the fact of existence of polio in ancient days (Fig. 1.1A).

Fig.1.1.A. An Egyption Stele representing polio victim

Fig. 1.1.B. Broken tulips (courtesy Fitzwilliam Museum, University of Cambridge)

An interesting phenomenon known as `tulipomania' was prevalent in France and Holland during mediveal periods. The plant concerned was a variety of tulip (virus diseased tulip) that produced beautiful flowers with different shades or striations than its healthy counterpart (Fig. 1.1B). People in those days were particularly crazy about `broken tulips'. Infact the craze was such that the cultivation and possession of these were regarded as status symbol. There are stories relating as to how brides possessing bulbs of such tulips were considered very much desirable by prospective bridegrooms.

Despite of serious and periodic appearances of these scourges, nothing was known about the actual nature of these mysterious agents until the late nineteenth century. Once the infectious diseases caused by bacteria, fungi, protozoa were discovered, people became more serious about knowing the causes for these diseases (caused by viruses). During this period, only real achievement made was development of vaccination for smallpox by Edward Jenner (1798). Two important dogmas of microbiology, the germ theory diseases of Louis Pasteur and Koch postulates prompted the scientists to investigate the so called viral diseases with all seriousness. Thus the scientific investigations began in the last quarter of the nineteenth century.

2. SCIENTIFIC INVESTIGATIONS

2.1 Plant viruses

Scientific investigations on viruses began with tobacco mosaic virus (TMV). Indeed, this virus served as a prototype model not only for plant viruses but for all viruses. Tobacco mosaic disease was causing a serious damage to tobacco crop in Holland and Germany. It was investigated from different perspectives in these two countries. Adolf Mayer (1843-1942), a German botanist, reported in 1886 that this disease could be transmitted by just applying the sap or juice of infected leaves onto the leaves of healthy individuals. Preoccupied with the notion that the disease is caused by bacteria, he was searching for bacteria in the infected plant sap. However, he could not see any bacteria in the clarified saps. Nevertheless, he recorded two important findings. In the first place, he noted that the intensity of infection remained unchanged even after partial purification of the sap. Secondly, he found that the capacity to infect was considerably destroyed when the clarified sap was heated and that above 55°C the loss of infectivity was almost total. A few years after a most remarkable discovery was made by a Russian botanist Dimitry Iwanowski (1864-1920). While investigating the tobacco mosaic disease he confirmed some of the observations made by Adolf Mayer. However, he strongly rejected Mayer's observation that the capacity to infect was lost when the sap was passed through a pad of filter papers. He demonstrated that infectivity was retained not only on filtration of the sap but even when the sap was passed through more sophisticated filters like chamberland-Roux filters. Based on his observations he called these agents as **filterable agents** which indirectly meant that these were smaller than known bacteria. Iwanowski, could not think about a new type of organism altogether. He simply hinted at the possible existence of bacteria smaller than they were known at that time. At about the same time, M.W. Beijerinck (1851-1931), a Dutch soil microbiologist who collaborated with Adolf Mayer also conducted some remarkable experiments on tobacco mosaic disease. Beijerinck also showed that the sap of infected tobacco plants could retain its infectivity after filtration through a Chamberland candle filter. He then extended these studies and have shown that the filtered sap could be diluted and then regain `its strength' after replication in growing tissue of the plant. The agent could reproduce itself but only in living tissue, in the cell free sap of the plant. This explained the failure to culture the pathogen outside its host, and it set the stage for discovery of an organism smaller than bacteria. Beijerinck called this agent a **contagium vivum fluidum**, or a contagious living liquid. This concept began a 25-year debate about the nature of viruses : were they liquids or particles. Finally, this concept was laid to rest when d'Herelle developed the plaque assay in 1937 and when the first electron micrographs of TMV were taken in 1939.

Adolf Mayer

Dimitry Iwanowski

M.W. Beijerinck

Handicapped by a suitable microscope which can provide sufficient resolution for the observation of filterable agents, scientists started probing the filterable agents from a different angle i.e. chemical constitution. The notion that viruses were proteins and so could be purified in the same way was first appreciated and applied by Vinson and Petre (1927-1931). They precipitated the infectious TMV agent from the crude sap of the infected plants using selected salts, acetone, or ethyl alcohol. At the same time, H.A. Purdy-Beale produced in rabbits antibodies directed against TMV that could neutralize the infectivity of this agent. This was taken as a further proof of the protein nature of viruses. In 1933, Max Schlesinger working on purified preparations of bacteriophage WLL showed that they were composed of proteins and contained phosphorus and deoxyribonucleic acid. This led to the suggestion that viruses were composed of nucleoproteins. Lastly, an American chemist W.M. Stanley (1935) was able to obtain TMV in purest crystalline form. It was a most interesting and remarkable achievement. Stanley was eventually awarded the Nobel Prize for his contribution. Once virus could be obtained in a pure crystalline form, further investigation became easy. Two British biochemists N.W. Pirie and F.C. Bawden (1936-37) analyzed these particles chemically and demonstrated that these were made up of proteins and nucleic acids (RNA). The contribution of these two scientists went a long way in enlightening the chemical nature of viruses. Dr. Pirie was later awarded the Nobel Prize for his achievements. The first electron micrographs of TMV were taken with a microscope built in Germany and they confirmed the rod shape of the virus particle. Based upon the x-ray diffraction patterns, Watson and Crick (1956) suggested that most simple viruses like TMV had to consist of one or few species of identical protein subunits. Frankel-Conrat (1956) demonstrated the reconstitution of TMV from disassociated subunits. Finally in between 1955-1957, the ultrastru-cture of TMV was elucidated by Frankin and Klug. Caspar and Klug (1962) elegantly described the geometrical principles of icosahedral virus structure for many of the isometric viruses. Thus, from 1929 to 1962, both structures and chemical compositions of many plant viruses were elucidated.

2.2 Bacteriophages

Scientific investigations on bacteriophages started rather late. Frederick W. Twort (1915) while attempting to grow vaccinia virus on nutrient agar plates observed the bacterial contaminated plates became `watery looking' (i.e., more transparent). Such bacterial colonies were no longer able to replicate when subcultured. Twort called this phenomenon glassy transform-ation, and he went on to show that infecting a normal colony of bacteria with the glassy transforming principle would kill the bacteria. At the same time, Felix d'Herelle, a Canadian medical bacteriologist, working at the Pasteur institute, Paris while cultivating the shigella dysentery bacillus observed clear circular spots where no bacteria grew. He called these *taches vierges* or **plaques**. D'Herelle made some interesting observations. He observed that when the bacteria were most plentiful in the feces of the patients the plaques appeared in the cultures. Further, he was able to demonstrate that the plaques appeared on fourth day of infection and killed the bacteria in the culture dish and interestingly, the patient began to improve on the fourth day after infection. D'Herelle named these viruses as **bacteriophages** (Greek, phage means to eat). In deviation from earlier notion, he reasoned that the appearance of plaques showed that the virus was particulate or `corpuscular'. D'Herelle also described cell lysis and release of infectious virus in clear and modern terms. Thus D'Herelle in many ways, was considered as one of the founders of modern virology. Following next two decades, D'Herelle focussed his efforts on the potential medical applications of phages. He hoped that bacterial diseases could be cured with phages. However, his hopes were proved to be false. In the next few years following the discovery, people became interested in understanding the replication processes and morpho-logy of phages. Emory Ellis and Max Delbruck (1939) designed the one-step growth curve experiments. This experiment defined the latent period when viral infectivity was lost. This became the experimental paradigm of the phage group. Shortly thereafter in 1942, Tom Anderson in association with Delbruck obtained the first electron photomicrographs of bacteriophages. The next 25 years (1950-1975) was a very fruitful period of virus research with bacteriophages. Hundreds of virologists produced thousands of publications that covered three major areas : (i) lytic infection of *E. coli* with the T-even phages (T_2, T_4); (ii) the nature of lysogeny, using lambda phage and (iii) the replication and properties of several unique phages such as ϕ x 174, the RNA phages, T_7 and Mu. Contributions from these investigations laid the foundations for modern molecular virology and biology (Table 1.1).

Table 1.1. Landmark discoveries in molecular biology through bacteriophages

Year	Discovery	Scientist(s)
1946	Genetic recombination	Delbruck
1947	Mutation and DNA repair	Luria
1952	Transduction	Zinder and Lederberg
1952	DNA-the genetic material	Hershey and Chase
1953	Temperate phages	Lwoff, Wollman
1955	Definition of a gene	Benzer
1958	Control of gene expression	Pardee, Jacob and Monod
1961	Discovery of mRNA	Brenner, Jacob and Monod
1961	Stop signals for translation	Campbell, Epstein and Bernstein
1964	Colinearily of the gene with the polypeptide chain	Sarabhai, Stretton and Brenner
1966	Pathways for macromolecular assembly	Edgar and Wood
1974	Vectors for rDNA technology	Murray and Murray, Thomas, Cameron and Davis
1976-1977	Genetic mapping of bacteriophage ϕ X 174	Frederick Sanger

Frederick Twort

Felix D'Herelle

2.3 Animal and human viruses

The first animal disease caused by virus to be investigated was foot and mouth disease of cattle. Dr. L. Loeffler and P. Frosch (1898) had suggested that filterable agents responsible for this disease were particles of new type. The first human virus to be identified, in 1901, was that responsible for yellow fever. Dr. Walter Reed and his associates especially Carlos Finley were able to successfully demonstrate that this disease was caused by filterable agents. They were also the first to establish that the filterable agents were carried from infected to healthy persons through the agents of insects. Rabies was another important human and animal disease whose causal agent was attributed to be filterable. In 1903, Dr. A. Negri demonstrated that the nerve cells of rabies infected dogs contained prominent crystalline bodies, now called `Negri bodies'. In 1909, Dr. Karl Landsteiner, the famous discoverer of human blood groups, and his colleagues demonstrated the filterable nature of agents responsible for poliomyelitis. In 1911 another important discovery was made. American virologist Dr. Peyton Rous had discovered that sarcoma (a type of cancer) of chicken was caused by filterable agents. Earlier, Ellerman and O. Bang (1908) also reported that the leukemia was caused by filterable agents. Since, then scientists began to probe the possibility of cancer of human beings to be caused by filterable agents. The contributions of Rous were recognised very late and he was awarded Nobel Prize in 1966.

Table 1.2. Landmark discoveries with animal viruses

Year	Discovery	Scientist(s)z
1949	Cultivation of poliovirus	Enders
1957	Interferon	Isaac and Lindermann
1957	Polio vaccination	Salk and Sabin
1960-65	Chemotherapeutic agents	Tam & Eggers, Wagnor, Kaufmann
1964-75	Reverse transcriptase	Temin & Baltimore
1971	Oncoviruses in primates	Schidolovski and Ahmed, Gallow
1973-80	Human oncoviruses	Sabin and Tam, Dress
1979	Genetic mapping of HBV	Galibert
1979	Cloning of HBV genes in *E. coli*	Saninsky
1983	HPV causes cerevical cancer	Harald Hausen
1984	AIDS as viral disease	Montagnier
1987	Development of antiretroviral drug	Border S *et al.*
1989	Oncoviruses	Bhishop and Varmus
2002	Development of papillomavirus vaccine (cerevical cancer)	Frazer I *et al.*
2003	Discovery of SARS Coronavirus, the first pandemic of 21st century	Urbani C *et al.*
2005	Sequencing of II influenza virus (1918) and the virus reconstructed	Taubenberger J *et al.*
2008	Discovery of human papillomavirus causing cervical cancer	Harald zur Hausen
2020	Discovery of Hepatitis C Virus	James Alter, Michael Haughton and Charles Rice

Inspite of these early breakthroughs with animal viruses, the pace of discovery was slow, because of the dangers and difficulties associated with experimental manipulation of human viruses. Consequently, agents of some important human diseases were not identified for many years. The virus that causes human influenza was first identified in 1933, by Wilson Smith, Chistopher Andrews and Patrick Laidlaw, because they found a host suitable for its propagation. At that time, people realised that unless the disease causing viruses could be isolated and grown in the pure form in laboratory conditions correct identification and precautionary measures like vaccine could not be developed. Efforts were, therefore, made in this direction. Dr. John Enders (1949) achieved a breakthrough by the cultivation of poliovirus in laboratory. He utilized the newly developed tissue culture techniques for this purpose.

In the later period i.e. after 1960s the investigations took a different direction. Virologists began to use viruses to probe questions central to understanding all life processes. Because viruses replicate in and are dependent on their host cells, they must use the rules, signals and regulatory pathways of the host. Virologists began to make contributions in all facets of biology. Secondly, during this period (1970 to the present), the recombinant DNA revolution began, and both bacteriophages and animal viruses played a critical role in this revolution. Viruses were used to probe many diverse and difficult questions in biology (Table 1.2).

Further reading

Lustig, A. and Levine, AJ. (1992). One hundred years of virology. *J. Virol.* 66 : 4629.

Van Helvort, T. (1993). Research styles in virus studies in the twentieth century : controversies and the formation of consensus. Doctoral dissertation.

Hughes, SS. (1977). The Virus : A history of the concept. London : Heinemann

Fenner, F. and Gibbs, AJ. (eds) (1988). Portraits of viruses. A history of virology. Basel : Karger.

Knipe, DM. and Howley, PM. (eds) (2001). Fields Virology 4th edn. Philadelphia : Lippincott-Raven.

Biswas, SB and Biswas, A. (1996). An Introduction to viruses 4th edn. Vikas Publishing House Pvt Ltd.

Waterson, AP. and Wilkinson (1978). An Introduction to the History of Virology. Cambridge University Press, London, UK.

Murphy FA (2012). The Foundations of Virology : Discoverers and discoveries, inventors and inventions, developer and technologies. Infinity Publishing, USA

2

ORIGIN AND EVOLUTION OF VIRUSES

Viruses are a unique group of intracellular infectious agents. While exhibiting the characters of both living organisms and non living entities, they are distinctive in certain features. Viruses are simple, acellular and sub-microscopic agents. A majority of the viruses are made up of only two types of chemicals, proteins and nucleic acids (nucleoproteins). However, certain complex viruses contain lipids also. Viruses lack lipman system i.e. they donot have energy (ATP) generating mechanism. Thus, they are totally dependant on host for their energy requirements. They donot respond to external stimuli and can be crystallized like any other chemical. Viruses don't have the cell structure, and donot fit into the dogma of cell theory. However, they replicate and produce progeny maintaining the genetic continuity. Their biological status is still an enigma.

1. UNIQUE FEATURES OF VIRUSES

Viruses are set apart from microorganisms on the basis of certain characteristics peculiar to them.

i. All are potentially infectious.

ii. Presence of a single nucleic acid (RNA or DNA) forming the genome.

iii. Reproduction of genetic material with the help of host system only.

iv. Lack of lipman system i.e. absence of enzymes for energy generation.

v. They are synthesized *de novo* i.e. the components of a virus are synthesized independendly and then assembled into mature virus particles.

vi. Absence of ribosomes

vii. Absence of information for the synthesis of ribosomal proteins, ribosomal RNA and soluble RNA.

viii. They can be crystallized and can be precipitated or sedimented like proteins.

The viruses differ from microorganisms like bacteria, rickettsiae, mycoplasma in the following properties.

i. Viruses are minute (submicroscopic)

ii. They donot have cellular organization.

iii. They contain only one kind of nucleic acid, either RNA or DNA.

iv. They donot reproduce outside the living cell. Thus viruses are not independant organisms, but use living cells of host for their multiplication.

v. They donot possess ribosomes

vi. They donot divide by binary fission.

vii. They don't grow in inanimate (artificial) media like agar agar

viii. They are usually resistant to antibiotics.

ix. They are sensitive to interferons.

Some viruses have properties like those of plasmids. However, viruses differ from plasmids in the following ways (Hull, 2002).

i. Normal viruses have a particle with a structure designed to protect the genetic material in the extracellular environment and to facilitate entry into a new host cell

ii. Virus genomes are highly organized for specific virus functions of no known value to the host cell, whereas plasmids consist of genetic material often useful for survival of the cell.

iii. Viruses can cause death of cells of disease in the host organism but plasmids do not.

2. DEFINITION

Defining a virus is a difficult aspect. A definition is based on certain parameters of characteristics known at that time. However, priorities of selected parameters change from time to time as more knowledge accumulates. The first definition of new infectious agents, the viruses was presented entirely in negative terms. They could not be seen, could not be cultivated, cannot be retained by bacterial filters etc. Recently proposed definition take into account the physical, biochemical and biological characteristics of the viruses. Some important definitions proposed by eminent virologists are given below.

Lwoff (1957) - Strictly intracellular and potentially pathogenic entities with an infectious phase and

i. possessing only one type of nucleic acid.

ii. multiplying in the form of their genetic material

iii. Unable to grow and undergo binary fission.

iv. Devoid of a Lipman system.

Luria (1959) - Elements of genetic material that can determine, in the cells where they reproduce, the biosynthesis of a specific apparatus for their own transfer into other cells.

Luria and Darnell (1967) - Viruses are entities whose genomes are elements of nucleic acids that can replicate inside living cells utilizing the cellular synthetic machinery and causing the synthesis of specialized elements that can transfer the viral genome to other cells.

Hull (2002) - A virus is a set of one or more nucleic acid template molecules, normally encased in a protective coat or coats of protein or lipoprotein, that is able to organize its own replication only within suitable host cells. It can usually be horizontally transmitted between hosts. Within such cells, virus replication is (i) dependant on the hosts synthesizing machinery, (ii) organized from pools of the required material rather than by binary fission, (iii) located at sites that are not separated from the host cell contents by a lipoprotein bilayer membrane, and (iv) continually giving rise to variants through various kinds of change in the viral nucleic acid.

Flint *et al.* (2004) - i. A virus is a very small, infectious, obligate intracellular (molecular) parasite.

ii. The virus genome comprises either DNA or RNA.

iii. Within an appropriate host cell, the viral genome is replicated and directs the synthesis, by cellular systems, of other virion components.

iv. Progeny virions are formed by *de novo* assembly from newly synthesized components within the host cell.

v. A progeny virion assembled during the infectious cycle for transmission of the viral genome to the next host cell or organism, where its disassembly leads to the beginning of the next infectious cycle.

3. ORIGIN OF VIRUSES

The origin and subsequent evolution of viruses are shrouded in mystery, because of the lack of a fossil record. However, recent advances in the understanding of virus structure and reproduction have facilitated to arrive at certain speculations on virus origins. At present, there are three major hypotheses on the origin of viruses.

i. Retrogressive hypothesis : According to this hypothesis atleast some of the more complex enveloped viruses, such as the poxviruses and herpes viruses, arose from small cells, probably prokaryotic, that parasitized larger, more complex cells. These parasitic cells would become ever simpler and more dependent on their hosts, much like multicellular parasites have done. Several objection have been raised against this hypothesis. Firstly, it is argued that viruses are radically different from prokaryotes, and it is difficult to explain the mechanisms by which such a transformation might have occured. Secondly, absence of intermediate forms between prokaryotes and atleast the more complex enveloped viruses.

ii. Cellular origin hypothesis : According to this hypothesis, viruses represent cellular nucleic acids that have achieved independence and managed to escape to cell. Possibly few mutations could convert nucleic acids, which are only synthesized at specific times, into infectious nucleic acids whose replication could not be controlled. This proposal is supported by the observation that the nucleic acids of retroviruses and a number of other virions do contain sequences quite similar to those of normal cells, plasmids and transposons. The small, infectious RNAs called virioids have base sequences complementary to transposons, the regions around the boundary of mRNA introns, and portions of host DNA. This has led to the speculation that they have arisen from introns or transposons.

It is also suggested that viruses have arisen by way of both mechanisms. Because viruses differ so greately from one another, it seems likely that they have originated independently many times during the course of evolution.

iii. Independent entities hypothesis : This hypothesis postulates that viruses evolved, on a course parallel to that of cellular organisms, from the self-replicating molecules believed to have existed in the primitive prebiotic RNA world.

4. EVOLUTION OF VIRUSES

Like all other living organisms, viruses undergo evolutionary change. Their genomes are subject to mutations at the same rate as all other nucleic acids. From evolutionary point of view, viruses have an advantage in that many polymerases (notably in the RNA viruses) lack proof-reading mechani-sms, as a result mistakes are not corrected. Thus mutants accumulate more rapidly. Viruses can exchange genetic information by recombination or reassortment of genome segments. Recombination creates new combinations of mutations that are essential for survival under selective pressures. Recombination occurs when the polymerase changes template (copy choice) during replication or when nucleic acid segments are broken and rejoined. Viruses that incorporate their genomes into host genomes (transduction) play an essential role virus evolution. Information can be exchanged in a variety of unexpected ways during viral infections. For example, a single host can be coinfected by many different viruses during its life time. These coinfections have profound influence on a virus evolution.

Table 2.1. Timeline of newly emerging viruses.

Virus	Year of emergence	Natural reservoir
West Nile virus	1937	Birds
Hantaan virus	1951	Mouse
Ebola virus	1976	Bats
HIV	1983	Bats
Avian Flu H5N1	1988	Birds
Sin Nombre Virus	1993	Mouse
Nipah Virus	1999	Bats
SARS – Coronavirus	2002	Bats
H1N1	2009	Pigs
SFTS	2009	Ticks
MERS – Coronavirus	2012	Bats
Avian Flu H7N9	2013	Avians
SARS – CoV_2	2019	Bats

Molecular biology of viruses of eukaryotes and their host cells differs considerably from those of bacteriophages and their prokaryotic host cells. It is not possible to grow bacteriophages in eukaryotic cells or vice versa. Thus, it is probable that phages and eukaryotic viruses have evolved independently and diverged at a very early stage. Virus evolution is also closely linked with their host. As viruses are absolutely dependent on their hosts, viral evolution tends to take one of two general pathways. In one pathway, viruses coevolve with their hosts so that both share a common fate. For instance, as the host prospers, so does the virus. However, in the absence of no other host, the viruse faces a problem because it can be eliminated with antiviral measures or by the loss of host. In the second pathway, viruses occupy broader niches and have multiple host species. Then, when one host species is compromised, the virus can replicate in another. In general, DNA viruses tend to follow the first pathway while many RNA viruses follow the second. Both pathways provide excellent strategies for virus survival.

One important fall out of virus evolution is the emergence of new viruses and viral diseases. An emerging virus is one that either has newly appeared in the population or is rapidly expanding its range of hosts with a corresponding increase in detectable disease. Some examples of emerging viruses are given in Table 2.1. Each example represents a story of contemporary virus evolution. Recent emergence of severe acute respiratory syndrome (SARS), bird flue can be added to the above list.

Further reading

Domingo E, Webster RG and Holland JJ (1999). Origin and evolution of viruses. New York Academic Press.

Koonin EV (ed.) (1992). Evolution of viral genomes. Seminars in virology 3(5): 311-417.

Reanny D (1984). The molecular evolution of viruses. In: The Microbe 1984. I Viruses, Vol. 36, The society for general microbiology symposium, pp. 175-196.

Morse, SS (ed.) (1993). Emerging viruses. Oxford University Press, Oxford, UK.

Becker, Y (1996). A short introduction to the origin and molecular evolution of viruses. Virus Genes 11 : 73-77.

Flint, SJ *et al.*, (2004). Principles of Virology, ASM Press, Washington, DC.

3

MORPHOLOGY AND CHEMICAL COMPOSITION

The mystery behind the contagium vivum fluidum or `filterable agents' could not be solved until late 1930's. Dearth of suitable instruments and techniques was the main drawback in the way of thorough investigation into the physical nature of the viruses. Three major major achievements within a short period of about ten years have cleared the clouds hovering the true physical nature of the viruses. These are : (i) the development of ultracen-trifuge by Svedberg (1925) and its use in isolating the viruses, (ii) the develo-pment of electron microscope by Knoll and Ruska (1931), and (iii) successful purification and crystallization of TMV by Stanley (1935). Observation of purified virus particles with the sophisticated techniques like electron micro-scopy and x-ray crystallography revealed a tremendous range in the morphology and architecture. Negative staining of virus particle with an electron-dense materials such as uranyl acetate, phosphotungstate, a technique which has been applied to viruses since middle decades of the twentieth century, provided the first rational basis for the classification of these agents based on morphology. Cryoelectron microscopy, that preserves the native structure and allows the direct visualization has increased the contrast between the virus and stain. An improved resolution upto 10°A was obtained when the cryoelectron microscopy was coupled with computerized mathematical analysis and three dimensional reconstruction. This approach has yielded a wealth of information about the external and internal structures of many viruses.

1. MORPHOLOGY

Morphology refers to size, shape, symmetry, architecture, physical components and their arrangement with respect to each other. Size, shapes and important features of representatives of plant, animal, bacterial, fungal viruses are presented in Table 3.1.

Table 3.1. Shapes and sizes of some important viruses

Virus	Shape	Size (nm)
Animal viruses		
Adenoviruses	Spheroidal	70-90
Herpesviruses	Spheroidal	100-150
Poxviruses	Brick shaped	230x300
Retrovirus	Spherical, enveloped	80-100
Picorna viruses	Spheroidal	20-30
Rhabdoviruses	Bullet shaped	75 x 130 – 230
Toga viruses	Spheroidal	40 - 60
Plant viruses		
Potyviruses	Flexuous filaments	13 x 700-900
Beet yellow	Flexuous rods	10 x 1250
Brome mosaic	Spheroidal	25
Cauliflower mosaic	Spheroidal	50
Cucumber mosaic	Spheroidal	30
Lettuce necrotic yellow	Bullet shaped	66 x 227
Potato virus X	Rods	13 x 480 – 580
Southern bean mosaic	Spheroidal	30
Tobacco mosaic	Rods	18 x 300
Tomato spotted wilt	Spheroidal	70-80
Geminivirus	Geminate	18-30
Bacteriophages		
Coliphage fd	Filamentous	6 x 100
Coliphage Lambda	Spheroidal head and rod like tail	54 x 150 x 10
Coliphage T_2, T_4, T_6	Spheroidal head and rod like tail	65 x 95; 25 x 110
Coliphage T_1, T_3, T_5	Spheroidal head and rod like tail	50; 65 x 170
Coliphage ϕ x 174	Spheroidal	27
Cyanophage SM	Spheroidal	88
Fungal viruses		
P. chrysogenum virus	Spheroidal	35
Pyricularia oryza virus	Spheroidal	25-32

Viral particles are complex assemblies of macromolecules exquisitely designed for protection and delivery of viral genomes. The terminology commonly used in the description of morphology of a virus is presented in Table 3.2.

Table 3.2. Terminology used in the description of virus morphology.

Term	Definition
Virion	Complete infectious virus particle
Envelope	Outer lipid bilayer, many a times carrying glycoproteins.
Nucleocapsid	Assembly of protein-coat and nucleic acid
Capsid or protein coat	Protein shell surrounding the nucleic acid genome
Capsomere	Morphological unit of capsid
Structural unit or protomer	Unit-from which capsomers are built
Spikes	External projections of viral capsides and envelopes

i. Envelope

A majority of animal viruses and a few plant and bacterial viruses contain an outer envelope. Such viruses are referred to as **enveloped** and without the envelope **nonenveloped** (Fig 3.1). Envelope is formed by a viral glycoprotein containing membrane derived from the host cell, but they vary considerably in size, morphology, complexity and the nature of the membrane. The foundation of all viral envelopes is a lipid bilayer acquired from the host cell during assembly. This membrane consists of a bilayer of phospholipids. Embedded in the membrane are viral proteins, the great majority of which are glycoproteins that carry covalently linked sugar chains or oligosaccharides. The envelopes of some complex viruses, including orthomyxoviruses and poxviruses contain integral membrane proteins that lack large external domains.

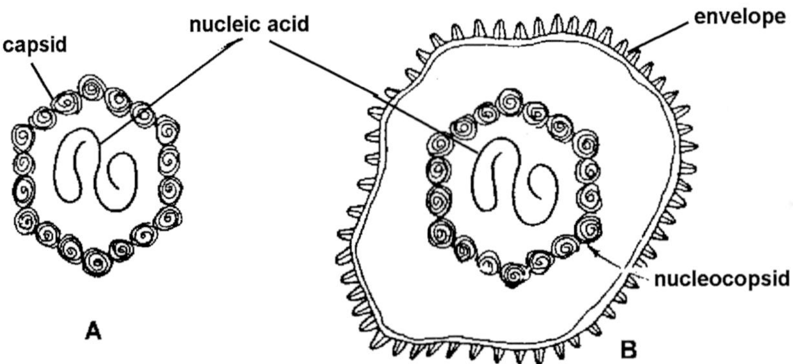

Fig. 3.1. Morphology of viruses. **A.** Nonenveloped, **B.** Enveloped.

ii. Capsid and nucleocapsid

A proteinaceous coat enclosing and protecting the genome is known as capsid. It is usually single layered but in some viruses (e.g., reoviruses) it is double layered. Generally, capsid and nucleocapsid words are used with different meanings. In nonenveloped viruses, protein coat encasing and protecting the genome is known as **capsid**, whereas in enveloped viruses, the inner capsid in association with nucleic acid is referred to as **nucleocapsid** (e.g., influenza virus, HIV). The morphological unit of the capsid is known as **capsomere**. The structural units of the capsomers are known as **protomers**. A capsomere may be formed from a single protein or many similar or dissimilar proteins (e.g., poliovirus). These proteins are referred to as subunits or protein units. The capsid may be made up of only type of capsomers (e.g., TMV) or it may be made up of more than one type of capsomers (e.g., Adenovirus). In later case, the capsomers are antigenically different. The shape and symmetry of virus is determined by the arrangement of the capsomers. Symmetry of viruses is described in next section of this chapter.

iii. Nucleic acid

Nucleic acid is the most important component of a virus. It is specific to a given virus and determines host specificity, pathogenicity and above all hereditary characters. It should be noted here that genomic nucleic acid in any given virus is either DNA or RNA. No virus, so far, is reported to contain the both. This trait distinguishes viruses from the known living organisms all of which possess both the types of nucleic acids. Viruses exhibit a bewildering diversity with respect to nucleic acids (Table 3.3). All known types of DNA/RNA are described from viruses. Infact, much of the molecular biology knowledge is gleaned from the study of viruses.

Viruses possessing DNA are referred to as **deoxyviruses**, while those possessing RNA as **riboviruses**. Apart from double stranded DNA (dsDNA) and single stranded RNA (ssRNA), many viruses are known to contain, exceptionally, single stranded DNA and double stranded RNA. These, inturn, may be linear or circular. Basically, thus, eight types of nucleic acids are found in viruses. In ssRNA containing viruses, the genome may be **positive strand** or **negative strand**. The positive strand has the same sense as that of mRNA. On the other hand, negative strand is complementary to mRNA.

In many cases, the genomic nucleic acid is found in a single contiguous molecule. However, in some viruses the genome is segmented, which may be found with in a single capsid (eg. orthomyxoviruses, reoviruses) or they may be found enclosed by different capsids (e.g. Alfamovirus, Furovirus and Ilarvirus). In any case, all these make a complete genome. Genomic nucleic acid size also varies considerably in different viruses. The number of bases or base pairs may range from hundred to several thousands. Similarly, the number of genes that make the genome may be as few as one as in the case of satellite tobacco necrosis virus to several hundreds (Table 3.4).

Table 3.3. Diversity of nucleic acids of different viruses

Strandedness	Nucleic acid configuration	Examples
Single-Stranded DNA		
	Linear	Parvoviruses
	Circular	ϕ X174, M13, fd phages Geminiviruses
Double-Stranded DNA		
	Linear	Herpesviruses, adenoviruses, Vaccina virus, T coli-phages, lambda phage, and other bacteriophages
	Closed circular	Papovaviruses (polyoma, SV-40), PM2 phage, cauliflower mosaic
Single-Stranded RNA		
	Linear, positive sense	Picornavirus (polio), togaviruses, RNA bacteriophages, TMV and most plant viruses
	Linear, negative sense	Rhabdoviruses (rabies), paramyxo- viruses (mumps, measles)
	Linear, segmented, positive sense	Brome mosaic virus (individual segments in separate virions)
	Linear, segmented, dip-loid (two identical single strands), positive strand	Retroviruses (Rous sarcoma virus, HIV)
	Linear, segmented, negative sense	Bunyaviruses (TSWV), Orthomyxoviruses (influenza)
Double-Stranded RNA		
	Linear, double stranded, segmented	Reoviruses, wound-tumor virus of plants, cytoplasmic polyhedrosis virus of insects, phage ϕ 6, many mycoviruses

Table 3.4. Genome size and number of genes of some important viruses

Virus	Genome size (Kbp or Kb)	Number of genes (approx.)
Bacteriophages		
T_4 phage	200.0	650
λ phage	48.0	160
φ x 174 phage	5.836	11
Mu phage	37.2	—
MS_2, Qβ	3.6	3
Plant viruses		
TMV	7.4	20
CaMV	14.0	46
Potex viruses	6.0	20
Geminivirus	2.7-3.0	6
Potyviruses	10.0	35
Turnip yellow mosaic virus	6.0	10.0
Wound tumor virus	24.0	80.0
Animal viruses		
Adenovirus	35.0	120
Poliovirus	6.0	20
Rous sarcoma virus	30.0	100
SV40	6.0	20
Herpes simplex virus	90.0	300
Influenza virus	6.0	20

iv. Other components

Some virus particles contain only the nucleic acid genome, capsid proteins or protein plus lipid membrane necessary for protection and delivery into a host cell. However, many viruses contain additional viral proteins or other components, which are generally present at much lower concentration but play essential roles in establishing an efficient infection.

Many viruses synthesize and store enzymes necessary for synthesis of nucleic acids. Such enzymes generally catalyze important synthetic reactions such as synthesis of viral mRNA from an RNA template or of viral DNA from an RNA template. More complex virions may also contain additonal viral proteins, that donot function as enzymes but are nevertheless important for an efficient infectious cycle. The best characterized protein of this class is VP16 protein of herpes simplex virus type I. This protein activates transcription of viral immediate early genes to initiate the viral program of gene expression. Some important viral enzymes and proteins and their functions are presented in Table 3.5.

Table 3.5. Some virion enzymes and proteins necessary for viral replication

Virus	Enzyme/Protein	Function(s)
Viral enzymes		
Rhabdovirus	L	RNA-dependant RNA polymerase
Influenza A virus	P proteins	RNA-dependent RNA polymerase Initiation of mRNA synthesis
Reovirus type I	λ_2	Guanyl transferase
	λ_3	dsRNA-dependent RNA polymerase
HIV	RT	Reverse transcriptase; Proviral DNA synthesis
	IN	Integrase
	PR	Protease
Herpes simplex virus type I	VP24	Protease
	VP18.8	Protein kinase
Vaccinia virus	RPO proteins	DNA-dependent RNA polymerase
	Poly(A) polymerase	Synthesis of poly (A)
	DNA topoisomerase	Sequence-specific nicking of viral DNA
Proteins		
HIV type I	Vpr, Vif	Required for efficient infection
HSV type I	VP16	Structural protein, activation of immediate early gene transcription
Cytomegalo virus	pp65	Inhibition of presentation of immediate early proteins as antigens
Vaccinia virus	VETF	Binds to early promoters
	RAP94	Transcriptional factor

Some of the viruses consist of special components in their structure, these are polyamines (T_2, T_4 phages). The polyamines are putrescine, spermidine and spermine. Polyamines are also reported in Herpes virus and Influenze virus. Traces of polyamines are identified in Turnip yellow mosaic virus and Broad bean mottle virus. Inorganic divalent metal ions were also found in certain viruses. E.g. Ca^{+2}, Mg^{+2} (TMV, Southern bean mosaic virus). Some plant viruses such as Tobacco streak virus (TSV) have a zinc finger binding domain that specifically binds an atom of zinc in a protein involved in nucleic acid binding.

2. SYMMETRY

The observation that certain viruses, particularly TMV, assemble *in vitro* from dissociated components led Crick and Watson (1956) to propose that virus shells would be highly symmetric objects. Further, they predicted that the shell of the viral particles must be composed of a small number of different kinds of molecules, perhaps of only one kind. Also, they reasoned that the shells assembly is based on the principle of minimization of energy, so that the subunits would fall into place spontaneously to form a stable structure. With these restrictions, there are only few possible arrangements in which subunits can be arranged to form a stable shell. A combination of data obtained from various physical and chemical investigations on a large number of viruses have proved the predictions of Crick and Watson correct. Today, we are aware that two basic types of symmetry are utilized in the construction of virus particles from protein subunits: i. **Cubic** and ii. **Helical**. A few complex viruses like bacteriophages of the T series exhibit a form which is a combination of two aforesaid symmetries. These viruses are said to be exhibit a **binal (or complex) symmetry** (Table 3.6).

i. Cubic symmetry

The regular icosahedron is the geometrical structure underlying the arrangement of capsomers in viruses with cubic symmetry (Fig. 3.2). This was strikingly shown by double shadowing of Tipula iridiscent virus.

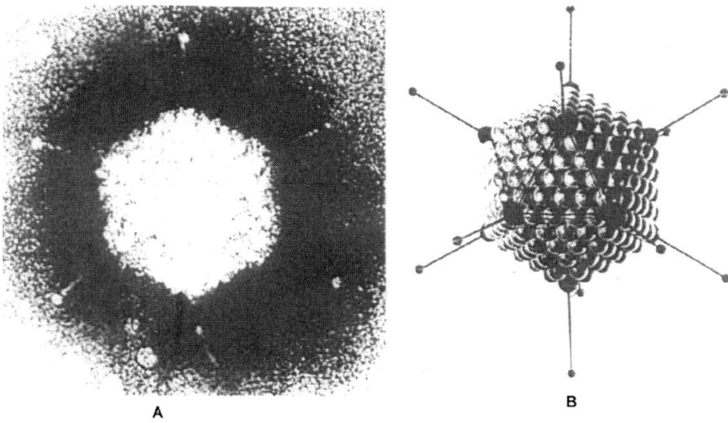

Fig. 3.2. Structure of Adenovirus **A.** Electron photomicrograph, **B.** Computer model.

A regular icosahedron is a solid with 20 faces (each of which is an equilateral triangle), 12 vertexes, and 30 edges. A regular icosahedron can be related into three equivalent positions. This is called a rotational axis of threefold symmetry. Similarly, if an icosahedron is viewed down on axis passing through one of the vertexes, there are five equivalent positions into which it can be rotated. This is called a fivefold symmetry axis. There is a twofold axis passing through the centre of the edges. The icosahedron has six fivefold axes, 10 threefold axes, and 15 twofold axes of rotational symmetry (Fig.3.3).

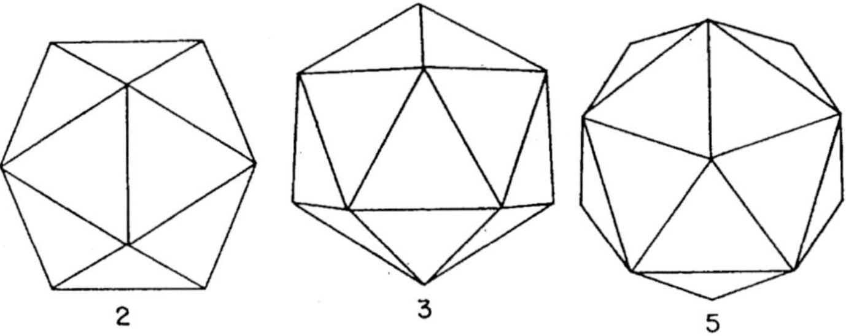

Fig. 3.3. The three symmetries of the regular icosahedron. The views are along a two fold, a three fold, and a fivefold axis of rotation symmetry.

Theoritically possible number of structural units or capsomers is given by the formula.

$$C = 10 x (n-1)^2 + 2$$

where C is the number of structural units on the virion, x = 1 or 3, and n is the number of units on an edge of the virion. Two series of possible forms are of interest here. One series is that generated when x (which represents on integer) is equal to one, and the other is that generated when x is equal to three. For example, the number capsomers in adenovirus capsid (252) can obtained as follows :

$$n = 6; x = 1$$

$$C = 10 x (n-1)^2 + 2$$

$$= 10 \times 1 (6-1)^2 + 2$$

$$= 10 \times 25 + 2 = 252$$

There is a great variety of virions with cubic symmetry, even though the possibilities are limited by the requirement of geometry. There are some virions with the theoretically minimum number of 12 capsomers and others with 812 capsomers.

ii. Helical symmetry

Helical symmetry is exemplified by a spiral staircase in which the identical rings are systematically arranged at about its central pole or axis. Many plant viruses and bacteriophages, and nucleocapsids of some enveloped animal viruses are rod-like or filamentous structures with helical symmetry. Helical symmetry is described by the number of structural units per turn of the helix, μ, the axial rise per unit, p, and the pitch of the helix, P, given by the formula

$$P = \mu \times p$$

A characteristic feature of a helical structure is that any volume can be enclosed simply by varying the length of the helix. Such a structure is said to be open. The best studied helical capsid is that of tobacco mosaic virus. The virus particles are rod shaped structures with the subunits of proteins arranged in a helical pattern (Fig. 3.4). It has been calculated that there are about 134 in one TMV protein coat. Utilizing x-ray diffraction analyses it has been shown that there are 2130 subunits forming the protein coat of TMV (Casper and Klug, 1960). Detailed analysis has shown that there are 49 subunits distributed over three turns. In an average, these three turns form one, structural unit which is repeated every 6.9 nm in the axial direction. In other words each succeeding ring is at a distance of 2.3 nm is in the same direction (i.e. pitch).

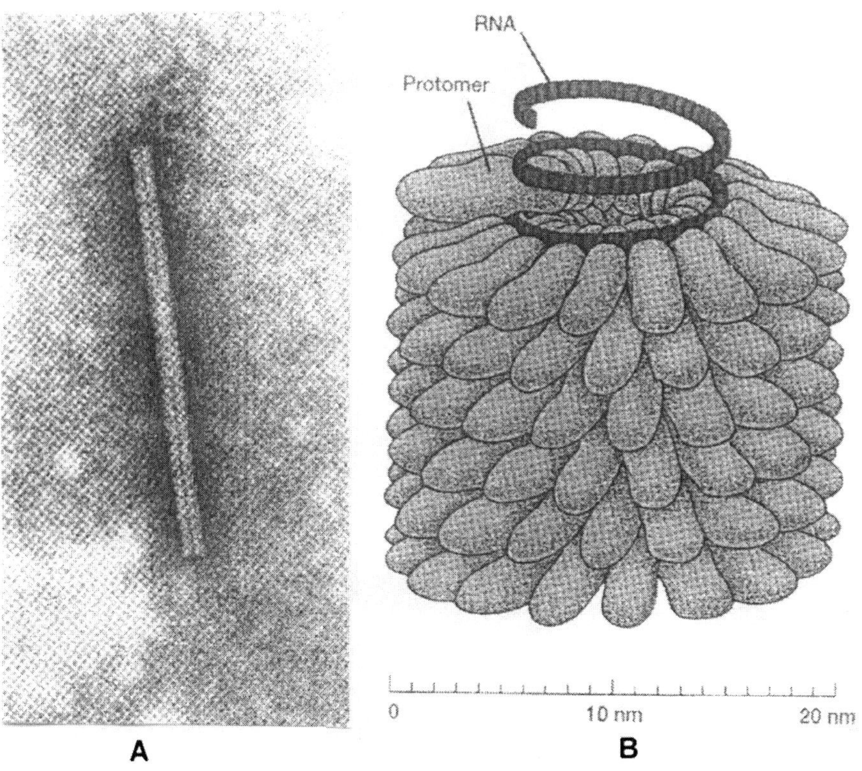

Fig. 3.4. Structure of TMV. **A.** Rod shaped as seen under electron microscope **B.** Helical arrangement of capsomers in the capsid.

Table 3.6. Symmetries of some important viruses

Virus	Symmetry
Bacteriophages	
T, λ, μ Phages	Binal
ϕ X 174	Cuboidal
Inovirus, Rudivirus	Helical
Plant viruses	
TMV, potyvirus, potexvirus	Helical
Caulimovirus, phytoreovirus, Bunyavirus, Comovirus, Bromovirus	Cuboidal
Animal viruses	
Poliovirus, Togavirus, Hepadnavirus, Iridovirus Parvoviridae, Circoviridae	Cuboidal
Herpesvirus, HIV, Bornavirus	Cuboidal (nucleocapsid)
Rhabdovirus, Influenzavirus, Filovirus	Helical (nucleocapsid)

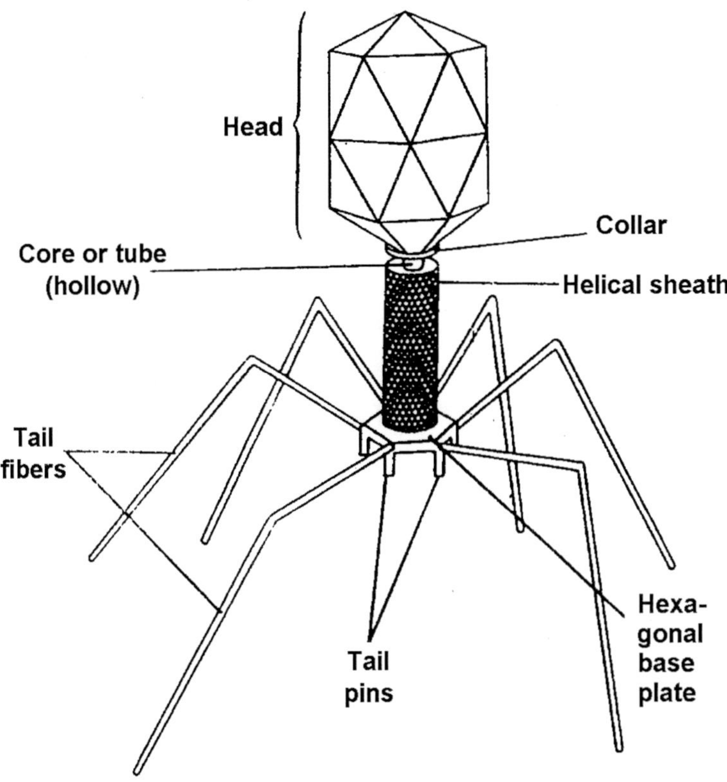

Fig. 3.5. Strutcure of bacteriophage T_4

iii. Binal symmetry

Many of the bacterial viruses contain a large head and a tail. Among these viruses, the structure and symmetry of Teven phages (T_2, T_4 and T_6) are known in great detail. Head and tail phages combine both types of symmetries i.e. cubic and helical. The polygonal head shows cubic symmetry whereas elongated tail is built upon helical symmetry (Fig. 3.5). The head of a T-even phage particle measures approximately 65 x 95 mu. In cross section, the head is approximately hexagonal and is capped by a hexagonal pyramid. The head is formed from a large number of subunits, each of which is a single protein of about 80,000 daltons molecular weight. The tail of the phage is a remarkable organ. It is elongated and is composed of several parts situated around a central core. The tail core is roughly 95 nm long and 8 nm in diameter. It extends from the head of the phage to the tail plate and has a lengthwise axial opening through which the DNA emerges from the head of the phage. Surrounding the core is the tail sheath, composed of about 200 subunits arranged helically or circularly.

Further reading

Baker, TS and Johnson, JE (1997). Principles of virus structure determination, P 38-79. In W Chiu, RM Burnett and RL Garcea (ed.) Structural Biology of Viruses. Oxford University Press, New York.

Baker, TS, Oslon, NH and Fuller, SD (1999). Adding third dimension to the virus life cycles. Three dimensional reconstruction of icosahedral viruses. *Microbiol Mol. Biol. Rev.* 63 : 862-922.

Brown, F and Wilson, G (1984). Topley and Wilson's Principles of Bacteriology, Virology and Immunity, 7th edn. Vol. 4. Virology, Baltimore: Williams and Wilkins.

Casjens, S 1985. Virus Structure and Assembly. Boston : Jones and Bartlett.

Cheng, RH, Kuhn, RJ, Olson, NH *et al.* (1995). Nucleocapsid and glycoprotein organization in an enveloped virus. *Cell* 80 : 621-630.

Flint, SJ *et al.* (2004). Principles of Virology, ASM press, Washington, DC.

Harrison, SC (2001). Principles of virus structure. In Fields Virology 4th Edn. (Edited by Knipe D.M. *et al.*) Lippincott Williams and Wilkins.

Rux, JJ and Burnett, RM (1998). Spherical Viruses *Curr. Opin. Struct. Biol.* 1 : 142-149.

4

ISOLATION AND CULTIVATION

The principal dogma of the microbiology stipulates that any microorganism needs to be obtained in pure and cultured form for all types of future investigations and viruses are no exception to this dogma. However, viruses are unique and differ in many respects with other microorganisms. Firstly, in nature they are never found alone. In accordance their mode of life they occur in the presence or in association with some living cell or other. Their total dependence on their host makes it extremely difficult for one to study viruses. Secondly, unlike other microorganisms they cannot be cultured on a defined medium. A specific living cell or tissue or a host is required for their cultivation. These tissues and cells would act as the medium for the virus in the laboratory conditions. The very size of the virus poses many problems during isolation and purification. Viruses are almost equal to the size of macromolecules and organelles of the cell. Therefore one requires the knowledge of not only virus but also cell biology of the host. Further, a prudence of the investigator is also necessary to select a technique that differentiates and separates the virus from host cell components. Many a times, more than one type techniques need to be employed for the complete purification of the viruses. Inspite of all these constraints, virologists have made spectacular achievements in isolation, purification and characterization that have paved the way for advancement of virology.

1. ISOLATION AND CULTIVATION OF ANIMAL VIRUSES

Isolation

The presence of a virus inside an infected host can be visualized by effects on host organism. Effects on hosts include through a wide range of symptoms, on organs and systems such as skin, mucous membrane, digestive, respiratory or nervous systems; or dysfunction of immune system, liver and heart. Effects on cultured cells include a variety of morphologic and cytological changes in infected cells collectively termed as **cytopathic effects**. In addition, the following immunobased techniques are also used for the detection of viral antigens in cell cultures and clinical specimens: (i) immunofluorescence, (ii) immunoenzyme staining, (iii) enzyme immunoassay (ELISA) and iv. radio immunoassay.

Once, the presence of a virus has been established, it is often purified either by limiting serial dilution or by plaque purification. Viruses are isolated from an affected host by harvesting excreted or secreted material like blood or tissue in a suitable cold buffer. Tissues and organs are homogenized. On proper homogenization, the tissues get completely disrupted and an uniform suspension is obtained. This suspension is then centrifuged, first at low speed (1400 x g) and then at high speed (7000 x g) for 20 minutes during each time. The clear suspension obtained is then passed through a bacterial filter to get rid of most of the larger suspended particles. The filtrate is then passed through a membrane filter to retain the virus particles. The filtered particles are then resuspended in a small amount of a suitable buffer. The virus preparation at this stage contains some cell proteins and liquids as impurities. Such a preparation is used as inoculum for cultivation of the virus.

Cultivation

Animal and human viruses are propagated by three different methods.

i. Propagation on cell culture

ii. Propagation on organized tissues

iii. Propagation on live animals.

i. Propagation on cell culture

Cell culture is the most common method for the propagation of viruses. Eversince the demonstration of propagation of polivirus on non-neuronal cells by Enders, Weller and Robbins (1949) protocols for propatation of different viruses have been standardized. To develop a cell culture, the simplified procedure is as follows (Fig. 4.1). Selected tissue is surgically removed from an anesthesized animal, minced and homogenized. Collagenase is added to breakdown extra culture collagen which does not attach intact cells. The cells are then purified by filteration through a coarse mesh to remove large fragments, and later concentrated by low-speed centrifiguation. The pelleted cells are washed in buffered media containing sera and then subjected to differential low-speed centrifugation to separate all types based on sedimentation rates. Different fractions are plated onto culture medium containing essential aminoacids, vitamins, antibiotics and serum and incubated under proper conditions. Cells grow as loose clumps which can be separated with mild trypsin treatment (trypsinization) and the individual cell types are inoculated into a fresh medium. Epithelioid and fibroblastic cells attach to the plastic and form a monolayer, whereas blood cells such as lymphocytes settle but do not adhere. Commonly used cell lines usually double in 24 to 48 h in such media. The development of monolayer cultures has proved to be a great advantage in the cultivation of viruses. Dulbecco and his associates in the USA demonstrated that monolayers could also be developed on agar containing the necessary nutrients. Virus particles grown on such monolayers were found to be extremely uniform in growth.

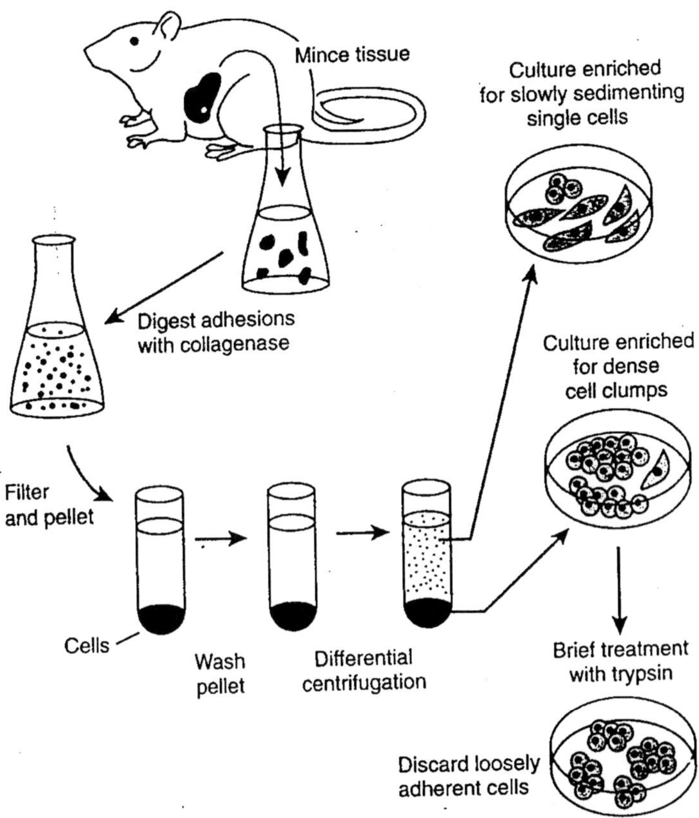

Fig. 4.1. Protocol for generating a primary cell culture.

There are three kinds of cell cultures (Table 4.1). The cell cultures prepared directly from animal tissues as described above are known as **primary cell cultures**. Such cultures consist of several cell types and have a limited life span, usually less than 5 or 6 cell divisions. Most commonly used primary cell cultures are developed from monkey kidney, human embryonic amnion, kidney or foreskin and embryos of chicken and mouse. Primary cell cultures are used for vaccine production or laboratory experiments. Cell lines developed from single cells of primary cell culture are called **diploid cell strains**. They consist of a homogenous population of a single cell type and can divide upto 100 times before dying. These cell strains possess diploid chromosome number. The most widely used diploid cells are those established from human embryos such as the WI-38 strain. **Continuous cell lines** consist of immortal cells that can be propagated indefinitely in culture. These lines are derived from tumor tissue or by treating a primary cell culture with a mutagenic chemical or tumor virus. Such cell lines often differ from the parent types in morphology, cytology and biochemistry. Examples of commonly used continuous cell lines include HE p-2 and HeLa cells, derived from human carcinomas. Continuous cell lines provide a uniform population of cells that can be infected synchronously for growth curve analyses.

Table 4.1. Some cell cultures in common use for viruses cultivation

A. Primary cell culture

1. Rhesus monkey kidney cell culture
2. Human amnion cell culture
3. Chick embryo fibroblast cell culture

B. Diploid cell strains

1. WI-38 - Human embryonic lung cell strain
2. HL-8 - Rhesus embryo cell strain

C. Continuous cell lines

1. HeLa - Human cercinoma of cervix cell line
2. HEP-2 - Human epithelioma of Larynx cell line
3. KB - Human carcinoma of nasopharynx cell line
4. McCoy - Human synovial carcinoma cell lines
5. Detroit-6 - Sternal narrow cell line
6. Chang/C/I/L/K - Human Conjunctive (C), Intestine (I), Liver (L), Kidney (K), cell lines
7. Vero - Vervet monkey kidney cell line
8. Baby hamster kidney cell line

The cell cultures can also be maintained in a state of **suspension culture** by shaking the culture flasks continuously on a mechanical shaker or by spinning through a magnetic stirrer. The advantage of suspension culture is that a large number of cells can be grown in a relatively small volume. Until recently, it was believed that viruses could not replicate in any cell free extract. However, recently it has been demonstrated that infectious poliovirus could be synthesized in a cell free extract of human cells incubated with viral RNA.

ii. Propagation in organized tissues

Propagation of viruses in organized tissues, like chick embryo is practiced since long back. Infact, before the advent of cell culture many viruses were propagated in embryonated chicken eggs. Embryonated eggs offer many advantages: they offer different membranes during the development in which different viruses can be propagated, perfect aseptic conditions can be maintained, easily available in large number and finally handling and maintenance is easy. Consequently, organized tissue structures are still considered important in culturing certain virus types.

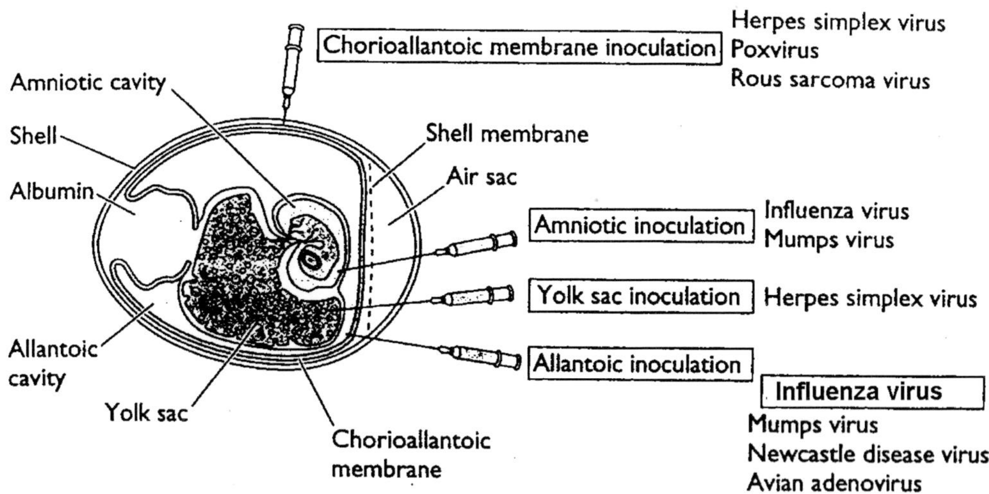

Fig. 4.2. Virus cultivation using an embryonated egg
(Adapted from F. Fenner *et al.,* 1974)

The procedure adopted for cultivation of viruses in chick embryonic membranes is very simple. Depending upon the virus and embryonic membrane required, 5-16 days old embryonated incubated eggs are selected. They are externally sterilized by swabbing. Then a hole is drilled in the shell and virus is injected into the site appropriate for its replication (Fig. 4.2). After inoculation, the open area of the shell is sealed with a wax or cellophane membrane. The sealed eggs are then incubated for several hours as in hatching. The virus particles infect the membrane at random and create pock marked appearance against the transparent background. These indicate viral lesions. Virus propagation in chick embryos is now used for several viruses. The high yield of this virus from embryonated chicken eggs has led to their widespread use in research laboratories and for vaccine production.

iii. Propagation on live animals

In early days of virus cultivation, when freezers and cell cultures were not available, viruses were used to be maintained by continuous passage of virus from animal to animal. This practice is continued even today. The advantages of this type of cultivation are : inoculation and maintenance are easy, chances of contamination are meagre. Further, for viruses which cannot be propagated in cell cultures, this is the only method available. For example hepatitis C virus replicates only in chimpanzees. Similarly, Norwalk virus, an enterovirus cannot be grown in cell culture. However, there are also certain disadvantages such as maintenance of large animals is cost prohibitive and some times, chances of getting infection while cultivating disease causing viruses may be there. For example, Herpes B virus can be cultivated in monkeys but can be dangerous for human beings as it causes encephalitis. Similarly, problems of selection of particular virus mutants by host may arise. Despite these drawbacks, experimental infection of laboratory animals will continue to be obligatory for investigations of viral pathogenesis, development of viral vaccines and antiviral drugs.

In recent years, small animals like mice and guinea pigs are extensively used. Maintenance of strains of murine leukemia virus (MLV) in selected mice strains is a standing example.

2. ISOLATION AND CULTIVATION OF PLANT VIRUSES

2.1 Isolation

Isolation, propagation and purification of plant viruses are essential steps for investigations on host range, physical and chemical properties, electron microscopy and molecular aspects including genomic organization. Effective isolation procedures have now been developed for many plant viruses. Generally, systemically infected plants give higher yields of the virus. Similarly, herbaceous plants are more suitable for virus extraction. Isolation and propagation protocols begin with preparation of crude sap extract of virus. Systemically infected leaves from young plants, showing typical symptoms are chopped into smaller pieces, ground in a pestle and motar or homogenized in a waring blender and extracted with distilled water or suitable buffer. The resultant pulp is squeezed through muslin cloth to extract the sap. The sap is then centrifuged at low speed (5000 rpm) to remove the remaining cell residues. All these operations must be carried out at 4°C. After this, the virus is sedimented by centrifuging at high speed (35,000 rpm). The supernatant is removed and the virus containing pellet is dissolved in a buffer solution. Centrifugation again at low speed is done for sedimenting non-virus materials. By repeating the low speed and high speed centrifugation (differential centrifugation), the virus may become free from non-virus components. This semi-purified virus can be used for either propagation or further purification.

2.2 Cultivation

Plant viruses are propagated through two methods : (i) propagation on hosts and, (ii) on tissue culture.

i. Propagation hosts

The hosts in which the virus is multiplied for purification are known as propagation hosts. The choice of host plant for propagating a virus is critical importance for its successful isolation. A propagation host should possess the following characteristics: (i) easy to grow under glass-house conditions, (ii) should support high titre of the virus, (iii) as far as possible should be free from inhibitors and (iv) the host constituents should be easily separable from the virus during purification. In practice, species of *Cucumis, Nicotiana, Petunia, Phaseolus,* peas and *Vigna* have been found suitable for the propagation of a large number of viruses. If the infection is mixed, a selective or differential host is required to eliminate the contamination. For many viruses, the concentration rises to a peak after a few days or weeks and then falls quite rapidly.

ii. Tissue culture

P.R. White (1934) was the first person to propagate plant viruses in tissue culture. He could successfully cultivate tobacco mosaic virus in excised tomato root tips. The method developed by White was rather simple. In this method, stem segments of systemically virus infected tomato plant were selected. Washed thoroughly and were suspended by threads in conical flasks containing little water. After a period of time the root tips developing from the stems are excised and transferred to flask containing a suitable medium. The tips were allowed to grow for about a week. Afterwards, the surviving cultures were cut into pieces of about 1 cm length and were then subcultured in fresh medium as before. Viruses were found to be multiplying in these subcultured tips. Using this technique, it has been found possible to culture such viruses as tobacco necrosis virus and wound tumor virus.

The above method envisages that viruses are normally cultivated on tissues obtained from already and systemically infected plants. However, the healthy tissues of hosts can be cultured first and later inoculated with virus to be propagated. A further progress in this direction is the use of isolated protoplasts for culturing protoplasts. The advantage of this method is that protoplasts can be uniformly and synchronously cultivated.

3. ISOLATION AND PROPAGATION OF BACTERIAL VIRUSES

Of the three important viruses, isolation and cultivation of bacterioph-ages is relatively easy. The greatest advantage is that all the steps involved in isolation and cultivation are carried out in a plate or flask within the laboratory. Bacteriophages exist in a variety of habitats such as soil, intesti-nal contents, sewage, and some insects like cockroaches and flies. However, they occur in low concentration and their isolation from their natural habitats is not easy. Addition of any enriched susceptible host cell culture to the sample increases the number of phage particles. The principle involved is on solid media phages produce clear, lysed areas in bacterial lawns or if, sufficiently diluted, small holes called **plaques** each of them corresponding to a single viable phage (Fig. 6.1A). In liquid media, phages sometimes cause complete clearing of bacterial cultures.

Phages are propagated by three types of techniques : (i) in liquid media inoculated with host bacteria, (ii) on agar surfaces with lawn of bacteria, and (iii) in agar double layers consisting of normal bottom agar covered with a mixture of soft agar (0.3 – 0.9%), phages and bacteria. Phages are harvested after a suitable incubation time, generally 3h for liquid cultures and 18 hours for solid media. Phages from agar cultures are extracted with buffer or nutrient broth. Phage suspensions, or lysates, are sterilized by filteration through membrane filters (0.45 μm pore size). Phage stocks are usually preserved at –70°C in 50% glycerol. They may also be stored in liquid nitrogen.

Further reading

Davis, BD *et al.* (1990). Microbiology, 4th ed. Philadelphia : JB Lippincott.

Flint, SJ *et al.* (2004). Principles of Virology, ASM press, Washington, DC

Freshney, RI (1994). Culture of animal cells 3rd Edn. New York : Wiley – Liss.

Habel, K and Salzman NP (eds.) (1969). Fundamental Techniques in Virology. New York : Academic Press.

Hull, R (2002). Matthew's Plant Virology. Academic Press New York.

Knipe, DM *et al.* (eds.) (2001). Fields Virology Vol. I, Lippincott Williams and Wilkins.

Kouchler, RJ (1977). Biochemical methods in cell culture and virology. Stroudsbury, PA : Dowden, Hutchinson and Ross.

Mahy, BWJ, Kangro HO Virology methods manual, San Diego, Academic Press.

Maramorosch, K and Kaprowski, H (eds.) (1967). Methods in Virology. New York : Academic Press.

5

PURIFICATION AND CHARACTERIZATION

Crude viral extracts prepared from either natural host or propagated host do not qualify for further investigations. In the crude extract, the virus is mixed with a variety of cell constituents that lie in the same broad size range as the virus and that may have similar properties. A highly purified virus is always essential for basic characterization and identification, for studying many physical and biological properties and also for production of anitserum and vaccines. To achieve this, they must be isolated carefully first and then suitably purified. Separation of virus particles from host components and their concentration *in vitro* is referred to as purification. However, it should be borne in mind that the purification procedures donot disrupt the physical, chemical, biological and infectivity properties of the virus. Needless to emphasize, the methods adopted for purification should be virus-friendly. Temperature, pH and salt concentrations are crucial factors to be considered during isolation and purification. A range of techniques are now available for purification of viruses. Some viruses can be purified by few steps or even by a single step. However, some viruses could only be purified with elaborative and cumbersome methods. Purification of plant, animal and bacterial viruses can be carried with similar methods, but they differ in minor details.

1. CLARIFICATION OF THE EXTRACT

The first step in virus isolation and purification is usually aimed at to remove much of the macromolecular host material, leaving the virus in solution. The extraction medium may be designed to precipitate ribosomes, or to disintegrate them. The extract may be subjected to some treatments such as heating to 50-60°C for a few minutes, freezing and thawing, acidification to a pH of less than 5 or addition of K_2HPO_4 to coagulate much of host material. The treatment has to be chosen on a case-by-case basis to avoid the damage to virus. The treated extract is then subjected to centrifugation at low speed (5000-10000 g for 10-20 minutes). This treatment sediments cell debris and coagulated host material. With the butanol-chloroform system, centrifugation separates the two phases: virus in the aqueous phase and denatured protein at the interface. However, it should be noted that organic solvents should not be used for the purification of enveloped viruses.

W.M. Stanley (1904-1971); Nobel Prize, 1946

2. SEMI-PURIFICATION AND CONCENTRATION

i. High-speed sedimentation

Centrifugation of the above clarified extract at high speed for a sufficient time will sediment the virus. This serves the double purpose of concentrating the virus as well as separating the low molecular-weight materials. However, high-speed sedimentation is a harsh process that may damage some particles (e.g. some reovirus). Some viruses form characteristic aggregates when the pellets are redissolved. The virus particles in these aggregates may be tenaciously bound together. The problem is further complicated, if host membranes are involved in binding. In such cases, addition of non-ionic detergent releases virus particles out of aggregates.

ii. Precipitation

Precipitation is the most effective method of purification since it preferentially precipitates the virus. It can be achieved by physical or chemical means. The suspension or extract containing virus and obtained by clarification and high speed centrifugation can be used for precipitation. Certain plant viruses can be precipitated in a single phase polyethylene glycol (PEG) system. However, host DNA molecules are likely to be precipitated. The exact conditions for precipitation depend on pH, ionic strength and concentration of macromolecules. PEG precipitation is applicable to many viruses including the unstable ones. Other chemical agents like ethanol, ammonium sulphate have also been found to be effective in the precipitation of viruses. Methanol, protamine and yeast nucleic acid have been found to be effective precipitants of animal viruses.

Precipitation can also be achieved by adjusting the hydrogen ion concentration (pH) of the suspension. Proteins, being ampholytes, each protein can be precipitated at a particular pH point. This pH is called **isoelectric point**. The isoelectric point can be achieved by adding more cations (Mg^{++}, Ca^{++}, H^{+} etc.) and anions, (Cl^{-}, Po_4^{---}) and thus particular protein could be precipitated out. Since viruses have an outer protein coat, they can also be precipitated out in this manner. Alfalfa mosaic virus is one of the first viruses to be precipitated by this method.

The precipitates obtained once could be redissolved in some suitable buffer solution and reprecipitated. In this manner, more and more purified preparations could be obtained. The advantage of this precipitation is that expensive ultracentrifuges can be dispensed with. However, differential centrifugation is often used as a second step in purification procedures.

iii. Dialysis

Dialysis is a separation process using membranes. It is based on the principle that solute molecules that are small enough to pass through the pores of the membrane will diffuse from the side with higher concentration to the side with lower concentration. Dialysis through synthetic membranes can be used to separate low-molecular weight materials from an initial extract. It is usually employed to remove salts, following salt precipitation or crystallization, or following density gradient centrifugation in salt or sucrose solutions.

3. ULTRA-PURIFICATION

The virus preparations made through above described procedures still contain some low-and high molecular weight host materials. Further purification steps are needed to remove these host impurities. Here onwards, the procedures to be used will depend on mainly on the stability of the virus, the degree of preparation and the purpose for which the preparation is required.

i. Density-gradient centrifugation

Density gradient centrifugation, developed by Brakke (1951, 1960), is one of the most useful procedures for further purification with out pelleting. This technique has proved to be highly versatile and has been widely used in the fields of virology and molecular biology. Two commonly used versions of density gradient centrifugation are: **rate-zonal centrifugation and equilibrium density gradient centrifugation.** In these methods, viral suspension is mixed with an appropriate gradient concentrations of sucrose and/or calcium chloride solution of decreasing concentration into 14 M sodium chloride solution taken in centrifuge tubes. The mixture is then allowed to equilibrate at 4°C for 12 hours for achieving a stable linear gradient. The mixture is subsequently centrifuged at very high speed (40,000-1,50,000 g) for varying periods (several hours) depending upon the virus. During centrifugation, the virus and other host contaminants move along gradient at different rates to form layers in the tube depending upon their shape, size and sedimentation coefficient. Different molecules aggregate at different regions in the centrifuge tube forming distinct bands or layers. In equilibrium density gradient centrifugation, the components of identifical densities form bands with stabilized position in the centrifuged fluid. This is because the bands are formed in the regions of the solution with densities in equilibrium with that of virus particles assembled at that region. These bands are stable and would not be disturbed even on prolonged centrifu-gation. This method is, therefore, also known as **pycnotic density gradient centrifugation**.

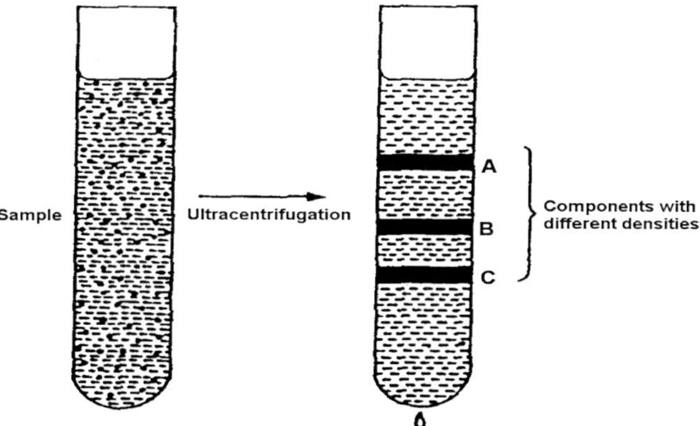

Fig. 5.1. Density gradient centrifugation

The number of bands formed indicates the number of types of particles in the suspension (Fig. 5.1). The bands can be visualized against a dark background as light scattering zones. These layers can then be separately collected by puncturing the tube at the bottom and collecting the drips into series of test tubes or drawing out the contents of bonds separately by a syringe. Because successive fractions from a gradient can be collected, a variety of procedures can be used to identify, the virus, non-infectious virus like components and host materials. These include infectivity, UV absorption spectra and exami-nation under the electron microscope. The resolution offered by gradient centrifugation is very high, so much so, the particles differing in shape as little as 0.05 g cm^{-3} can easily be separated.

A number of viruses were purified and characterized by density gradient centrifugation. Shope papilloma virus and poliovirus were purified by utilizing the rate-zonal centrifugation. Similarly ϕ x 174 phage and TMV strain particles were characterized by using density gradient centrifugation.

ii. Gel chromatography

Gel chromatography using columns containing Sephadex and Sepharose may also be used for purification of viruses that are unstable to the pelleting involved in high-speed centrifugation. In this, a virus containing suspension is allowed to run with buffer through the column (Fig. 5.2). As the mixture passes, molecules which are too large to enter the pores of the gel move down quickly whereas movement of smaller molecules which enter pores, is slowed down or retarded. The fractions are monitored by UV spectrophotometer and the virus is eluted. The virus fraction may also be eluted using an automatic fraction collector. This method has been successfully used to produce purified prepa-ration of certain viruses. However, the technique has not been widely used.

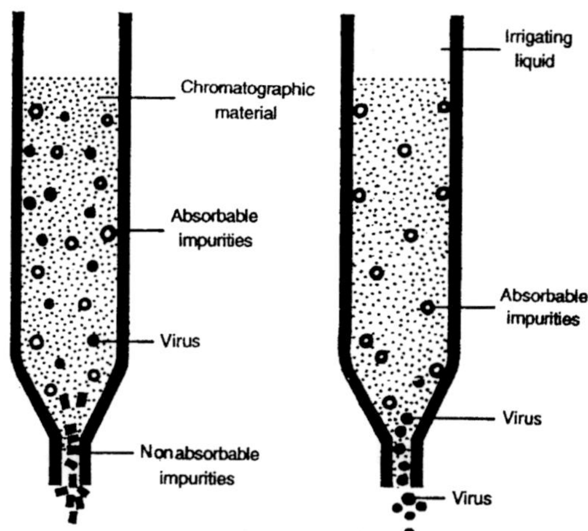

Fig. 5.2. Purification of virus preparation by chromatography

iii. Immunoaffinity columns

This technique combines the chromatography and serology. In this, monoclonal antibodies generated against the virus are bound to a solid support matrix such as agarose to form a column that will specifically bind the virus from a solution passed through the column. The virus can then be eluted by lowering the pH. Immunoaffinity columns use is only selective and justified only in special circumstances.

iv. Chromatography

Chromatographic techniques have occasionally been used for purification of viruses from partially purified preparation. For example, LNYV was purified by using a column of calcium phosphate gel in phosphate buffer. Similarly, OBDV was purified using cellulose column chromato-graphy. Two electrophoretic forms of CPMV were separated using a fast protein liquid chromatography.

Table 5.1: Different purification methods of viruses

S.NO	Method
01.	**Clarification of extracts**
02.	**Semi purification and concentration**
	(i) High - speed sedimentation
	(ii) Precipitation
	(iii) Dialysis
03.	**Ultra purification**
	(i) Density – gradient centrifugation
	(ii) Gel chromatography
	(iii) Immuno affinity columns
	(iv) Chromatography

4. CRITERIA FOR PURITY OF VIRUS PREPARATION

After going through the laborious and time consuming schedule of virus purification, the question that aries in the mind of an investigator is about its purity. The purity or homogeneity has to be judged with respect to (i) physicochemical composition and, (ii) biological characteristics. Chemically, a preparation is said to be `purest' when it is available in the crystallized state. An extremely purified sample shows certain characteristics of its own. For example, a purified suspension of TMV shows a brilliant shine. Physical criteria like size, shape, solubility and molecular weight can also be used for judging the purity. Spectrophotometry, electrophoresis, electron microscopy and serology or different combinations of these have been found useful in determining the homogeneity of a sample.

Biological purity refers to the capacity of the preparations to infect a host. As the samples are subjected to rigorous and harsh methods of purification, they may loose the infectivity. Hence, the biological activity should be monitored at every step of the purification. Generally, it has been observed that the infectivity tends to get deminished with successive stages of purification.

5. STORAGE AND PRESERVATION

Once isolated and purified, the viral sample need to be stored for the period to the maximum extent possible with out the loss of viability. This is necessary for various reasons. It must be remembered here that viral particles, due to their very nature, are amenable to fluctuations in the environmental conditions. Therefore, the role of environmental factors must be understood and suitable remedial conditions should be maintained. As for as temperature is concerned, 4°C is found to be ideal for preserving the particles for a day or two, but much lower temperature such as frozen CO_2 (dry ice)- 70°C and more preferably liquid nitrogen – 196°C, are required for long term preservation. Purified plant virus preparations can be stored at -20°C by adding a drop of sodium azide and glycerol or thymol to prevent growth of microbes and stabilize the virus. Liquid preparations can be stored at low temperature by the addition of an equal volume of glycerol. The enveloped viruses are more heat labile than nonenveloped viruses, therefore, they tend to lose infectivity even if stored at low temperature. Normal practice in laboratories is to freeze the viral suspensions very rapidly in a medium containing protective proteins, followed by storage at –70°C or –196°C. Lyophilization followed by preservation of the dried powder at –20°C is also recommended.

6. CHARACTERIZATION

Once the viral preparations are purified, the next logical step is characterization. Characterization implies probing of various morphological and chemical characteristics of the virus. The different techniques employed for this purpose include: electron microscopy, x-ray crystallography, sedimentation analysis and various physico-chemical procedures (Fig. 5.3).

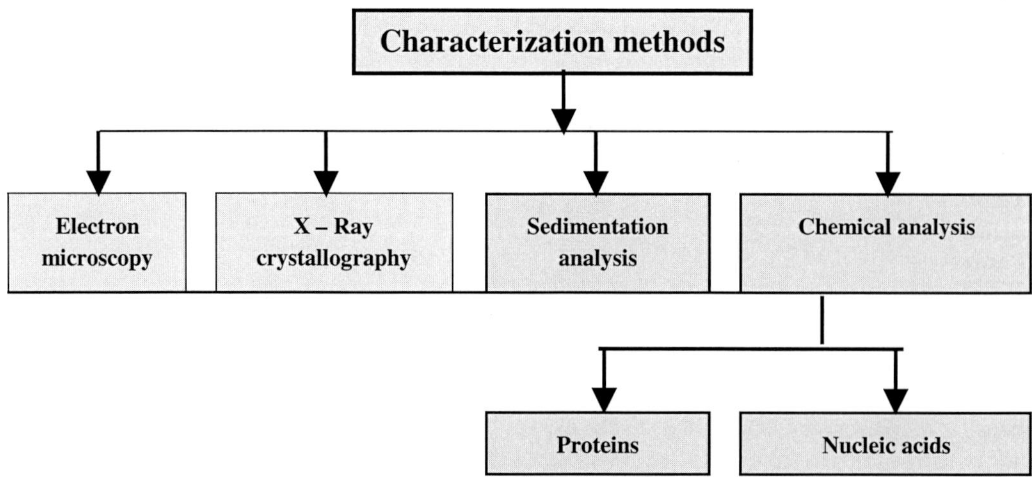

Fig 5.3: Methods of characterization of viruses

i. Electron microscopy

Knowledge of the size, shape and any surface features of virus particles is a basic requirement for virus identification. Electron microscopy can provide this information quickly and reliably. The tobacco mosaic virus particles are one of the first viruses to be observed under electron microscope (Williams and Wykoff, 1945). Since then electron microscopy and related technologies have improved commendably. A large number of viruses have been observed and their morphological details were obtained using these techniques. The scope of this book does not allow the detailed description of these technique, but nevertheless utilities of these techniques are mentioned here (Table 5.2). The most commonly employed electron microscopy related techniques are metal shadowing and positive and negative staining.

Metal shadowing technique was first employed by Williams and Wykoff. In this method vapours of a metal, like gold were projected at an angle onto the membrane to which virus particles were adsorbed during observation. The vapourized metal forms a thin coat over the particles. Because of this coating, the particles become opaque to the electron beams and their contours become more clear. The net result is that a shadow is formed on the side opposite to the direction of metal vapours. Thus, morphology and surface details can be deduced very clearly from a study of the uniformly shadowed particles.

Negative staining was first developed by Brenner and Horne (1959). In this technique, virus particles are treated with electron opaque chemicals like osmium tetroxide and sodium phosphotungstate in such a way that background is stained and some areas of the particles are also stained. As a result the virus particles are clearly seen. This method provides the information only on surface details. Positive staining utilizes the chemicals like uranyl salts and ferretin which get selectively absorbed by nucleic acids and proteins respectively. This technique, thus highlights only the viral particles but not the background and hence the name positive staining. This technique is most suitable for studying the internal details.

Table 5.2. Electron microscopy based techniques and their applications

Technique	Application
Metal shadowing	Morphology, surface details
Negative staining	Surface details
Positive staining	Internal structure
Freeze drying	Size and shape
Carbon replica	Surface details
Thin sectioning	Internal details

ii. X-ray crystallography

X-ray crystallography technique has been extensively and successfully used in physical laboratories for elucidation of the internal architecture of a crystal. This technique has now been extended and extensively used to elaborate the architecture of virus crystals. Our present knowledge of the nature of the alignment of different viral components are mainly due to x-ray crystallography studies. Its continued use in future is likely to unravel many more unknown features of viral morphology.

iii. Sedimentation analysis

Sedimentation is the inherent property of the particles in suspension to settle down when subjected to gravitation. The sedimentation constant is the velocity of the suspended particles in a centrifugal field. This value is generally denoted as `S' or Svedberg unit after its discoverer. Some physical properties of the viruses like molecular weight of virus particles or their components can be obtained through the analysis of sedimentation properties. For instance, the molecular weights of the particles can be calculated from their `S' values.

iv. Chemical analysis

A highly purified viral sample is a prerequisite for accurate chemical analysis of the virus. Chemical analysis can be carried out using a wide range of physico-chemical techniques. These analyses reveal the major chemical constituents of different groups of viruses. In addition, chemistry of individual viral components like nucleic acid and coat proteins have also been worked out in detail.

Proteins : Proteins are found distributed in different locations of the viral particles. However, it is essential to separate and purify these proteins from remaining chemical constituents. Dissociation of proteins can be achieved through treatment with acid or alkali or by any suitable detergent. The degree of effectiveness varies according to method as well as the nature of the virus. Keeping in view the draw backs of these methods, such as protein denaturation, in recent years techniques like polyacrylamide gel electrophoresis (PAGE) and western blotting in combination are widely used for analyses of viral proteins.

The protein samples prepared from a virus can be analyzed for different end uses that include: (i) types of aminoacids (ii) amino acid sequence, (iii) protein end groups *i.e.* C-terminal and the N-terminal, (iv) protein subunit, (v) electrophoretic band patterns, (vi) peptide fingerprinting, and (vii) antisera production.

Nucleic acids : Nucleic acids, whether RNA or DNA make the most important constituent of a virus. Characterization of a virus mostly depends on the properties of its nucleic acid. Several isolation and purification methods are now available to obtain purified nucleic acid preparation, some of which are : (i) hot salt method, (ii) phenol method, (iii) detergent method, and (iv) alkaline method.

Once the viral nucleic acid is obtained in pure form it can be subjected to different analyses. Today, molecular biology offers a bewildering range of techniques for the analysis of nucleic acids. Basically, there are five approaches to use nucleic acids for characterization of viruses:

i. The type, number and ratio of bases and nucleotides e.g. base ratio (AT/GC), nucleotide ratio, polynucleotide end group, nucleotide sequence analysis.
ii. The type and molecular sizes of nucleic acids. e.g. electrophoretic isolation of dsRNA, polymorphism in heteroduplex RNAs.
iii. Cleavage patterns of viral DNA or cDNA e.g. restriction fragment length polymorphism (RFLP) analysis.
iv. Hybridization between nucleic acids e.g., Dot-blot hybridization, tissue print hybridization, *in situ* hybridization.
v. Polymerase chain reaction e.g. random amplified polymorphic DNA (RAPD), RT-PCR.

The secondary and tertiary structure of viral nucleic acids can be analyzed by using methods such as spectrophotometry, sedimentation, optical rotation, viscosity and x-ray diffraction patterns.

In addition to proteins and nucleic acids other chemical constituents like carbohydrates and lipids can also be purified and their chemistry can be studied. This data can also employed for characterization of viruses.

Further reading

Brakke, MK (1967). Density gradient centrifugation. In Methods in Virology (eds. Maramorosch, K and Koprowski, H.), Vol II : 93-118. Academic Press, New York.

Flint, SJ *et al.* (2004). Principles of Virology ASM Press, Washington, DC.

Francki, RIB (1972). Purification of Viruses. In: Principles and techniques in plant virology (eds. Kado CI and Agarwal HD), 292-235 Van Nostrand Reinhold: Princeton

Hamilton, *et al.* (1981). Guidelines for the identification and characterization of plant viruses *J. Gen. Virol.,* 54 : 223-241.

Hull, R (2002). Matthews Plant Virology 4th ed. Academic Press, New York.

Mahy, BWJ and Kangro, HO. Virology Methods Manual, San Diego, Academic Press.

Steere, RL (1959). The purification of plant viruses. *Adv. Virus Res.* 6 : 1-73.

6

DETECTION AND ASSAY

Many a time, when the suspected sample or infected specimen given to an investigator, some preliminary investigations are required as to assess whether the sample or specimen does really contain a virus, or if present whether in biologically active form and at what concentration it is present. Assay procedures are targeted to address these questions. Information obtained from assay studies can be applied to different situations. For example, if the virus in the sample or specimen is biologically inactive it can be concluded that it cannot cause the infection. Similarly, if the titre is below the level of inoculum potential, the same conclusion can be drawn. The titre of the virus in the vaccine (live or attenuated) is a crucial factor for its effectiveness. Assay of viruses is also important in the prognosis and monitoring the progress of the diseases like HIV/AIDS, where periodical determination of viral load is required.

Viruses can be assayed by different methods. Approaches for detection and assay of viruses fall into three categores: (i) assays based on infectivity, (ii) assays based on serology and (iii) assays based on nucleic acids (Table 6.1). Each method has its own merits and demerits. Thus, a clear understanding of the nature and efficiency of different quantitative assays is required to make effective use of data obtained from any assay.

1. ASSAYS BASED ON INFECTIVITY

i. The plaque assay

The plaque assay is the most elegant, the most quantitative, and most useful assay for both bacterial and animal viruses. Developed originally for the study of bacteriophages by d'Herelle in the early 1900s, was modified and applied to animal viruses by Renato Dulbecco (1952) (awarded Nobel prize in 1975). In this assay, monolayers of cultured cells are incubated with a preparation of virus to allow the virus to adsorb to cells. After the removal of unadsorbed virions, the cells are covered with nutrient medium containing a supplement that results in the formation of a gel, most commonly agar. When the original infected cells release new progeny viruses, the semi-solid medium prevents the formation of secondary plaques through diffusion of virus from the original site of infection to new sites, ensuring that each plaque that develops in the assay originated from a single infectious particle in the starting inoculum. A virus particle able to initiate a productive infection is termed a **plaque forming unit (PFU)** (Fig. 6.1). If the infected cells are damaged, the plaque can be distinguished from the surrounding monolayer. The plaque assay can be used to determine titres only of viruses that cause visible cell damage. To enhance the contrast between the plaque and the surrounding monolayer, the cells are usually stained with a vital dye, neutral red or crystal violet. Living cells absorb dye and the plaques appear clear against a red (neutral red), or purple (crystal violet) back ground of healthy cells. Usually, there is a linear relationship between the number of infectious virus particles and the plaque count.

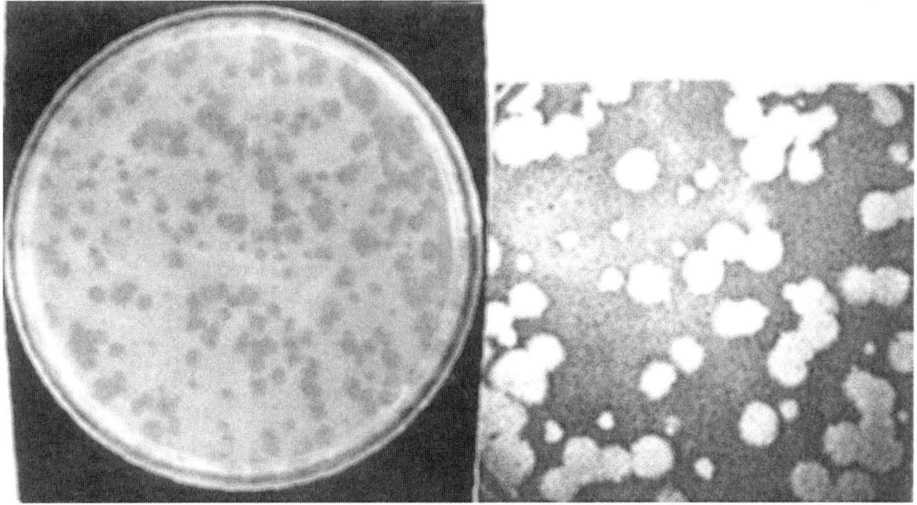

Fig. 6.1. Viral plaques. **A.** Phage plaques, **B.** Animal virus plaques.

An important benefit of the plaque assay is that it measures infectivity but not the exact number of virus particles in the sample. Apart from its utility as a quantitative assay, the plaque assay also provides a way to detect genetic variants of virus that form different types of plaques.

ii. Pock assay

Chorioallantoic membrane of developing chick can be used for assay of certain viruses. In this assay, fertilized eggs are incubated for 2 weeks, then carefully opened to expose the membrane. The virus suspension is then placed on the membrane, the egg resealed, and virus pocks (or plaques) allowed to develop (Fig. 6.2). Although cumbersome, this assay can be used to quantify viruses in a fashion similar to a plaque assay. The pock assay was commonly used before the adoptation of the plaque assay to animal virology, but now it has largely been replaced by other assays and is used only for specialized purposes.

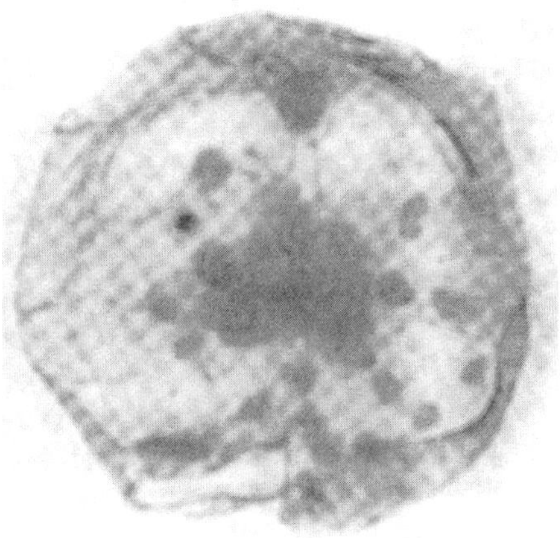

Fig. 6.2. Cowpox induced pock formation on the chorioallantoic membrane of chick embryos

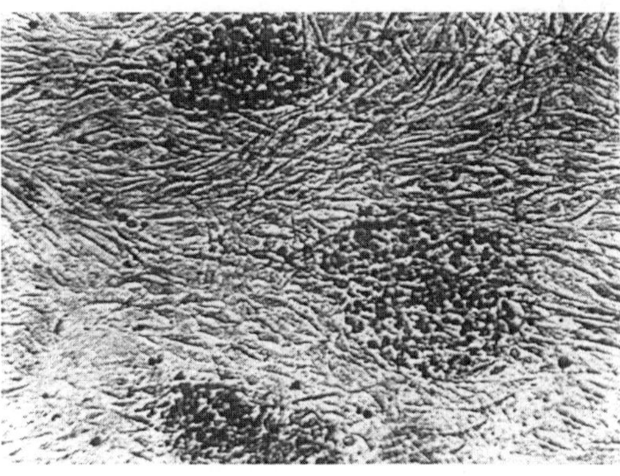

Fig. 6.3. Focus assay: Transformed cell foci of Rous sarcoma virus on a monolayer.

iii. The focus assay

The focus assay is useful for determining the titres of some viruses (e.g., retroviruses) that normally transform the cells rather than killing them. They can be quantified by taking the advantage of transformation cytopathology. For example, Rous sarcoma virus (RSV) transformed cells lose contact inhibition and therefore grow in foci, or piles of transformed cells on top of contact- inhibited cell monolayer (Fig. 6.3). Dense foci of transformed cells stain more darkly than cells in a monolayer and thus can be quantified by treating an infected monolayer with an appropriate stain. Otherwise, the focus assay is similar to plaque assay in principle. Infectivity is expressed in focus forming units per milliliter.

iv. Infection centre assay

This assay is used to determine the fraction of cells in a culture infected with a virus. Monolayers of cells infected with a virus are suspended before progeny viruses are produced. Dilutions of a known number of infected cells are then plated on monolayers of susceptible cells and covered with an agar overlay. The number of plaques indicate the number of cells infected in the original culture. The fraction of infected cells in the culture can therefore be easily determined. One principal use of infectious centre assay is to measure the proportion of infected cells in persistently infected cultures.

Table 6.1: Different assay methods of viruses

S. No.	Method
1. Assays based on infectivity	
(i)	Plaque assay
(ii)	Pock assay
(iii)	The focus assay
(iv)	Infection centre assay
(v)	Dilution end point assay
2. Assays based on physical components of virus	
(i)	Electron microscopy
(ii)	Hemagglutination
(iii)	Viral enzyme activity
3. Assays based on serology	
(i)	Virus neutralization
(ii)	Hemagglutination inhibition
(iii)	Complement fixation
(iv)	Immuno staining
(v)	Immuno precipitation or Immuno blotting
(vi)	Enzyme – linked immuno sorbent assay (ELISA)
4. Assays based on viral nucleic acids	
(i)	Hybridization techniques
(ii)	PCR techniques

v. Dilution end point assay

In the assay of some viruses, there comes a point in the dilution of any virus stock below which a desired property is not observed. If a virus stock is diluted far enough, and then a small measured sample is taken, chances are that there will be no infectious virus present. For instance, 1 ml of diluted sample may not contain one PFU. Because the virus stocks can be diluted to the level that any given aliquot will usually have no PFUs. Thus one can measure infection by dilution instead of by titration. This type of endpoint dilution method is often called a **quantal assay** because it is a statistical analysis rather than quantitative one. In this type of assay, a known number of subjects (animals, cell culture, wells etc.) are infected with increasing dilutions of virus and then scored for illness, death or cytopathogenicity. By plotting a log dilution against percentage of infected subjects, one can estimate a virus dilution that results in half the aliquots in that dilution containing virus and half not. In an assay of a disease in animals, this end point is called ID_{50} (median infectious dose), or LD_{50} (median lethal dose). For measurement of gross cytopathology in tissue culture wells, it is called $TCID_{50}$ (median tissue culture infectious dose).

2. ASSAYS BASED ON PHYSICAL COMPONENTS OF VIRUS PARTICLES

Very often, the number of virus particles and infectious units are not equal. Physical assays which are more rapid and easier to carry out, provde the information on the actual number of virus particles present in the sample or specimen.

i. Electron microscopy

Almost all viruses (except members of *Poxviridae*) are too small to be observed by light microscopy. However, they can be seen readily by electron microscopy. Thus, counting of viral particles can be done by direct observation under electron microscope. One widely used procedure is **latex droplet method** (Backus and Williams, 1950). In this method, a known volume of the suspension is spread on a cellophan film after being mixed with known number of polystyrene latex droplets. The film is dried and observed under the microscope. Technique like metal shadowing is employed to increase the resolution of the particles. With this treatment, virus particles alone get coated with the latex and are seen very clearly. Larger free latex particles are also seen visible side by side. Both the types of particles are enumerated and their relative proportions are calculated. Since, the concentration of the latex particles sprayed as well as the original concentration of the latex particles per ml is known, it is easy to calculate the total concentration of virus particles (Fig. 6.4).

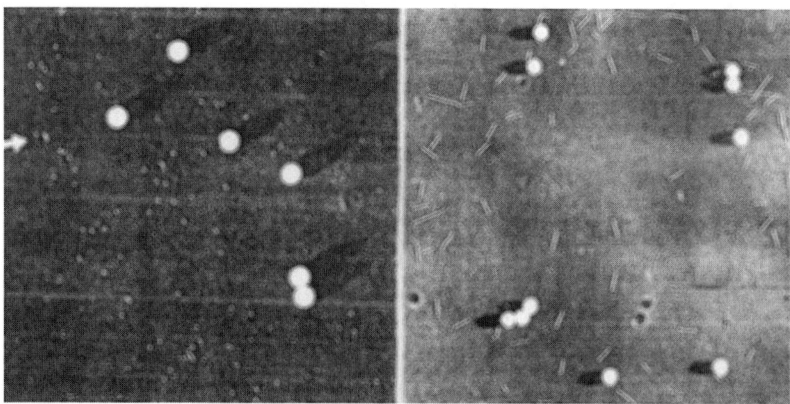

Fig. 6.4. Counting of virions mixed with polystyrene latex particles. **A.** Poliovirus, **B.** TMV (Courtesy virus laboratory, University of California and Symington *et al.*, 1962)

ii. Hemagglutination

Many viruses possess proteins that can bind to erythrocytes or RBC. Through this property they can link multiple cells, resulting in a lattice. This phenomenon is known as **hemagglutination** and the agents responsible are called **hemagglutinins**. For example, influenza virus contains an envelope glycoprotein called hemagglutinin (HA) which binds to N-acetyl neuraminic acid-containing glycoproteins on erythrocytes. A virus particle attaches simultaneously to two red blood cells and bridges them (Hirst, 1941). The most commonly used hemagglutinin assay procedure is the pattern method. This is based on the principle that when a suspension of blood cells and the virus suspension is mixed and left undisturbed in a tube or well, after some time, the nonaggregated cells tumble to the bottom of the well and form a sharp dot or button while agglutinated form a lattice that coats the wells.

In practice, the assay is carried out by an end point procedure (Fig. 6.5). Two fold dilutions of the virus sample are taken in replicates and mixed with a standard dilution of red blood cells. The last dilution showing complete hemagglutination is regarded as the end point. This value can be taken as an estimate of the viral population. The titre estimated by this method has, however, one limitation in that it is not sensitive for determining small numbers of particles. Nevertheless, because the assay is rapid, it is of often used as a quick indicator for the relative quantities of virus particles.

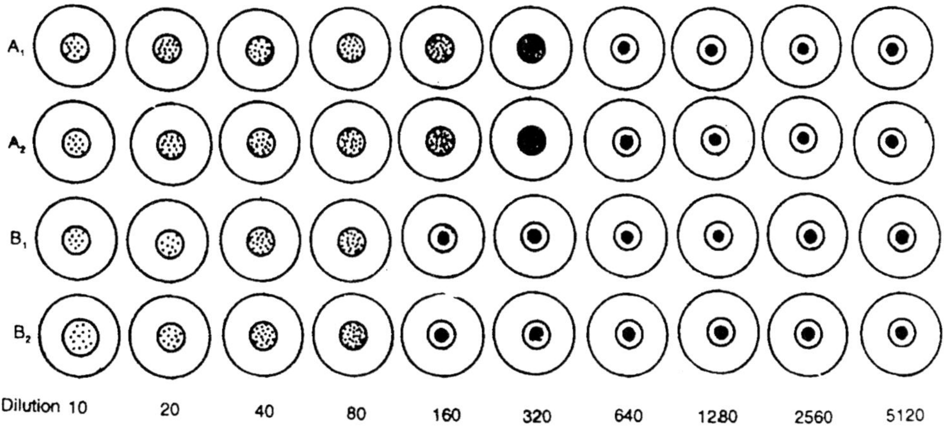

Fig. 6.5. Counting of viruses by haemagglutination assay. The method depicted here is the patternmethod which is one of the several in use. Two samples A and B were diluted serially by using two-fold steps (10,20,40,80 and so on). 0.5 ml of each dilution was mixed with an equal volume of red blood cells suspension (10^7 / ml). Each mixture was placed in a cup drilled in a clean plastic plate and left for 30 minutes at room temperature. Each assay was done in duplicate. Sample A caused complete haemagglutination until dilution 320. Sample B did the same until dilution 80. This meant that these dilution had just enough viruses to agglutinate the blood cells present. Now

the number of red blood cells per ml of the reaction mixture was 10^7. Since one virus particle is capable of agglutinating two blood cells cimultaneously, the number of viruses required for agglutinating 10^7 cells would 5 x 10^6. Therefore, sample A had 320 x 5 x 10^6 virus particles and sample B, 80 x 50 x 10^6 particles. However, the population are expressed simply in haemagglutinin units. Accordingly, samples, A and B had 320 and 80 units respectively.

iii. Viral enzyme activity

Some animal virus particles contain polymerases which can be assayed by mixing permeabilized particles with radioactively labeled precursors and calculating the incorporation of radioactivity into nucleic acid. Such type of assays are most frequently used for retroviruses, many of which donot transform cells or form plaques. For example, the reverse transcriptase of virus particles is assayed by mixing cell culture supernatants with a mild detergent, a poly (rc) template and an oligo (dG) primer, and [α- ^{32}P] dGTP. If reverse transcriptase is present, a radioactive product will be produced by priming on the template, which can be detected by precipitation or bound to filter and quantified. This assay is not useful for enumerating the number of infectious or noninfectious particles. Nevertheless, since the amount of enzymatic activity is proportional to particle number, the assay does allow rapid tracking of virus production in the course of an infection.

3. ASSAYS BASED ON SEROLOGY

Antigen antibody interactions form the basis for serological assay methods. Assays have been designed to detect the presence of viral antigens with defined antibodies, or antibodies against a virus with the use of a defined antigen. Some of the serological techniques used for viral assay are:

i. Virus neutralization

Neutralization tests are used to determine the degree of infectivity of a virus or to detect new serotypes of viruses. Normally, the infectivity of a known type of virus would be neutralized by specific antibodies effective against it. Neutralization assays are usually conducted by mixing dilutions of serum or monoclonal antibody with virus. After incubation, assayed for remaining infectivity. Cultured cells, embryonated eggs or animals are used to detect the cytopathogenicity. The end point is defined as the highest dilution of antiserum that inhibits the development of cytopathic effect in cultured cells, by at least 50%. Different serotypes of the same virus often exhibit different dilution end points.

ii. Hemagglutination inhibition (HI)

The principle of this method is that antibodies against viral proteins with hemagglutination activity can block the ability of the virus to bind to red blood cells. In this assay, dilutions of antiserum are incubated with virus, and erythrocytes are added. After incubation, the HI titre is read as the highest dilution of antiserum that inhibits hemagglutination. This test is sensitive, simple, inexpensive and rapid. It can be used to detect antibodies to viral hemagglutinin in animal and human sera.

iii. Complement fixation

Complement fixation assay can be used to detect the presence of antibodies against a virus in the serum. The interaction of viral antigen and antibody can cause complement fixation, which leads to membrane lysis. Generally, RBC are used to demonstrate membrane lysis. In this assay for complement fixing antibodies, samples of patient sera are heated to inactivate endogenous complement and then incubated with preparations of viral antigen and a standardized quantity of guinea pig complement. If there is antigen-antibody interaction, complement fixation will occur, which is detected by adding sheep RBC that have been coated with rabbit anti-RBC antibodies. In case of complement fixation, the RBC will remain intact, if the complement is not fixed, the RBC will be lysed by the action of free complement. Complement fixation assays are not as sensitive as neutrali-zation or HI test. Nevertheless, it is often the first assay performed on sera from infected patients to identify the group to which infecting virus belongs.

iv. Immunostaining

The presence of either viruses or their protein components can be visualized with the help of antibodies which bind to antigens of viruses. This binding can be detected with the help of indicator molecules such as fluorescent dyes (fluorescein or rhodamine) or enzymes such as alkaline phosphatase, horse radish peroxidase and B-galactosidase. There are two types of stainings: (i) direct immunostaining, and (ii) indirect immuno-staining. In direct immunostaining, an antibody directed against viral antigen is directly coupled to an indicator. In indirect immunostaining a second antibody is coupled to the indicator. This is more sensitive than direct immunostaining because multiples of second antibody bind to the first antibody and thus amplify the signal from indicator. The reaction mixtures are observed under fluorescent microscope in case of fluorescent dyes or in light microscope for coloured product in the case of enzymes.

Immunostaining technique offers wide applications in both research and diagnosis. It is used in the research laboratory in locating the specific prote-ins, monitoring the expression of viral proteins, determining the effects of mutation on protein expression and investigating the sites of virus replica-tion in animal hosts. In clinical laboratories immunofluorescence assays with nasal swabs or washes are routinely utilized to diagnose infections with respiratory syncytial virus, influenza virus, measles virus and adenovirus.

v. Immunoprecipitation or immunoblotting

Immunoprecipitation assays are based on the principle of interaction of specific antiviral antibodies with viral proteins in solubilized extracts of infected cells or tissues. In recent years, immunoprecipitation in liquid state is largely replaced by precipition reactions in agarose gel. Immuno precipitation can establish many properties of viral proteins, such as protein types present in virions as well as infected cells.

Immunoblotting, also known as western blot analysis, depends on the reaction of an antibody with a viral protein. However, it is conducted on a different format, immobilized on a thin membrane. First, proteins are fractionated by electrophoresis and then transferred to a thin synthetic membrane that has a strong affinity for proteins. Later proteins bound to the membrane (nitrocellulose) are detected by immunostaining. The main advantage of this technique is that it does not require labeling proteins and thus can be conducted with tissues and organs as well as with cultured cells.

vi. Enzyme – linked immunosorbent assay (ELISA)

ELISA is a very popular technique today. It was first described simulta-neously and independently by Van Weeman and Schuurs, and Engvall and Perlman in 1971. This technique is a very sensitive method for detection and quantification of antigens and antibodies and combines the specificity of enzymatic and immunological reactions. In this, a suitable enzyme is used as a label. Ideally, the enzyme substrate is colourless and the product is coloured so that it can be easily detected and measured. Both, Ag and Ab can be enzyme labeled. ELISA is a very versatile technique. Several versions of ELISA are now available to meet the different types of requirements.

ELISA shows high sensitivity and high specificity. It requires small amounts of reagents and is rapid. Both qualitative and quantitative detection and measurement and easy visual detection are possible. The results are reproducible. It can be automated and portable kits are available. Both monoclonal and polyclonal Abs can be used. A large number of samples can be screened at a time. ELISA kits are now widely used in the epidemiological studies of plant and animal viral diseases. Preliminary diagnosis of HIV/AIDS at field level is routinely carried out using ELISA kits.

4. ASSAYS BASED ON VIRAL NUCLEIC ACIDS

The advent of DNA based markers, revolutionized the detection and characterization of viruses. With these marker techniques, it has now become possible to detect viral DNA in host cell. Hybridization techniques such as Southern blot, Northern blot, *in situ* hybridization are widely employed in detection and quantification of viruses. In Southern blot hybridization analysis, DNA preparations are digested with restriction endonucleases, fractionated by gel electrophoresis, denatured within a gel and transferred to a membrane to which they bind strongly. The bound DNAs are then detected by hybridization with an isotopically labeled viral nucleic acid sequence. Northern blot hybridization analysis is used to detect RNA in a similar manner. Nucleic acid hybridization probes can also be used to locate viral genomes or transcripts in sections of cells, tissues by a technique known as *in situ* hybridization.

The polymerase chain reaction (PCR) has also found wide application in the detection of viral nucleic acids. Specific oligonucleotides are used to amplify viral DNA sequences from infected cells or tissues. PCR reaction techniques have also allowed the identification of new viruses associated with human diseases.

5. COMPARISON OF DIFFERENT ASSAY PROCEDURES

The various methods of assay have different sensitivities and measure different properties. The focal assay methods are most satisfactory for their high efficiency combined with simplicity, reproducibility and economy. Assays based on infectivity are also more sensitive than those based on physical and chemical properties. Chemical and physical technique titrate not only infectious but also non-infectious virions. These methods, therefore, can be useful for studies requiring measurement of the total number of viral particles. Hemagglutination and immunologic methods can also titrate only soluble components.

The ratio of the number of viral particles, as determined by electron microscopy, to the number of infectious units measures the efficiency of infection, which varies widely among different viruses, and even for the same virus assayed in different hosts.

Further reading

Davis, BD *et al.* (1990). Microbiology 4th ed. Philadelphia; JB Lippincott.

Flint, SJ *et al.* (2004). Principles of Virology ASM Press, Washington, DC

Habel, K and Salzman NP (eds.) (1969). Fundamental techniques in virology. New York : Academic Press.

Hull, R (2002). Matthew's Plant Virology, Academic Press, New York.

Kouchler, RJ (1977). Biochemical methods in cell culture and virology. Stroudsberg, PA : Dowden, Hutchinson and Ross.

Knipe, DM *et al.* (eds.) (2001). Fields Virology Vol. I. Lippincott – Williams & Wilkins.

Mahy, BWJ and Kangro, HO (1996). Virology Methods Manual, San Diego, Academic Press.

Maramorosch, K and Kaprowski, H (eds.) (1967). Methods in Virology. New York : Academic Press.

7

TAXONOMY

Viral taxonomy, like the taxonomy of other living organisms, deals with three aspects viz., identification, nomenclature and classification of the viruses. The discipline of virus taxonomy is relatively young about 50 years old. The need for viral taxonomy was felt a long before enough was known about the viruses themselves to form the basis for a rational system. The problems with viral taxonomy are many mainly due to their submicroscopic structure, host diversity, biological properties, need for sophisticated instrumentation and ultra modern techniques, changing priorities of characters for classification etc. Critical issues and uncompromising debates and controversies surfaced when attempts were made to equate the virus taxonomy on par with other groups of microorganisms. Equally, if not more serious, concern is about the application of species concept for viruses. The time tested species concept of plants and animals and other microbial groups cannot be applied here for obvious reasons. Another critical issue was whether the classification of viruses should consider virus properties in a monothetical hierarchial fashion, or in a polythetical heirarchial fashion. Finally, much talked about and pampered phylogenetic arrangement of the taxa of viruses appears to be a far cry at this juncture. Thus, viral taxonomy is still an evolving field and this chapter endeavours to summarize the present state of art including important historical land marks that shaped the present system of viral taxonomy.

The number of distinct viruses that have been described cannot be ascertained with any precision, but the total must be several thousands. There are over two thousand descriptions of viruses infecting bacteria alone. Recently, **International Committee on Taxonomy of Viruses (ICTV)** has identified 1550 well characterized species, 2404 tentative species grouped in 243 genera 69 families and 3 orders.

1. EARLY SYSTEMS OF CLASSIFICATION

In the early 1900s, viruses were classified as distinct from other organisms by virtue of their ability to pass through unglazed porcelain filters known to retain the smallest of bacteria. As increasing numbers of filterable agents became recognized, they were distinguished from each other by the only measurable properties available, namely the disease or symptoms caused in an infected organism. Thus, animal viruses that caused liver diseases were grouped together as hepatitis viruses, and viruses that caused mottling in plants were grouped together as mosaic viruses. Johnson (1927), for the first time stressed the need for using some criteria other than disease symptoms and host for identifying viruses. He suggested that a virus should be named by adding the word "virus" and a number to the common name for the host in which it was first found e.g., tobacco virus I for TMV.

Host based classification schemes have also been proposed at one stage. Thus, the viruses infecting plants, animals and bacteria sometimes are placed into three separate classification schemes. There are viruses, however, that infect nearly every species of plant or animal including algae, mushrooms, many different insects, bacteria, and most vertebrates and higher plants. Many of these viruses have great species specificity, so that following this scheme would require a separate classification of viruses for each species of plant or animal.

Bergey's Manual of Determinative Bacteriology (7th edition) considered viruses to be bacteria like microorganisms and an ordinal rank was granted for them, the order **virales,** (Holmes 1948). This order was further sub-divided into three sub-orders, namely the **phaginae** (infecting bacteria), **phytophaginae** (infecting plants) and **zoophaginae** (infecting animals). In subsequent editions, however, viruses were excluded from any such consideration. In 1930s an explosion of technology spawned a description of the physical properties of many viruses, providing numerous new characteristics for distinguishing viruses one from another. The technologies included procedures for purification of viruses, biochemical characterization of purified virions, serology and electron microscopy. By the 1960s, a great wealth of information on different aspects of viruses became available. Thus a necessity was felt that an organized effort is required to classify and name the viruses.

Cooper (1961) originally proposed that viruses might best be classified in terms of their chemical composition. First, he placed all animal viruses into two major divisions, depending on whether they contained RNA or DNA as their nucleic acid. The next subdivision depended on whether they contained lipid. Finally, he considered their size and serological reactions to place them into different groups. He intended this classification scheme to apply only to the animal viruses. Nevertheless, Cooper's approach opened up new vistas for the classification of viruses. Following the Cooper's lead many classification schemes have followed that divided the viruses into those with RNA and those with DNA (Lwoff *et al.,* 1962; Hamparian *et al.,* 1963; Becker, 1966; Bellet, 1967 and Gibbs, 1969).

i. The LHT System

Among all these, the classification scheme proposed by Andrew Lwoff and his colleagues, Horne and Tournier (Lwoff *et al.,* 1962) (the LHT system) merits somewhat detailed explanation on many accounts. It was the first comprehensive scheme for the classification of all viruses (bacterial, plant and animal) under the classical Linnaen heirarchial system consisting of phylum, class, order, family, genus, species. A major deviation of LHT system from earlier systems was that viruses should be grouped according to their shared properties rather than the properties of the cells or organisms they infect. A second deviation was a focus on the nucleic acid genome as the primary criterion for classification. This system forms a basis for all modern classifications. LHT system delineated four characteristics to be used in the classification of all viruses.

1. Nature of the nucleic acid in the virion (DNA or RNA).; 2. Symmetry of the capsid; 3. Presence or absence of envelope; 4. Dimensions of virion and capsid

Andrew Michel Lwoff (1902-1994), Nobel Prize, 1965

LHT scheme of virus classification upto family level is presented in Fig. 7.1. The classification at generic and subgeneric levels is based on such determinants as (i) primary structure of nucleic acid (base composition, base sequence, homology), (ii) molecular weight, antigenic properties of the capsid, (iii) organization of the nucleocapsid, (iv) specificity for host, (v) mode of development, (vi) virulence and (vii) clinical effects.

The LHT system, at that time, was hailed as the best system and evoked a considerable interest among contemporary virologists. The merits of this classification are: (i) it was the first attempt to classify viruses as a whole, (ii) it is based on the structure and composition of the

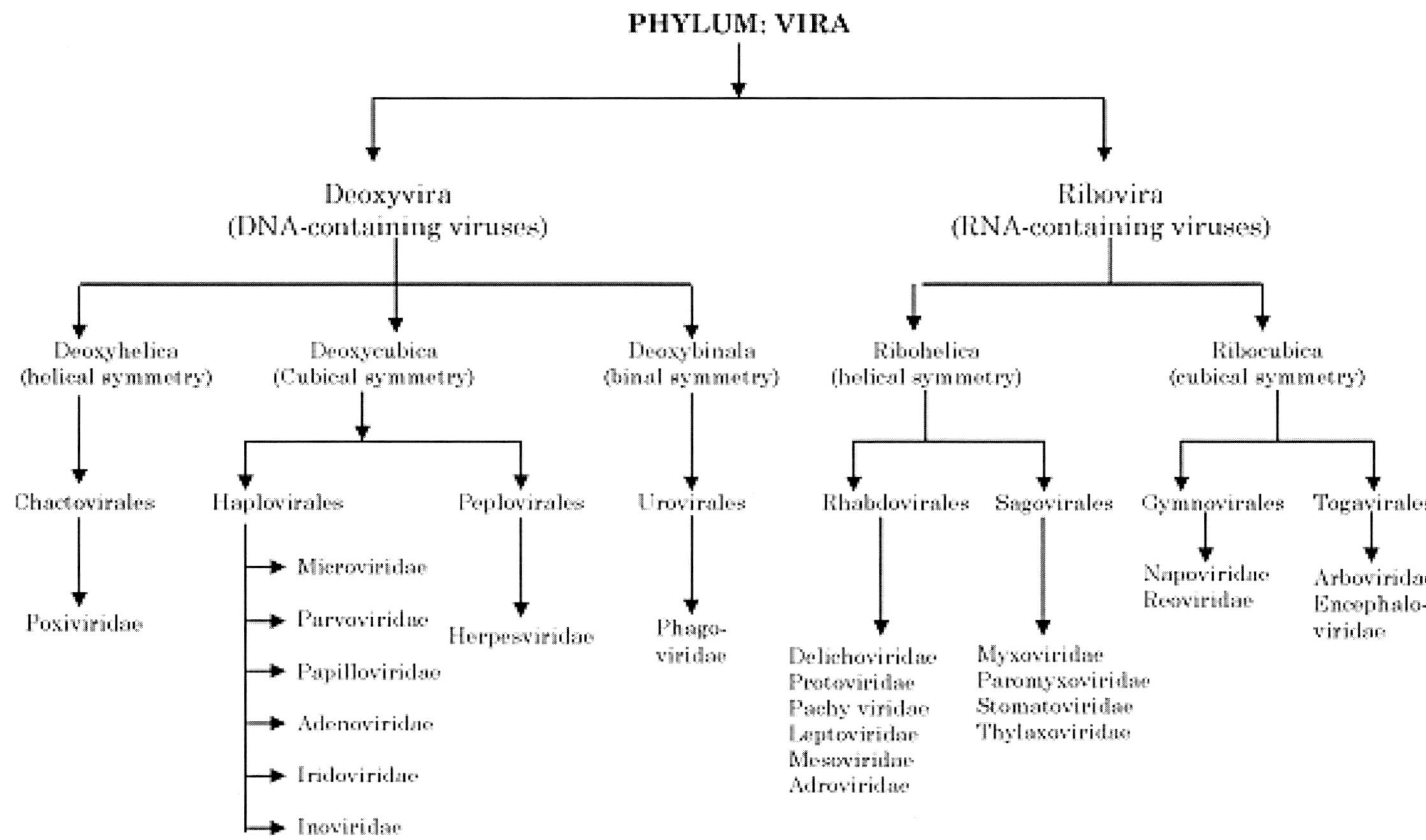

Fig. 7.1. The LHT scheme of classification of viruses

viruses and (iii) it attempted to classify on the basis of correlation amongst characters. Nevertheless, the scheme has also been widely criticized. Wildey (1962) opined that viruses were a heterogenous assemblage of entities and, therefore, any classification of viruses in a heirarchial pattern is likely to be artificial. Gibbs (1969) dubbed the system as arbitrary as it gave more weightage to a few chemical and physical characters whose selection is not based on proper evaluation. Bellet (1967) felt that the LHT system was artificial as it did not attempt to take affinities amongst different groups into consideration.

Despite the criticisms, the LHT system has been drawing keen attention in recent years. Not withstanding its limitations, till to date, the system remains the comprehensive. The Provisional Committee on Nomenclature Viruses (PCNV) (1965), an adhoc body constituted by IUMS has recommended this classification at that time. On any account, it is a significant contribution towards virus taxonomy.

ii. Bellet's system

Bellet (1967) proposed another system for classifying viruses. It was based on two physico-chemical characteristics namely the molecular weight and the percentage of G+C content of the viral nucleic acids. Additionally, antigenic and serological reactions and phenotypic properties were also taken into consideration. In this system, the viruses were divided into two major subgroups namely, the viruses with single stranded and those with double stranded nucleic acids. The coefficient similarities of the various virus groups belonging to these major subgroups were then computed. More closely related subgroups were then clamped together. However, this system restricts itself mainly to animal viruses and can, therefore, be only of limited use.

iii. Gibb's system

Gibbs (1969) proposed a system of classification for plant viruses based on (i) the shape of the capsid i.e., whether isometric or anisometric, (ii) the mode of transmission, (iii) the type of vector, (iv) the symptoms of infection, and (v) the nature of the accessory particles. Accordingly, he grouped some 135 known plant viruses into 6 broad groups or clusters. These broad groups were further categorized into subgroups showing a higher degree of inclusiveness. Gibb's system helped in categorizing plant viruses on the basis of information available at that time. Despite the criticisms, some of the groups formulated by Gibb's have been accepted as stable taxonomic groups by ICNV (Gibbs, 1987).

As for as identification of viruses is concerned, Gibbs and his associates (1966) proposed that viruses could be technically identified on the basis of certain approved parameters. They suggested that the following eight characters could be used for the purpose.

i. The nucleic acid type

ii. The number of strands in a nucleic acid

iii. Nucleic acid molecular weight
iv. Percentage of nucleic acid in a virion
v. The form of the particle
vi. The form of the nucleic acid
vii. The host and
viii. The vector

They further suggested that these parameters be defined by abbreviations and that the abbreviations can be used in describing a virus in the form of formulae. They renamed such formulae as **cryptograms**. The cryptogram besides giving the reader an immediate summary of a virus specific characteristic also helps to relate the particular virus to other with similar properties. An example of a cryptogram and an explanation of its use is given in Table 7.1. However, it is not now being used in plant virology.

Table 7.1. Key to plant virus cryptograms

Examples of a cryptogram				
Tobacco mosaic virus (TMV) =	R/1 (1st)	2/5 (2nd)	E/E (3rd)	S/O (4th)

Key to terms

1st term :	type of nucleic acid / number of nucleic acid strands			
	R = RNA	1 = single		
	D = DNA	2 = double		
2nd term :	Molecular weight of nucleic acid in millions/% of nucleic acid in infective particle			
3rd term :	outline of particle shape / outline of nuclear capsid			
	S = spherical (isometric) E = elongate (rod shaped) B = bacilliform			
4th term :	type of host infected / type of vector			
	B = bacterium	Ap	=	aphid
	F = fungus	Au	=	leafhopper
	I = invertebrate	Cl	=	beetle
	S = seed plant (angiosperm)	Fu	=	fungus
		Ne	=	nematode
		Th	=	thrip
		W	=	whitefly
		O	=	spreads without vector
		Se	=	seed transmitted

2. INTERNATIONAL COMMITTEE ON TAXONOMY OF VIRUSES (ICTV)

The requirement for a sound classification system for viruses based on well defined characteristics, became imperative during 1960s as more and more information on individual viruses has accumulated. During the 9th International Congress for Microbiology in Moscow in 1966, the International Union of Microbiological Societies (IUMS) formally established a permanent **International Committee on Nomenclature of Viruses (ICNV)**. The ICNV initially was constituted so that most countries would be represented by atleast one member. Several meetings of this committee and a general meeting was held during the congress, some of these were stormy. One group led by Andrew Lwoff wished to adopt an all embracing taxonomic scheme for the classification of viruses into subphyla, classes, orders, suborders and families. This proposal was strongly opposed by a group of English virologists (A.J. Gibbs, B.D. Harrison, D.H. Watson and P. Wildy). There was also conflict over matters of nomenclature. One group suggested a list of conserved genera with type species having latinized binomial names. Even this proposal was opposed by the English group. Finally, the heirarchial classification proposed originally by Lwoff was ignored by ICNV.

By the end of the congress, the ICNV had adopted a code of nomencla-ture and elaborated a set of rules that laid the basis for progress in the years ahead. Important features were that the code of bacterial nomenclature would not be applied to viruses; that nomenclature would be international and applied to all viruses; that the law of priority would not be observed; that no persons name would be used; and that an effort would be made towards a latinized binomial nomenclature. Thanks to Dr. P. Wildy, the first president, under whose leadership the turbulent period passed, an unanimity was reached and a constructive work began.

In 1973, the ICNV changed its name to confirm more closely its actual function and became the **International Committee for Taxonomy of Viruses (ICTV).** The statutes define objectives of the ICTV these are :

i. To develop and internationally agreed taxonomy for viruses.

ii. To develop internationally agreed names for virus taxa.

iii. To communicate taxonomic decitions to the international community of virologists.

iv. To maintain an index of agreed names of virus taxa and other mobile genetic elements (MGE) that are part of virosphere (hereafter termed as "viruses".)

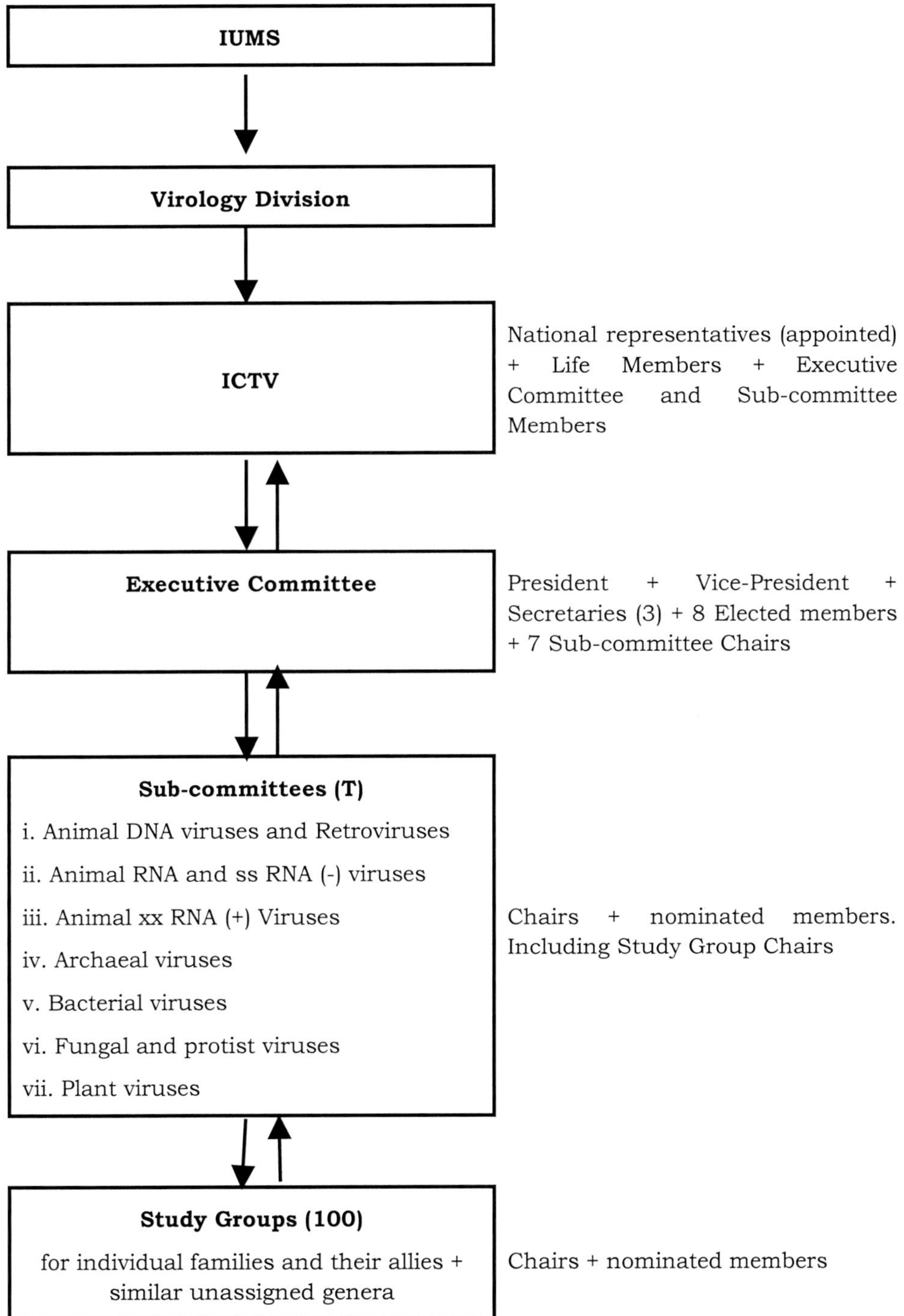

Fig. 7.2. Structure of the ICTV and its position in the organization of IUMS. The compositions of the ICTV Committees and Groups are indicated to the right.

The composition of ICTV is rather complex (Fig. 7.2). The committee operates through subcommittees and their specialized study groups. Five of these subcommittees address viruses affecting specific group of organisms viz., vertebrates, invertebrates, fungi, bacteria and plants. Recently a sixth subcommittee was constituted for virus to facilitate the management and distribution of virological data (ICTVdB) (Buchen and Osmond, 1997). Subsequently these sub-committees set up study groups composed of specialists from different countries working on particular family or proposed family of viruses. The objective for the study groups was to bring forward taxonomic proposals for discussion by the subcommittee. If the subcommittee approved a proposal, it will be forwarded to the executive committee. If approved by this committee, a taxonomic proposal will be finally put to the vote at a plenary meeting of the full ICTV. Individual virologists or groups of virologists are welcome to publish taxonomic proposals in **Archives of Virology**, the official journal of virology section of IUMS. This procedure although a lengthy and cumbersome, produces a stable taxonomic outcome, derived from the cooperative and coordinated efforts of panels of specialists rather than from the initiative of single individual, as is common to other scientific disciplines (Martelli, 1992). Summary of the officially approved viral taxonomy and nomenclature is updated and published every three years by the ICTV and provided to virologists with a complete taxonomic guide within a single volume. So far, nine reports were brought out with the latest in 2011. The current, 10th report (http://ictv.global/report) is being published on line, and is replacing the previous hard copy edition with a completely open access, continuously updated publication. The ICTV report has been made an authoritative source of data concerning virology.

The ICTV database of virus taxonomy

Once ratified, the newly approved taxonomy is entered in the ICTV taxonomic database. The database utilizes a relation schema built using Microsoft SQL Server. The tables store a controlled vocabulary used for specific fields where the possible data values remain fairly constant. These include taxonomy level (viz, species, genus, subfamily, family and order), possible taxonomic changes from the previous release (new, split, merge, move, merge, rename, and abolish), and virus genome type (e.g. dsDNA, ssDNA, dsRNA, ssRNA(+), ssRNA(–), etc.) that forms part of the annotation available for the member viruses of each species. Each record in the taxonomy table stores data on a single taxon including its name, numerical id, parent taxon, release number and date, most recent taxonomic change, filename of the proposal establishing the taxon, and a pointer to its previous record if a change has occurred in taxon data (such as moving a species to another genus). With this information, a complete history of every taxon can be reconstructed from the database.

Data stored in the taxonomy database can be accessed visually from the ICTV web site (http://ictv.global/virusTaxonomy.asp) using the taxonomy browser: a hierarchical series of expandable, tabular rows, with each row containing information on a single taxon. Each taxon is presented in its proper position in the taxonomic hierarchy. Each row is labeled with the number of lower-level taxa belonging to each higher-level taxon, and a link is provided to the complete history of each taxon. A search box is available to search the current, or current and historical taxonomy releases for any taxon that includes the text entered in the search box fach search result is linked to the identified taxon in the hierarchical taxonomy browser.

The complete taxonomy for any one release (the Master Species List, MSL) is also available for download as an Excel spreadsheet (http://ictv.global/msl.asp). These spreadsheets are available for every release back to 2008.

3. THE INTERNATIONAL CODE OF NOMENCLATURE

One of the objectives of the ICTV is to establish internationally agreed names for taxonomic groups of viruses. Sustained efforts were being made for the last 30 years to develop a uniform code of nomenclature. International Congress of Microbiology held in Mascow,1966 had adopted a code of nomenclature and elaborated a set of rules. Subsequently, this code was amended in 1981(Mathews, 1982,1983). The rules of this nomenclature are listed here as they form the basis for understanding the changes brought out subsequently.

i. The code of bacterial nomenclature shall not be applied to viruses.

ii. The nomenclature shall be international.

iii. An effort will be made towards latinized nomenclature.

iv. Existing latinized names shall be retained wherever possible.

v. The law of priority shall not be observed.

vi. No person name shall be used.

vii. A virus species is a concept that will normally be represented by a cluster of strains from a variety of sources, or a population of strains from a particular source, which have in common a set or pattern of correlating stable properties that separate the cluster from other cluster of strains.

viii. The genus name and species epithet together with strain designation must give an unambiguous identification of the virus.

ix. The species epithet must follow the genus name and be placed before the designation of strain, variant or serotype.

x. Artificially created laboratory hybrids between different viruses will not be given taxonomic consideration.

xi. The approval by ICTV of newly proposed species names and type species will proceed in two stages. In the first stage provisional approval may be given. Provisionally approved proposals will be published in an ICTV report. In the second stage after 3-years waiting period, the proposals may receive the definite approval of the ICTV.

xii. The genus is a group of species sharing certain common characters.

xiii. The ending of the name of a viral genus is___________ *virus*. E.g., Adenovirus, Parvovirus

xiv. The family is a group of genera with common characters, and the ending of the name of a viral family is ________ *viridae*. E.g., Poxvirdae, Retroviridae, Microviridae

xv. Approval of new family must be linked to approval of a type genus; approval of new genus must be linked to approval of a type species.

Recently, the Executive Committee of the ICTV has developed an International code of nomenclature based on adhoc rules (Murphy *et al.*, 1995; Mayo, 1996). This code has the formal approval of the Virology Division of the IUMS. The main principles of the rules are that a new name should be distinctive (Rule, 14), easy to remember (Rule 12), be free of association with any individual's name (Rule 11) and avoid absurdity or offence in any language (Rule, 19). Rule 18 is devised to exclude names that seem to convey meaning which might exclude legitimate members of the taxon or which would seem to include viruses that are classified in different taxa. Issues related to naming of taxa are regulated by rules 8 to 20 of the code. Some more salient features of the recent code of nomenclature are

1. Four principal taxa officially recognized by ICTV are : species, genus, family and order.

   Intermediate taxa, such as sub-family are used in some instances but only when their use solves difficult problem. There is no official recognition for viruses below the level of species. The suffixes for different taxa are

Order	-	*virales* (E.g., Caudovirales)
Family	-	*viridae* (E.g., Geminiviridae, Myoviridae)
Sub family	-	*virinae* (E.g., Parvovirinae, Densovirinae)
Genus	-	*virus* (E.g., Geminivirus, Rhabdovirus)

2. Taxon names when used formally are captalized and italicized.
3. A virus species is a polythetic class of viruses that constitutes a replicating lineage and occupies a particular ecological niche (van Regenmortel, 1990).
4. Species names are to be printed in italics and have the first letter of the first word capitalized. Other words are not capitalized unless they are proper nouns (Rule 3. 40).

   E.g., Tobacco mosaic virus, Poliovirus,

   Murray River encephalitis virus.
5. The naming of viruses, now virus species has been followed in different traditions in different branches of virology.
   a. Many bacterial viruses have names consisting of combinations of letter and number codes. E.g., T_4, λ, ϕX174, Mu, PM_2, M13.
   b. Apart from descriptive words, number series are also used in some fields of vertebrate virology. E.g., Picornavirus, SV40.
   c. Plant virus names are usually of the form 'host name' plus 'symptom name' plus virus e.g., tobacco mosaic virus. The current rules forbid the 'host name' plus 'virus' (Rule 23).

Note : *Over the years, several changes have been suggested in the nomenclature and classification systems of viruses. The detailed International Code of Virus Classification and Nomenclature (ICVCN) released in March, 2021 is presented in Annexure IIA.*

4. CRITERIA USED IN VIRUS CLASSIFICATION

The taxonomic method adopted for use in virology is polythetic, meaning that any given virus group is described using a collection of individual properties. The description of a virus group is nonsystematic in that there exists no fixed list of properties that must be considered for all viruses, and no strict formula for the ordered consideration of properties. Instead, a set of properties describing given virus is simply compared with other viruses described in a similar fashion to formulate rational groupings. Viruses exhibit a great diversity in terms of morphology, genome structure, mode of infection, host range, tissue tropism, pathogenesis etc. Each of these properties can be used to place viruses into groups.

I. Virion properties

A. Morphological properties of virions.

i. size, ii. shape, iii. presence or absence of envelope, iv. structure and symmetry of capsid.

B. Physical properties of virions.

i. Molecular mass, ii. buoyant density, iii. sedimentation coefficient, iv. pH stability, v. thermal stability, vi. cation (Mg^{2+}, Mn^{2+}, Ca^{2+}) stability, vii. solvent stability, viii. detergent stability, ix. radiation stability.

C. Properties of the genome.

i. Types of nucleic acid (DNA or RNA), ii. strandedness (ds or ss), iii. linear or circular, iv. sense : positive or negative or ambisense, v. number of segments, vi. size of genome or genome segments, vii. presence or absence and type of 5' terminal cap, viii. presence or absence of 5' terminal covalently-linked polypeptide, ix. presence or absence of 3' terminal poly(A) tract, x. nucleotide sequence comparison.

D. Properties of proteins

i. Number, ii. size, iii. functional activity (especially virion transcriptase, virion reverse transcriptase, virion hemagglutinin, virion neuraminidase, virion fusion protein, iv. amino acid sequence comparison.

E. Lipids

i. Presence or absence, ii. nature

F. Carbohydrates

i. Presence or absence, ii. nature

II. Genome organization and replication

i. Genome organization, ii. strategy of replication of nucleic acid, iii. characteristics of transcription, iv. characteristics of translation and post-translational processing, v. sites of accumulation of virion proteins, site of assembly, site of maturation and release, vi. cytopathology, inclusion body formation.

III. Antigenic properties

i. Serological relationships, ii. mapping epitopes.

IV. Biological properties

i. Host range, natural and experimental, ii. pathogenicity association with disease, iii. tissue tropisms, iv. mode of transmission in nature, v. vector relationships, vii. geographic distribution.

5. MODERN SYSTEMS OF CLASSIFICATION

Baltimore scheme

Classifications based on phenetic characters present certain drawbacks. For example, many viruses with similar morphology have quite different mechanisms of replication and may cause a wide range of different diseases in their respective hosts. Further, they do not provide a good basis for unifying discussion of virus replication process.

David Baltimore (1938-); Nobel Prize,1975

During the past 35 years or so there has been an enormous increase in knowledge of the molecular biology of viruses and cells. Among the most significant advances has been the elucidation of pathways by which information encoded in genomes is expressed. It is a well known fact that genes of cells are encoded in their DNA but that cells express this information via messenger RNAs (mRNAs) which are translated in the cytoplasm by ribosomes and associated machinery. Viruses do not contain the molecules necessary to translate mRNA and relay on the host cell to provide these. They must, therefore, synthesize mRNAs that are recognized by the host cell ribosomes. The central role of the translational machinery and of the importance of mRNA molecules in the programming of viral protein synthesis attracted the attention of **David**

Baltimore, a Nobel laureate, and inspired him to propose a classification, popularly known as Baltimore scheme, based on the fundamental importance of mRNA in the replication cycle of the virus. At this stage, it is important to have clear idea about positive or negative sense of mRNA. By convention, an mRNA is designated as positive (or plus) sense RNA. Strands of viral DNA and RNA that are complementary to the mRNA are designated as negative (or minus) sense and those that have the same sequence are termed as positive sense (Fig. 7.3). In this way, Baltimore divided all viruses, irrespective of their host, into 7 classes and each is commonly referred to by the nature of the virus genomes in that class. (Fig. 7.4)

A	mRNA	5'... GAC UCG AGC ... 3'
B	+ strand DNA	5' ... GAC TCG AGC ... 3'
C	- strand DNA	3' ... CTG AGC TCG ... 5'
D	+ strand RNA	5' ... GAC UCG AGC ... 3'
E	- strand RNA	3' ... CUG AGC UCG ... 5'

Fig. 7.3. How to designate DNA or RNA genome as plus (+) or minus (-).

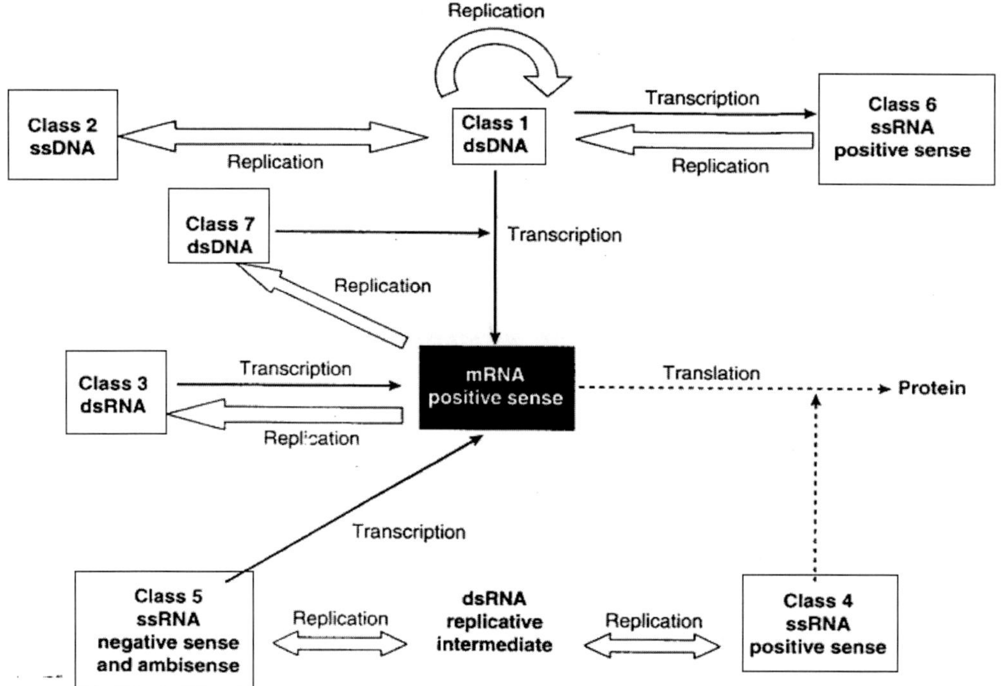

Fig. 7.4. The revised Baltimore scheme : solid arrows transcription; open arrows, replication; dashed arrows, translation.

Class I : Viruses with double-stranded (ds) DNA genomes

Viruses included in this class possess ds DNA. mRNAs may come from either strand. Transcription can occur using a process, similar to that found in the host cells.

Class II : Viruses with single-stranded (ss) DNA genomes

It includes viruses that have single-stranded (ss) DNA. The DNA can be positive or negative sense. The DNA is converted to double-stranded replicative form (Rf dsDNA) before the synthesis of mRNA can proceed.

Class III : Viruses with dsRNA genomes

Most viruses of this type have segmented genomes. Like dsDNA viruses, mRNA is synthesized from one strand only. RNA dependent RNA polymerase is encoded by the virus genome, packaged in the viron, and carried into the cell by the virus to initiate the infectious process.

Class IV : Viruses with ssRNA genomes with positive (+) sense

These viruses synthesize a complimentary strand, generating atleast a partially dsRNA intermediate. The enzyme required for RNA synthesis is virus encoded. However, it is synthesized after infection is initiated.

Class V : Viruses with ssRNA genomes with negative (-) sense and ambisense

Synthesis of mRNA occurs directly from the genomic RNA that requires a novel virus encoded enzyme. Some viruses also use the newly synthesized 'antigenome' RNA strand as a template for production of an mRNA. These are referred to as 'ambisense' viruses.

Class VI : Viruses with ssRNA genomes with positive sense

In genomic structure these viruses similar to class 4, but generate a dsDNA intermediate as prelude to replication. This is carried out by a virus-coded enzyme, reverse transcriptase which is carried in the virion.

Class VII : Viruses with dsDNA

These viruses are termed **reversiviruses**, that have been transferred from Class 1 into a new class 7. This is based on their novel replication strategy via a positive-sense ssRNA intermediate and a reverse transcriptase.

A critical appraisal of the Baltimore scheme reveals its both strengths and weaknesses. A particular strength is that assignment to a class is based on fundamental, unchanging characteristics of virus. Once assigned to a class, certain predictions about the molecular processes of nucleic acid synthesis can be made. Infact, ICTV partially adopted this scheme together with other parameters to place viruses into families and genera. A major weakness is that, although it brings together viruses with similarities of replication mechanism, the scheme takes no account of their biological properties. For example, bacteriophage T_2 and variola virus are grouped together in class I, although they are totally dissimilar in structure and biology. Similarly, the identification of a positive-sense RNA genome is not sufficient to classify the virus unambiguously because viruses of classes 4 and 6 have similar forms of genome nucleic acids.

Realising these drawbacks, Dimmock *et al.* (2001) felt that a number of classification schemes must be used together for full understanding of the relationships between viruses. Further, on the ground that the principles by which viruses are classified are yet to be finalized and assignment of viruses to families or genera is not yet complete, for the sake of convenience, they have divided the viruses into six groups based on the host they infect.

1. Viruses of vertebrate and invertebrate animals.
2. Viruses that multiply in plants.
3. Viruses multiplying in algae, fungi and protozoa.
4. Viruses (phages) multiply in *Archaea*, bacteria, mycoplasma and spiroplasma.
5. Statellite viruses and satellite nucleic acids of animals, plants, fungi and bacteria.
6. Viroids (unclassified as make no mRNA).

The groups were further classified upto family and genus level based on Baltimore scheme.

6. ICTV CLASSIFICATION OF VIRUSES (2020 Release)

Approved in EC meeting 52 (Online); Email ratification March 2021

Realms – 6; Kingdoms – 10; Phyla – 17; subphyla – 2; Classes –39; Orders – 59 Sub orders – 8; Families – 189; Sub families 136; Genera – 2224; subgenera – 70; Species – 9110

Realm I : Adnaviria

i. Filamentous virion

ii. ds DNA genome

iii. Genome codes for dimeric capsid protein with the SIR2-like fold

Kingdom-1; Phylum-1; Class-1; Orders-2; Families-3; Genera-13; Species-31

Realm II: Daplodnaviria

i. ds DNA genome

ii. Genome encodes a major capsid protein containing the HK97 fold (HK97-MCP) and small terminase subunits required for DNA packing into capsids.

Kingdom-1; Phyla-2; Classes-2; Orders-2; Families-17; Subfamilies-17; Genera-950; Species-2944

Realm III: Monodnaviria

i. Genome ss DNA

ii. Genome encodes a rolling-circle replication endonuclease of the HUH superfamily (or derived from such a virus)

Kingdom-4; Phyla-5; Classes-8; Order-13; Family-17; Subfamilies-7; Genera-167; Species-1416

Realm IV: Riboviria (all RNA) ds, SS+, SS-

i. Genome RNA

ii. Genome codes for RNA dependant RNA polymerase (RdRp)

Kingdom-2; Phyla-6; Subphyla-2; Classes-21; Orders-29; Suborders-8; Families-111; Subfamilies-46; Genera-934; Subgenera-70; Species-3850

Realm V: Ribozyria

i. Includes Hepatitis D virus (HAD) and its immediate relatives

ii. Genome Ribozyme

Family-1; Genera-8; Species-15

Realm VI: Varidnaviria (various DNA viruses)

i. Genome DNA

ii. Genome encodes roll major capsid proteins (MCP) and containing the vertical jelly fold forming pseudo hexameric capsids

Kingdom-2; Phyla-3; Classes-6; Orders-12; Families-17; Subfamilies-4; Genera-63; Species-268

***The detailed classification is provided in Annexure I**

**** A complete taxonomy (the Maste Species List, MSL) is available at http://ictv.global/msl.asp.**

Further reading

Baltimore, D (1971). Expression of animal virus genomes. *Bacteriological Reviews,* 35: 235-241.

Bellet, AJD (1967). Preliminary classification of viruses based on quantitative comparisons of nucleic acids. *J. Virol.,* 1 : 245-259.

Dimmock, NJ, Easton, AJ and Leppard, KN (2001). Introduction to Modern Virology, 5th ed. Backwell Science.

Gibbs, AJ (1969). Notes on classification of viruses. *Advan. Vir. Res.,* 14 : 163-328.

Hamparaian, VV, MR Hilleman and A Ketler (1963). Contributions to characterization and classification of animal viruses. *Proc. Soc. Exptl. Biol. Med.,* 112 : 1040-1050.

Lwoff, A, Horne, R and Tournier, P (1962). A classification system of viruses. *Cold Spring Harbor Symposia Quant. Biol.,* 27 : 51-55.

Martelli, GP (1992). Classification and nomenclature of plant viruses : State of the art. *Plant Diseases,* 76(5) : 436-442.

Mathews, REF (1982). *Intervirology,* 17 : 1-199.

Mathews, REF (ed.) (1983). A critical appraisal of viral taxonomy, Boca Raton, Fla CRC Press, pp. 256.

Mayo, MA (1996). Recent revisions of the rules of viral classification and nomenclature. *Archives of Virology,* 141 : 2479-2484.

Mayo, MA and Pringle, CR (1997). Virus taxonomy. *J. Gen. Virol.,* 79 : 649-657.

Murphy, FA, Fauquet, CM, Bishop, DHL, Ghabrial, SA, Jarvis, AW, Martelli, GP, Mago, MA and Summer, MD (1995). Virus taxonomy, classification and nomenclature of viruses. Sixth Report of the International Committee on Taxonomy viruses, Springer, Wein New York, *Arch. Virol.* (Suppl. 10).

Pringle, CR (1999). Virus Taxonomy. *Arch. Virol.,* 144(2) : 421-429.

Van Regenmortel, MHV, Fauquet, CM, Bishop, DHL, Carstens, E, Ester, MK, Lemon, S, Maniloff, J, Mayo, MA, McGeoch, DJ, Pringle, CR and Wickner, R (1999). Virus taxonomy, classification and nomenclature of viruses. Seventh Report of the International Committee on Taxonomy of Viruses. Academic Press, New York.

Van Regenmortel, MHV (1990). Virus species a much overlooked but essential concept in virus classification. *Intervirology,* 31 : 241-254.

Gorbalenya *et al,* (2020). The new scope of virus taxonomy : partitioning the virosphere into 15 heirarchial ranks. *Nature Mircrobiology* 5 : 668-674

Lefkowitz Ed. *et al,* (2018) Virus taxonomy : the database of the international committee on Taxonomy of Viruses. *Nucleic Acids Research* 46 : D708-D717.

Van Regenmortel MHV (2016). Classes, taxa and categories in heirarchial virus classification : a review of current debates on definitions and names of virus species. *Bionomina* 10 : 1-21.

PART TWO

BACTERIOPHAGES

8

DISTRIBUTION AND CLASSIFICATION

Bacteriophages, or simply phages are viruses of prokaryotes including eubacteria and archaebacteria. They were discovered and described twice independently, first in 1915 by British pathologist **Frederick William Twort** and then in 1917 by the Canadian bacteriologist **Felix Hubert d'Herelle**. Since then a large number of phages, approximately 4600 isolates, have been described. They are known to infect 130 bacterial genera representing all major bacterial groups. Taxonomically, they belong to 1 order, 13 families and 30 genera (ICTV, 2005). All major morphological forms of viruses are found in phages. They include tailed, cuboidal filamentous and pleomorphic forms. Nucleic acid diversity is also greater than other viral groups. Replication cycles vary from simple multiplication to establishment of lysogeny i.e., equilibrium between phage and the host. Research on phages has contributed to the major advances in not only virology but also in emerging fields of biology like molecular biology, molecular genetics and recombinant DNA technology. Phages are now routinely used in phage typing of bacteria and genetic engineering. The high hopes that were initially pinned on phage therapy of bacterial diseases were later proved to be disappointing. Some phages usually create a problem in fermentation industry as they destroy the valuable cultures and fermentation process.

1. DISTRIBUTION AND ECOLOGY

Phages are known to infect more than 130 bacterial genera distributed in all major bacterial groups. They are found in aerobes, anaerobes, actino-mycetes, archaebacteria, cyanobacteria and other phototrophs, endospore formers; appendaged, budding, gliding and sheathered bacteria; spirochetes; mycoplasmas and chlamydias. Tailed phages predominate in all bacterial groups except in mycoplasmas. In archaebacteria, they have been reported in halobacteria and methanogens but not in extreme thermophiles. Siphoviridae members are frequent in actinomycetes, coryneforms, lactococci and streptococci. Phages belonging to myoviruses and podoviruses are common in enterobacteria, pseudomonads, bacilli and clostridia.

Felix d' Herelle

As obligate parasites, phages are distributed in the same habitats as their hosts. In nature, they occur in a wide variety of habitats ranging from Icelandic solfataras to fish sauce, fetal calf serum, and cooling towers of thermal power stations. They are found on the surfaces and in normal and pathological products of humans and animals, on plants and in food, soil, air and water especially sewage. Body cavities with large bacterial populations such as intestines, rumen are extremely rich in phages. According to the nature of habitat the phages may be acido-, alkali- halo-, psychro- or thermophilic. However, these properties are not linked to particular phage groups but rather appear as individual adaptations.

2. PHYSIOLOGY

Phages for their survival and replication establish two types of association with their host viz., the **lytic cycle** and the **lysogenic cycle**.

The lytic cycle, also called vegetative or productive, results in the production of new phages almost immediately after infection. Phages involved in lytic cycles are known as **virulent**. Lytic cycles consist of several steps and show considerable variation according to the type of phage. In lysogenic or temperate cycle, infection results in a special equilibrium between phage and host. The phage genome integrates into the host genome and persists in a latent state in the host cell, replicates more or less in synchrony with it and may be perpetuated indefinitely in this way. If this equilibrium is disturbed either spontaneously or after induction, progeny phages are produced as in lytic cycle. Both lytic and lysogenic cycles are described in detail in chapter 9.

3. TAXONOMY

As mentioned earlier, so far nearly 4600 isolates of phages have been recorded. Logically this number necessitated application of taxonomic principles for phages. Early attempts at classification by serology, host range and inactivation tests were found to be inadequate and premature. Today the phage classification is mainly based on morphology and nature of the nucleic acid of the virion. The International Committee on Taxonomy of Viruses (ICTV) currently recognizes one order, 13 families and 30 genera in the phages (Table 8.1 & Fig. 8.1). Families of the phages are based on criteria like nucleic acid, particle shape and the presence or absence of envelope. Nomenclature rules are similar to other viruses except that species are designated by the vernacular names of their best known members like T_4, λ, Mu, ϕX174.

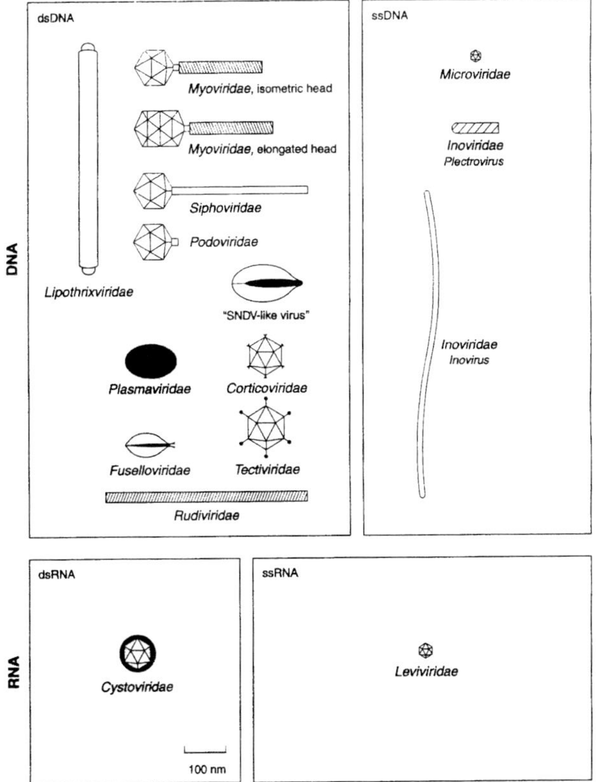

Fig. 8.1. Major bacteriophage families and genera.

Table 8.1. Classification of bacteriophages (ICTV, 2005)

Order	Family	Genus	Type species
Caudovirales	*Myoviridae*	*``T_4 - like viruses"*	*Enterobacteriophage T_4*
		"P_1 - like viruses"	*Enterobacteriophage P_1*
		"P_2 - like viruses"	*Enterobacteriophage P_2*
		"Mu - like viruses"	*Enterobacteriophage Mu*
		"SPO1 - like viruses"	*Bacillus phage SP01*
	Podoviridae	*"T_7 - like viruses"*	*Enterobacteriophage T_7*
		"29 - like viruses"	*Bacillus phage 29*
		"P_{22} - like viruses"	*Enterobacteriophage P_{22}*
	Siphoviridae	*"λ - like viruses"*	*Enterobacteriophage λ*
		"T_1 - like viruses"	*Enterobacteriophage T_1*
		"T_5 - like viruses"	*Enterobacteriophage T_5*
		"C_2 - like viruses"	*Lactococcus phage C_2*
		"L_5 - like viruses"	*Mycobacterium phage L_5*
—	*Corticoviridae*	*Corticovirus*	*Alteromonas phage PM_2*
—	*Cystoviridae*	*Cystovirus*	*Pseudomonas phage Y_6*
—	*Inoviridae*	*Inovirus*	*Enterobacteriophage M_{13}*
		Plectrovirus	*Acholeplasma phage MV-L51*
—	*Leviviridae*	*Levivirus*	*Enterobacteriophage M52*
		Allolevivirus	*Enterobacteriophage OB*
—	*Microviridae*	*Microvirus*	*Enterobacteriophage O x 174*
		Spiromicrovirus	*Spiroplasma phage 4*
		Bdellomicrovirus	*Bdellovibriophage MAC1*
		Chlamydiamicrovirus	*Chlamydis phage 1*
—	*Tectiviridae*	*Tectivirus*	*Enterobacteriophage PRDI*

Archaea

Order	Family	Genus	Type species
—	*Fuselloviridae*	*Fusellovirus*	*Sulfolobus virus SSV1*
—	*Guttaviridae*	*Guttavirus (was "Sulfolobus SNDN like viruses)*	*Sulfolobus neozealandicus droplet shaped virus (was sulfolobus virus SNDV)*
—	*Lipothrix viridae*	*Lipothrix virus*	*Thermoproteus virus 1*
Caudovirales	*Myoviridae*	*QH - like viruses*	*Halobacterium virus QH*
—	*Rudiviridae*	*Rudivirus*	*Sulfolobus virus SIRV1*
Caudovirales	*Siphoviridae*	*Y M1 - like viruses*	*Methanobacterium YM1*

Notwithstanding the taxonomy, for the convenience of identification, phages are usually divided into four morphological groups (Table 8.2).

i. Tailed phages
ii. Cubic phages
iii. Filamentous phages and
iv. Pleomorphic phages

i. Tailed phages

Tailed phages, the largest virus group, comprise approximately 96% of phages. They contain a single molecule of dsDNA and are strikingly charact-erized by a tubular protein tail. All the tailed phages have recently been given order rank and the name **Caudovirales**. They are classified into three families.

1. ***Myoviridae*** : It includes phages with long complex tails consisting of a core and a contractile sheath. Six genera namely T_4, PI, P_2, Mu, SPO1 and ϕH are included in this family.
2. ***Siphoviridae :*** It includes phages with long non-contractile and flexible tail. Six genera named after phages λ, T_1, T_5, L_1, C_2, ψM are included in this family.
3. ***Podoviridae*** : It includes phages with short tails. Three genera placed in this family are : T_7, P_{22} and ϕ 29.

Table 8.2. Main properties and frequency of phage families (from Ackermann, 2003)

Shape	**Nucleic acid***	**Family**	**Genus**	**Particulars**	**Example**	**No. of members[b]**
Tailed	DNA, ds, L	*Myoviridae*	6, see text	Tail contractile	T4	1128
		Siphoviridae	6, see text	Tail long, noncontractile	λ	2816
		Podoviridae	3, see text	Tail short	T7	646
Cubic	DNA, ss, C	*Microviridae*	*Microvirus*	Conspicuous capsomers	φX174	40
			Bdellomicrovirus		MAC-1	
			Chlamydiamicrovirus		Chp1	
			Spiromicrovirus		SV_4	
	DNA, ds, C,S	*Corticoviridae*	*Corticovirus*	Complex capsid, lipids	PM2	37
	DNA, ds, L	*Tectiviridae*	*Tectivirus*	Double capsid, pseudo-tail, lipids	PRD1	16
	RNA, ss, L	*Leviviridae*	*Levivirus*		MS2	37
			Allolevivirus		Qβ	
	RNA, ds, L	*Cystoviridae*	*Cystovirus*	Envelope, lipids	φ6	1
Filamentous	DNA, ss, C	*Inoviridae*	*Inovirus*	Long filaments	fd	53
			Plectrovirus	Short rods	L51	
	DNA, ds, L	*Lipothrixviridae*	*Lipothrixvirus*	Envelope, lipids	TTV1	4
	DNA, ds, L	*Rudiviridae*	*Rudivirus*	Stiff rods, no envelope, no lipids	S1RV1	2
Pleomorphic	DNA, ds, C, S	*Plasmaviridae*	*Plasmavirus*	Envelope, no capsid, lipids	MVL2	5?
	DNA, ds, C, S	*Fuselloviridae*	*Fusellovirus*	Lemon shaped, envelope, lipids	SSV1	2?

* L - linear; C - circular; S - segmented; ds - double streanded; ss - single stranded.

ii. Cubic phages

Cubic phages posses a tail-less capsid with cubic symmetry. They may possess DNA or RNA. Cubic phages are assorted into five families.

1. ***Microviridae :*** Members of this family possess ssDNA. Well known type genus *Microvirus* includes phage ϕX174.
2. ***Corticoviridae :*** Phages of this family possess circular dsDNA. Maritime phage PM2, a member of this family consists of two protein shells and a lipid bilayer sandwitched in between.
3. ***Tectiviridae :*** Phages of this family possess linear dsDNA and are characterized by a double capsid. Tectiviruses are found in widely different bacteria e.g. PRDI.
4. ***Leviviridae :*** Leviviruses possess linear ssRNA. Most of these are plasmid specific coliphages that adsorb to F or sex pili. Two genera namely *Levivirus* and *Allolevivirus* are differentiated, e.g., MS2 and Qβ phages.
5. ***Cystoviridae :*** These are enveloped phages with linear dsRNA. The RNA is multipartite and consists of three molecules. Virions possess their own RNA polymerase. The only genus, *Cystovirus* consists of a single species ϕ6.

iii. Filamentous phages

Filamentous phages are with helical symmetrical capsid. They may be long or short; rigid or flexous rods. Three families are differentiated in filamentous phages.

1. ***Inoviridae :*** Phages of this family possess circular ssDNA. The family includes two genera with very different host ranges and similarities in replication and morphogenesis. It includes two genera *Inovirus* and *Plectovirus*. Inovirus phages are mostly plasmid specific. The *Plectovirus* genus phages are short straight rods that occur in mycoplasmas only.
2. ***Lipothrixviridae :*** This family includes four viruses of the archaebacterial genus *Thermoproteus*. Particles are characterized by the combination of a lipoprotein envelope and rodlike shape. Genome is linear dsDNA.
3. ***Rudiviridae*** : This family includes two phages with linear dsDNA genome that infect archaebacteria, *Sulfolobus* and *Thermoproteus*.

iv. Pleomorphic phages

These phages, due to the absence of well defined capsid exist in pleomorphic state. They all possess circular supercoiled dsDNA. Two families are recognized in pleomorphic phages.

1. ***Plasmaviridae :*** The only one member of this family described from *Acholeplasma* is MVL2 or L2.
2. ***Fuselloviridae:*** This family has only one known member SSVI, from archaebacterium *Sulfolobus shibatae.* Particles are lemon-shaped with short spikes at one end. SSVI persists in bacterial cells as an integrated prophage.

4. ORIGIN AND EVOLUTION

Due to the seemingly wide difference between most phage families and the host range, phages are thought to be polyphyletic and originated at different times. Tailed phages seem to be phylogenetically related and represent the oldest phage group. Their wide distribution in eubacteria and archaebacteria suggests that they appeared even before their hosts diverged. Some phages might have been derived from cell constituents that acquired a coat and became independent. For example, leviviruses from mRNA and filamentous inoviruses from plasmids.

Further reading

Ackermann, HW, (1987). Bacteriophage taxonomy in 1987. *Microbiol. Sci.* 4 : 214-218.

Ackermann, HW, (1998). Tailed bacteriophages – The order *Caudovirales. Adv. Virus Res.* 51: 135-201.

Ackermann, HW, (2003). Bacteriophages: In "The Desk Encyclopedia of Microbiology (ed. Schaechter, M.), Elsevier Academic Press, pp. 135-146.

Ackermann, HW, and Du Bow, MS, (1987). Viruses of Prokaryotes, Vol. I & 2, CRC Press, Boca Raton, F1.

Calender, R (ed.) (1988). The Bacteriophages, Vol. 1 & 2, Plenum, New York.

Klaus, S., Kruger, D and Meyer, J (1992). Bakterien viren, Gustav Fischer Jena.

Maniloff, J and Ackermann, HW, (1998). Taxonomy of bacterial viruses : Establishment of tailed virus genera and the order *Caudovirales. Arch. Virol.,* 143 : 2051-2063.

Wilson, DR and Finla, BB (1998). Phage display applications, innovations and issues in phage and host biology. *Can. J. Microbiol.,* 44 : 313-329.

9

REPLICATION STRATEGIES

The process of reproduction in viruses is generally referred to as replication in order to differentiate the same process in true living organisms. Much of our current knowledge of replication of viruses was initially gathered from bacteriophages. Phage-bacteria system has been adopted as a technically suitable model system from the early days of virology. Replication of phages has been chiefly elaborated by Max Delbruck and S. Luria and their associates in the United States and by Andrew Lwoff's school in France. These workers mainly concentrated on the T-phages of the bacterium *E. coli,* particularly the T even phages. Some time after, other phages like lambda, ϕ x 174 and f_2 were also investigated for their replication details. Today, we know a wide range phages differing with each other in several aspects. The most important differences lie in host specificity, morphology and nucleic acid properties. Morphological forms include: tailless icosahedral phages, viruses with contractile tails, viruses with noncontractile tails, and filamentous phages. There are even a few phages with envelope. The genetic material may be either DNA or RNA; double stranded or single stranded. Naturally, phages have developed different mechanisms of replication, and along with replication variations in recombinations in their hosts and themselves. Thus it is not possible to generalize replication of different phages. Hence, strategies adapted by different phages are discussed under different headings. However, before going to individual replication cycles, details of one-step growth experiment will be elaborated.

1. THE ONE-STEP GROWTH CURVE EXPERIMENT

One step growth experiment developed by Ellis and Delbruck (1939) and Delbruck (1940) marked a major turning point in the study of viruses. This technique enabled the investigators to dissect in great detail the viral growth cycle.

Max Delbruck (Nobel Prize, 1969) performed the one-step growth experiment as follows: A known quantity of T-even phage was mixed with a known number of bacterial cells. The mixture was incubated for a short period of time, called the adsorption period, during which the virus became attached to the cells. The mixture was then diluted so that no further adsorption could take place i.e synchronized infection resulted. Samples taken at intervals after infection were assayed for virus content. The concentration of virus remained constant for a period of time, then suddenly increased to a new level and remained constant. As the dilution had a single bacterial cell, no further increase in the concentration of virus even incubated for longer time.

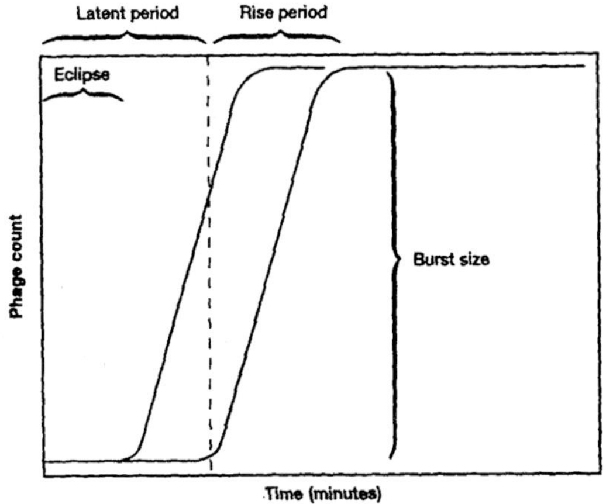

Fig. 9.1. The one step growth curve

Max Ludwig Henning Delbruck (1906-1981) (Nobel Prize, 1969)

Based on these experiments, the growth cycle, which lasts about 45 minutes at 37°C, is divided into several stages (Fig. 9.1). The first stage, lasting about 17 to 19 minutes is called the latent period because there is no increase in titre. The latent period ends when the titre begins to rise. The second period, the time of rising titre, lasts about 15 minutes. The one-step growth period then ends and the phage titre is at a constant new level that is higher then the initial titre. The ratio of final titre to initial titre is called the burst size and it represents average number of phage particles produced by each infected cell. Delbruck also performed further experiments to determine the factors affecting the shape and duration of the one-step growth curve. He found, for example, that if the temperature of incubation was lowered to 30°C instead of the usual 37°C, the latent period was greately lengthened. Thus, a lot of useful information about the replication of phages can be generated with one step growth curve experiments.

2. REPLICATION OF dsDNA PHAGES: THE LYTIC CYCLE

A phage life cycle that culminates with the host cell bursting and releasing virions is called **lytic cycle**. The phages which replicate through this process are called **virulent phages**. T-even phages with double stranded DNA (e.g. T_2, T_4 and T_6) which infect *E. coli* serve as the models for understanding the details of lytic cycle. Infact, much of our information concerning the multiplication of bacteriophages is based mainly on the behaviour of virulent T-even phages. We will consider the life cycle of T_4 phage in the following section. However, it must be remembered that with minor differences the life cycles of other T-even phages will also be similar.

Morphology

T-even phages, of *Myoviridae* family, are tailed phages with an anisometric hexagonal head and with elongated rigid tail (Fig. 9.2). They resemble lunar landing modules. Morphological details are given in chapter 3.

Genome

The genome of the T_4 consists of about 168000 bp of double stranded DNA containing glucosylated hydroxy methyl cytosine (HMC) residues. Mature DNA molecules, packaged into virions are linear and show what is called **terminal redundancy** that is, a base sequence is repeated at both ends of the molecule (Fig. 9.4). However, the DNA released from the head is highly branched and contains multiple covatently linked (head to tail) copies of the genome. These structures are called **concatemers**. Mature T_4 DNAs are cut during packaging of intracellular DNA at nearly random map positions. As a consequence, the ends of different individual genomes are almost randomly permuted over the circular genetic map. In other words, the sequence of genes in each T_4 virus of a population is the same but starts with a different gene at the 5'end. The genome is reported to contain approxi-mately 130 genes with known functions (Fig. 9.3). More than 100 additional genes were revealed by DNA sequencing, many of them are still in search of their functions. There are many overlapping coding regions and many open reading frames (ORF) are very small. Inspite of the large genome size only a few short regions are devoid of coding capacity. These observations suggest that apparently redundant and non-essential genes confer selective advantage.

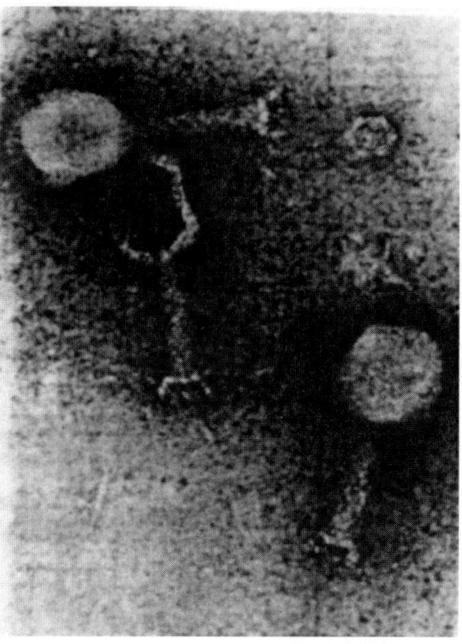

Fig. 9.2. Electron microphotograph of phage T_4

One-step growth curve experiments conducted by Delbruck (1940) clearly demonstrated that the life cycle passes through distinct phases and (Fig. 9.6) culminates in the production of progeny virions. These phases are

1. Adsorption
2. Penetration
3. Latent phase
4. Synthesis of nucleic acids and proteins
5. Lysis and release of virions.

1. Adsorption : The infective cycle begins with the coming together of a phage particle and a susceptible bacterial cell. This physical contact is brought about by active participation of either the cell or the phage particle or both or it may be brought about by simple diffusion. The attachment of the virion to the host bacterium is called **adsorption**. Adsorption is dynamic and specific process. The affinity required for the adsorption is determined by the chemical nature and of charge distribution on the host cell wall. Bacterial cell walls have specific receptor sites for the attachment of phages. The cell's surface has different receptors for different phages. A mutation causing the gain or loss of receptor activity for one phage does not necessarily result in gain or loss of receptors for another phage. After the tip of the phage tail has collided successfully to an appropriate receptor, events begin that make the adsorption irreversible (Fig. 9.5). The tail fibres are important for irreversible attachment.

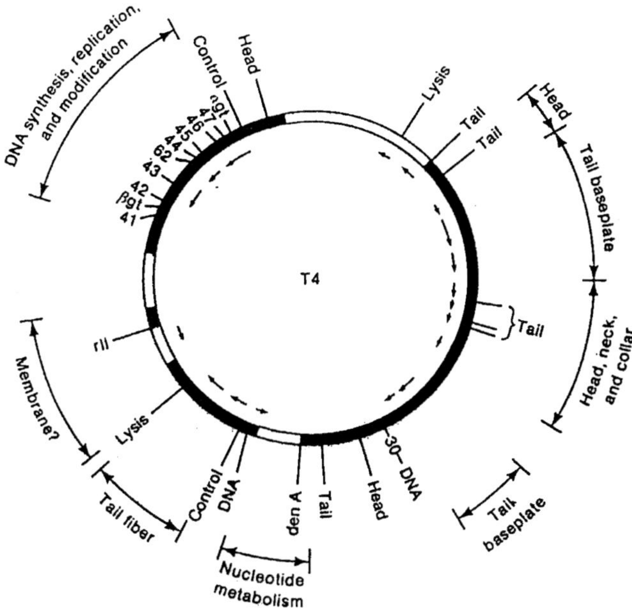

Fig. 9.3. Genetic map of phage T_4 showing some, but not all genes. The clustering of genes with related functions should be noted. (From Maloy, S.R. *et al.,* 1994 Microbial Genetics)

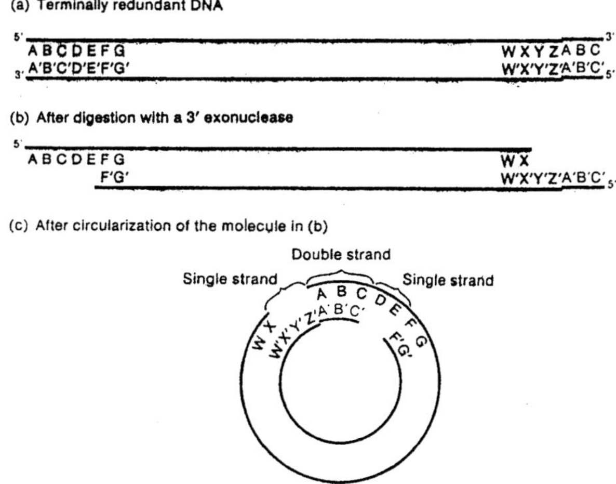

Fig. 9.4. A terminally redundant molecule and its identification by means of exonu-cleolytic digestion and circularization. A nonredundant DNA molecule cannot be circularized in this way (From Maloy, S.R. *et al.,* 1994 Microbial Genetics)

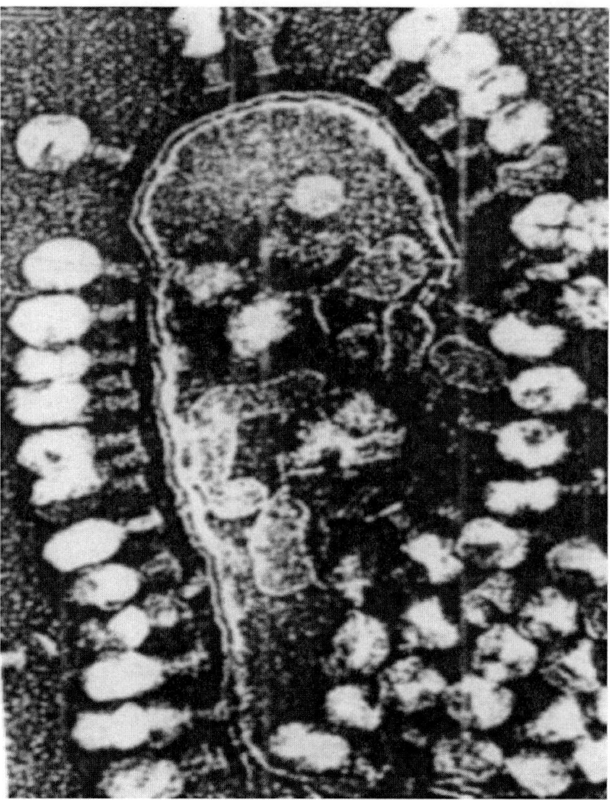

Fig. 9.5. Electron micrograph of *E. coli* infected with phage T_4.

An important enzyme, called **lysozyme**, is carried in the phage tail. It resembles the egg white lysozyme both in structure and function. This enzyme erodes a hole in cell wall by dissolving the peptidoglycan.

2. Penetration : Experiments performed by Hershey and Chase (1952) demonstrated that only the DNA of a phage is necessary to initiate a cycle of infection in a cell. Other minor proteins present in the phage head also enter the cell, but the major portion of the phage protein, the shell remains outside. The contraction of the tail sheath is part of the mechanism for injection of the DNA into the cell. The mechanism of contraction of the sheath is closely related to its structure. The contraction of the sheath pushes the tail core into the bacterium with a slight twisting motion. Energy for contraction is furnished by ATP contained in the sheath.

The tail core is hollow like a hypodermic needle and connects the head of the phage to the interior of the bacterial cell. The phage DNA is extruded through the opening of the tail core into the cell. Along with extruded DNA, several proteins contained within the phage head also enter into the bacterial cell. Entry of phage DNA into the bacterial cell initiates the next stage of the replicative cycle.

3. Latent period : The period between the entry of phage DNA and final lysis of host cell and release of progeny virions is referred to as **latent period**. It was so named because no perceptible virions could be observed. Experiments of Doermann (1952) have provided the evidence to support that the latent period could be divided into two periods. For approximately, the first half of the latent period, the infected bacterial cells contained no detectable phage. This period is called the **eclipse**. During the second half of the latent period, called the **maturation period**, completed infective phage particles accumulated within the infected cells before being detectable by the one-step growth experiment. Doermann (1952) suggested that eclipse period is the time of synthesis of viral components that will be assembled into finished phage particles later during the maturation stage.

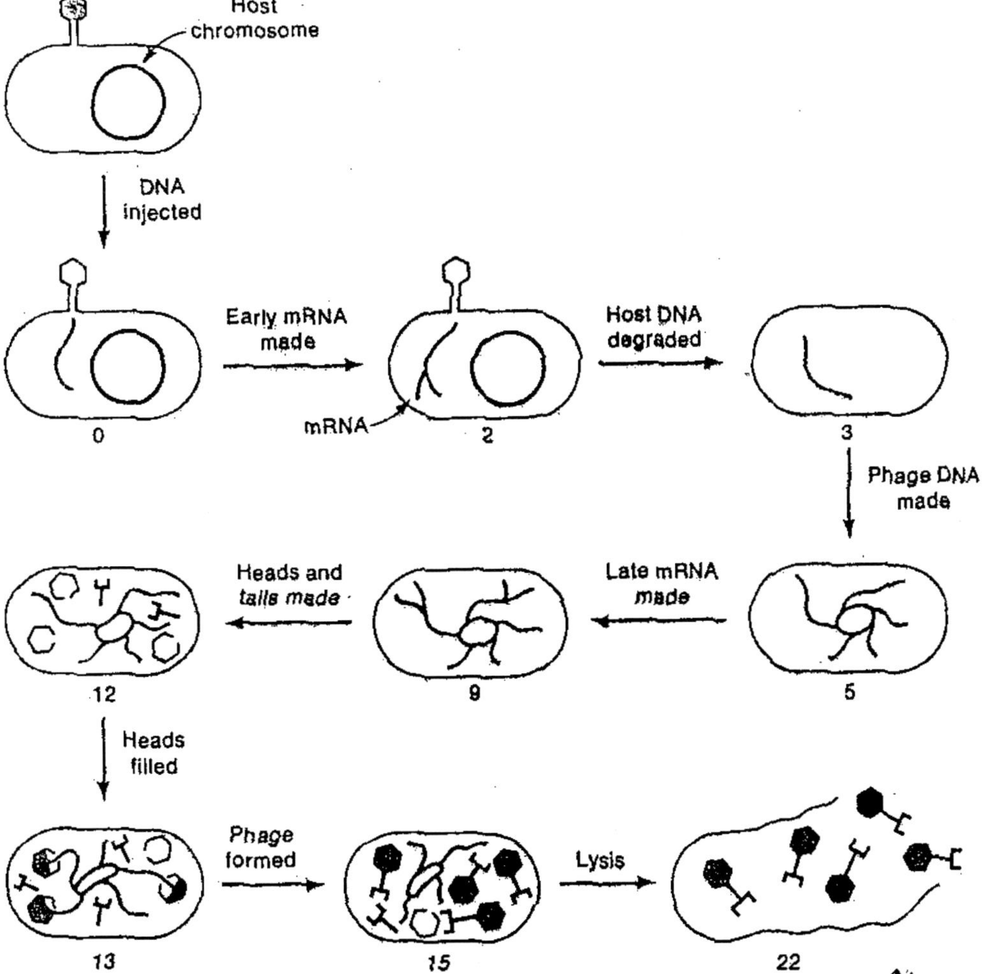

Fig. 9.6. A schematic diagram of the life cycle of phage T4. The numbers represent time after injection in minutes at 37°C. (From Maloy, S.R. *et al.,* 1994 Microbial Genetics)

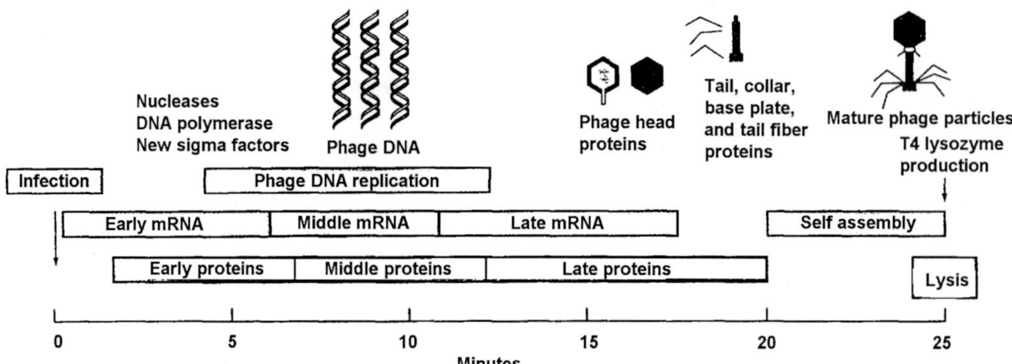

Fig. 9.7. Time course of events in phage T4 infection. (From Brock Biology of Microorganisms)

4. Synthesis of nucleic acids and proteins : When the phage DNA enters a bacterial cell, the cell embarks on a new set of bio-synthetic processes. Broadly speaking, the synthesis of host DNA, RNA is halted and the cell is forced to make viral constituents. The two immediate affects of entry of the viral nucleic acid are : (i) immunity against infection by phages of the same type and (ii) supression of all activities of the host cell. The first process is mediated by the formation of specific enzyme repressors which act against multiplication of phages of the same type. The mechanism behind the immunity is not very well understood. The phenomenon of supression, on the other hand, involves the cessation of synthesis of all cellular DNA, RNA and proteins. T-even phages inactivate host translation and transcription in many small steps by multiple redundant mechanisms. The temporal regulation of phage gene expression is exerted at many levels : transcript initiation, elongation and termination; stability conformation and recognition by ribosomes of the host. Different classes of phage genes are distinguished in terms of timing as : early (immediate early, IE), middle (delayed early, DE) or late. One distinguishing feature of IE genes is that they can be transcribed by host RNA polymerase without modification by phage proteins (Fig. 9.7). On the other- hand, expression of all other T-even genes requires phage protein synthesis.

i. Synthesis of early mRNA and proteins

The first set of promoters, early promoters, are similar to *E. coli* promoters. They are recognized by *E. coli* RNA polymerase at a time when the host DNA is still intact. *E. coli* RNA polymerase starts synthesizing phage mRNA within two minutes. This mRNA and all other early mRNAs direct the synthesis of protein factors and enzymes required to take over the host cell and manufacture viral nucleic acids. Some of the early virus specific enzymes degrade host DNA to nucleotides, thereby simultaneously halting host gene expression and also providing raw material for virus DNA synthesis. Within five minutes, the process progresses to the point that virus DNA synthesis commences. The expression of the T_4 genes is a sequentially and tightly regulated process. Even the organization of the genome appears suited for efficient control of the life cycle. As can be seen in Fig. 9.3, genes with related functions are clustered together, early and late gene clusters are separated, further they are transcribed in different directions. As a prelude to DNA synthesis, hydroxymethyl cytosine (HMC) is synthesized by two phage-encoded enzymes. Phage DNA replicates by **bidirectionally**. It is extremely complex process requiring atleast seven phage encoded proteins. After the DNA synthesis, HMC is glucosylated to protect T_4 DNA from attack of *E. coli* restriction enzymes.

ii. DNA replicatton

As explained earlier, T_4 phage DNA shows terminal redundancy. When a number of copies have been made, about 6-10 copies are joined together to form very long DNA strands called **concatemers**. However, again during assembly the concatemers are cleaved into individual units with different genes at either end, but with same amount of genome.

iii. Synthesis of late mRNA and late proteins

Late mRNA genes are expressed after DNA replication and about 9 minutes after penetration. They encode for three different types of proteins : (i) phage structural proteins, (ii) proteins that help phage assembly without becoming part of the virion structure, and (iii) proteins involved in cell lysis and phage release. All the proteins required for phage assembly are synthesized at the same time (Fig. 9.7).

iv. Assemble of proteins into capsid

The assembly of T-even phages is interesting as a model for the formation of biological structures containing many different component proteins and for their stabilization. Complementation studies have shown that the T_4 capsid is assembled through four independant subassembly lines, which produce the phage head, the fibreless tails and tail fibres, using mostly proteins belonging to late class (Fig. 9.8). The base plate is constructed from 15 gene products. After the base plate is finished, the tail tube is built on it and the sheath is assembled around the tube. The phage head is assembled through a series of sequential steps in which each protein molecule undergoes a conformational change as it is assembled, revealing the binding site that is recognized by next molecule.

Then the phage head spontaneously combines with the tail assembly. Tail fibers attach to the base plate after the head and tail have come together. Many of the steps in assembly occur spontaneously, however, some require special virus enzymes or host cell factors.

v. DNA package into head capsid

DNA is packed into the head through several mechanisms. After the T_4 head is filled with concatemeric DNA, the excess is cut off by an endonuclease that works only when the head is complete. Because this enzyme does not recognize DNA sequences, a headful of DNA is packaged. Its length is precisely determined by the size of the capsid. Usually, the head is packed full and contains about 2% more DNA than is needed for the full T_4 genome. The first complete T_4 particles appear at 37°C about 15 minutes after infection.

5. Lysis and release of phage particles: At the end of latent period the host cell is filled with newly synthesized progeny virions. The host wall ruptures or bursts and the virions are released out. Then the host cell is said to have undergone **lysis**. The lysis of *E. coli* takes place after about 15 minutes at 37°C. The average number of progeny virions released from infected host cell is referred to as **burst size** and it is approximately 300 virions for T_4 phage. Several T_4 genes are involved in the lysis process. One gene directs the synthesis of a lysozyme that attacks the peptidoglycan of the cell wall. Another phage protein dama-ges the bacterial cell membrane, allowing lysozyme to reach the cell wall.

3. REPLICATION OF dsDNA PHAGES : LYSOGENIC CYCLE

Many double stranded DNA (dsDNA) phages establish a different relationship with their host. After adsorption, penetration, the viral genome does not destroy the host while producing progeny virions. Instead, the viral genome remains within the host cell and replicates along with bacterial genome. This co-replication generates a clone of infected cells that may grow and divide for long periods while appearing perfectly normal. However, under appropriate environmental conditions each of these cells undergo lysis and produce progeny virions. This relationship between the phage and its host is called **lysogeny**. Bacteria which possess the potential to produce phage particles under some conditions are said to be **lysogenes** or **lysogenic**. The phages that enter into this type of relationship are **temperate phages**. The latent form of virus genome that remains within the host but does not destroy it is called **prophage**. In many temperate phages, the prophage is usually integrated into the bacterial genome but some temperate phages, like P1 the phage genome exists independantly. A lysogenized culture, either under spe-cific natural conditions or by artificial methods can be initiated to commence lytic cycle and produce progeny virions. This process is known as **induction**.

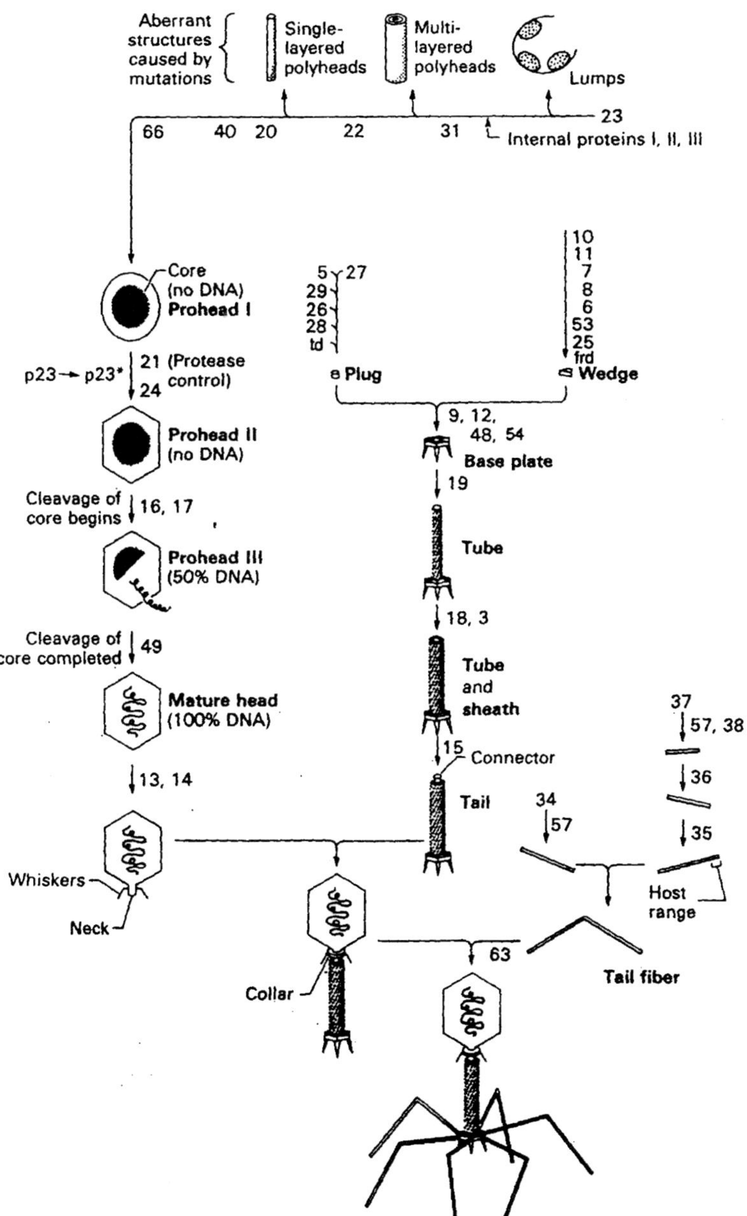

Fig. 9.8. Assembly of T_4 virions. Assembly occurs in three major subassemblies; head, tail and tail fibers. Numbers indicate the T_4 genes participating in a given step.

Life cycle of lambda (λ) phage

The lambda phage that infects K12 strain of *E. coli* is the best under stood temperate phage will serve as a model for understanding the lysogenic cycle. This phage has probably been studied longer and in greater detail than most other biological particles.

Structure

Lambda phage is a member of *Siphoviridae*. The virion consists of an icosahedral head 55 nm in diameter and a thin noncontractile tail of 15 x 135 nm with a thin fibre (23 nm long) at the end (Fig. 9.9). The main tail protein is 31 kD. The largest protein (135 kD) forms the tail fiber. The two major capsid proteins, 420 per virion, are 38 and 12 kD. These, as well as the tail components, are proteolytically cleaved as the virion matures. Lambda DNA, 47,000 bp long and of 32 x 10^6 daltons, is a double stranded linear molecule with cohesive ends. These cohesive ends are single stranded stretches, 12 nucleotides long, have complementary base sequences and can base pair with each other. Because of these cohesive ends, the linear genome cyclizes immediately after its release from head capsid (Fig. 9.10). *E. coli* DNA ligase then seals the breaks, forming a closed circle. The lambda genome has been carefully mapped and over 40 genes have been mapped. Most genes are clustered according to their function (Fig. 9.11). In the linear DNA, all the genes for proteins of the head or playing a role in its assembly are located in the first 18 per cent of its length and the tail-related functions in the next 18 per cent. The next 50 per cent is responsible for lysogenic functions and their regulation, followed by the genes for DNA replication and cell lysis.

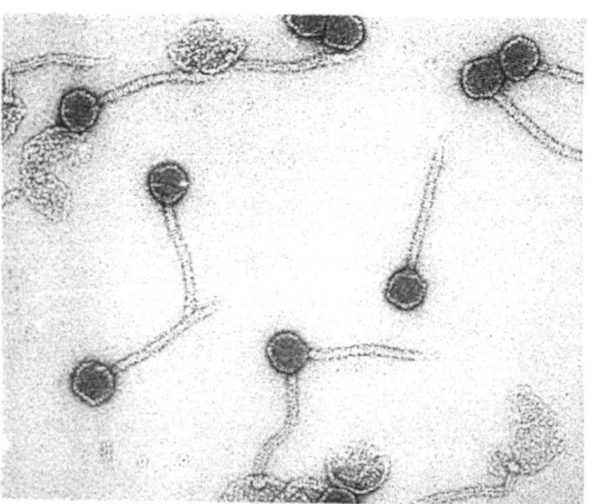
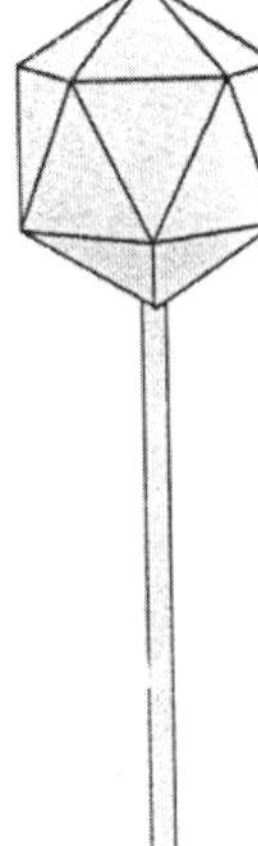

Fig. 9.9. Bacteriophage lambda

Lytic cycle

Lambda phage, a typical temperate phage, can reproduce using a normal lytic cycle. The virion attaches to a specific protein in the cell wall of *E. coli* and injects its DNA. Immediately, after entry, the phage DNA is converted into a covalent circle. If the cell is not a lambda lysogen (immune) expression of the phage genome begins. The first few steps in gene expression are same whether the final result is lysis or lysogeny. Immediately, certain parts of the phage DNA are transcribed by host RNA polymerase.

The polymerase binds to both a rightward promoter (PR) and a leftward promoter (PL) and begins to transcribe in both directions, copying different DNA strands. The first genes that are transcribed are *N* and *CIII* on the leftward and *Cro* and *CII* on the rightward. All these four protein products are regulatory in nature. The *N* product favours continuation of these particular transcription processes. *CII* product activates the genes for the structural and lysis proteins. The rightward transcription leads to products of O and P, which initiate replication of the phage DNA starting at site *ori*. Cro protein represses transcription of the early genes which are no longer required in the productive replication cycle.

Lambda DNA replication and virion assembly are similar to the processes already described for the T_4 phage. However, the difference is that at the beginning lambda DNA replicates by **bidirectional** mode of replication. But later on it switches on to the **rolling circle mechanism** to form long concatemers that are finally cleaved to give complete genomes.

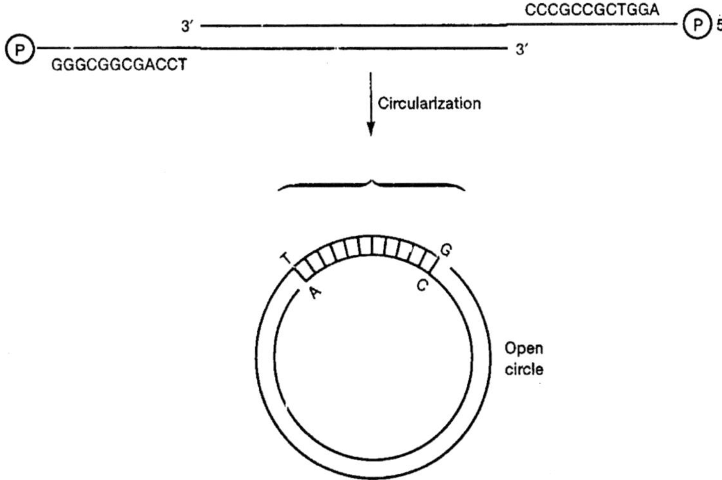

Fig. 9.10. **Lambda phage DNA.** A diagram of lambda phage DNA showing its 12 base, single-strained cohesive ends and the circularization their complementary base sequences make possible. (From Maloy, S.R. *et al.*, 1994, Microbial Genetics)

Lysogeny

The establishment of lysogeny and also immunity to superinfection can be attributed to the presence of the **lambda repressor** encoded by the *CI* gene. This repressor protein is 236 amino acid long and folds into a dumbbell shaped structure with globular domains at either end. With one domain it binds to DNA and with another domain it binds to another repressor forming a dimer. Infact, in dimer state lambda repressor is most active. If the host is a prospective lysogen the repressor is produced continuously, binds to the operators OL and OR, thereby blocking RNA polymerase activity. These events close the doors for lytic cycle and at the same time pave the way for lysogeny.

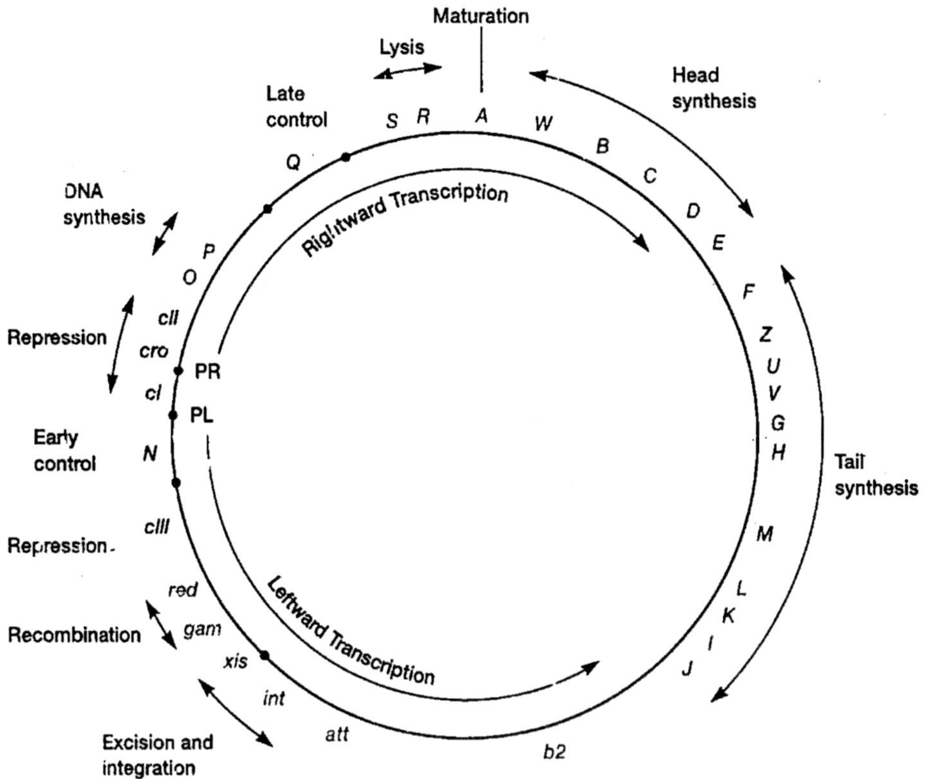

Fig. 9.11. The Lambda phage genome: The direction of transcription and location of leftward and rightward promoters (*PL* and *PR*) are indicated on the inside of the map. The positions of major regulatory sites are shown by dots on the map. Lambda DNA is double standard and transcription proceeds in opposite directions on opposite strands (From Maloy, S.R. *et al.,* 1994, Microbial Genetics)

The sequence of events leading to the initial synthesis of repressor and the establishment of lysogeny is well known. Immediately after the entry and circularization, *CII* and *CIII* genes are transcribed. These proteins accumulate and bind to the presite of the promoter for the *CII* gene that facilitates the RNA polymerase binding. As a part of this transcription and later translation lambda repressor, a product of *CI*, is synthesized rapidly. Now this, repressor binds to OR and OL and inhibits the mRNA synthesis and the production of the *CII* and *CIII* proteins. As a matter fact, as a part of this inhibition, production repressor should also be inhibited. However, the *CI* gene continues to be transcribed at a low rate mainly due to the activity of a second promoter (prm) that is activated by the repressor itself (Fig. 9.12). Thus, the lambda repressor is both a repressor and an activator. This control circuit in which lambda repressor stimulates its own synthesis ensures that lysogeny will normally be stable once it is established.

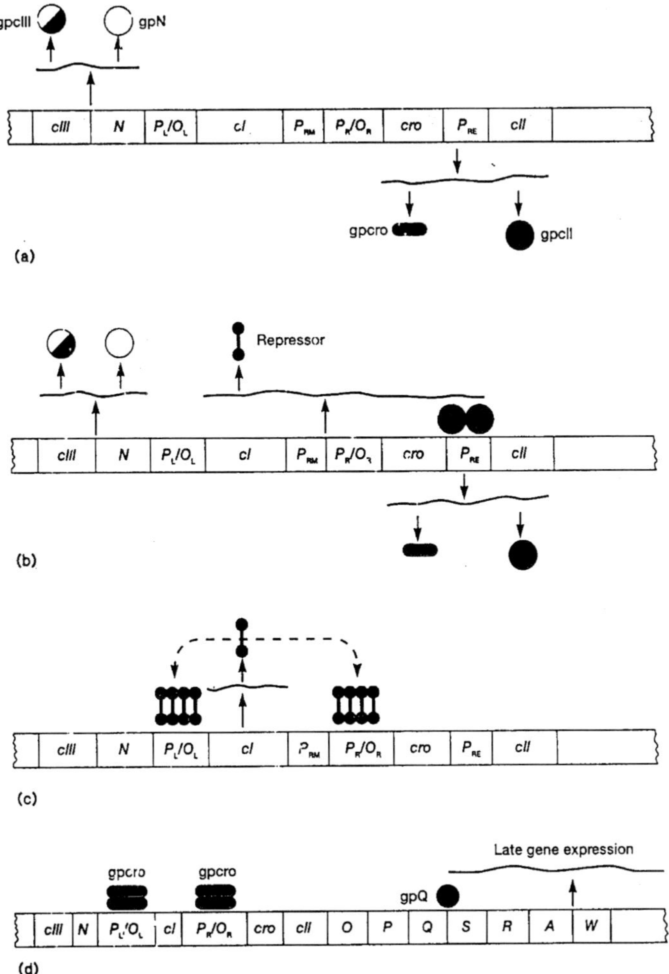

Fig. 9.12. Events involved in the choice between the establishment of lysogeny and continuation of the lytic cycle. (From Prescott *et al.,* 2005, Microbiology)

At the same time, the cro protein encoded by *cro* gene also accumulates. This protein binds to OR and OL and thus turns off the synthesis of repressor protein. Lambda repressor also tries to bind to the same promoters and inhibits cro protein. Nevertheless, under normal conditions (to the lysogen) the lambda repressor will be in an advantageous position because it has greater affinity for binding to the promoters than cro protein. As a result lysogeny will be established and continue integration. If lambda repressor binds to the promoters, the circular lambda DNA is inserted into the *E. coli* genome. Insertion (integration) is possible because the CII protein stimulates transcription of the *int* gene at the same time as that of CI gene. The *int* gene codes for the synthesis of an integrase enzyme, which is a site specific topoisomerase catalyzing recombination of the phage and bacterial attachment sites (the att site). Lambda DNA has an attachment site that can base pair with bacterial attachment site located between the galactose or gal operon and the biotin operon on the *E. coli* chromosome (Fig. 9.13). Once these two sites match up, the integrase enzyme, with the aid of special host protein catalyzes the physical exchange of viral and bacterial DNA strands. The circular lambda DNA is integrated into the *E. coli* DNA as a linear region next to the gal operon and is called a **prophage**. During integration, the linear order of the phage genes is changed or permuated.

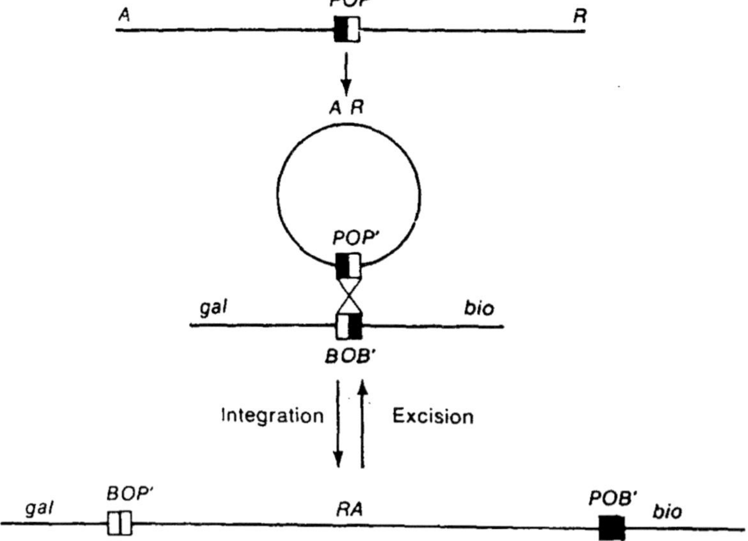

Fig. 9.13. The mechanism of prophage integration and excision of phage lambda. The phage attchment site has been denoted *POP'* in accord with subsequent findings. The bacterial attachement site is *BOB'*. The prophage is flanked by two new attachment sites denoted *BOP* and POB. (From Maloy, S.R. *et al.*, 1994; Microbial Genetics)

During cell growth, the lambda replication system prevents expression of the integrated lambda genes except for the gene *CI*, which codes for lambda repressor. During host DNA replication, the integrated lambda DNA is replicated along with the rest of the host genome and transmitted to progeny cells. When release from repression occurs, the lambda productive cycle (lytic) commences.

Induction

The lambda prophage will leave the *E. coli* genome and begin the production of new phages when the host faces a threat of survival. This process is known as **induction** and is always triggered by drop in lambda repressor level. Occasionally, the repressor will spontaneously decline and the lytic cycle commences. Environmental agents that induce lambda lysogens to produce phage are agents that damage DNA, such as UV radia-tion, x-rays, nutrional stress, high temperature and DNA damaging chemicals such as the nitrogen mustards. These agents by a cascade of events reduce the level of lambda repressor. When DNA damage occurs a host defense mechanism called the SOS response is brought into play. An array of 10-20 bacterial genes is turned on, some of which help bacterium survive radiation. However, one significant result of DNA damage is that a bacterial protein called RecA is turned into special kind of protease that participates in the destruction of lambda repressor. With the lambda repressor destroyed, the inhibition of expression of lambda lytic cycle genes is abolished. An early gene located next to the *int* gene, the *xis* gene, codes for the synthesis of an **excisionase** protein that binds to the integrase and enables it to reverse the integration process and free the prophage. The lytic cycle then proceeds normally.

Decision between the lytic and lysogenic cycle

Lambda serves as one of the best studied models of how a `decision' is taken at molecular level. As soon as the entry of DNA, the phage has to take a decision as to whether go for lytic or lysogenic cycle. The mechanism that determines the choice between the lytic and lysogenic cycle has not been fully worked out. However, it is clear that the lambda repressor plays a central role. This protein has three important functions : (i) it activates the the *pre* promoter, (ii) it activates *PI*, the *int* promoter, and (iii) it delays synthesis of late mRNA (which encodes phage heads, phage tails and the lytic system). For a lytic cycle to occur either the synthesis or the activity of CII protein needs to be reduced to the point that *pre* is not activated. If CII abundance or activity initiates the decision between lysis or lysogeny, these features must be responsive to various signals. Apart from CI and CII products, the CIII protein as well as host proteins like Hfl A, Hfl B, Him A and Him B also play an important in taking the decision. The complex decision phenomenon requires more study before it will be understood.

4. REPLICATION OF dsDNA PHAGES : BACTERIOPHAGE Mu

Mu is one of the most interesting bacteriophages. This phage, a coliphage, is temperate like lambda but has an unusual property of replicating as a transposable element. It was **Larry Taylor**, in 1963, who first reported the discovery of a virus which uses DNA transposition as lifestyle. He named this temperate coliphage Mu (for **Mutator**) because he found that its prophage was capable of randomly inserting in the bacterial chromosome, including within a gene, and thus Mu lysogens with a much higher frequency.

Structure

Bacteriophage Mu, a member of *Myoviridae*, is large dsDNA virus with an icosahedral head, a helical tail and six tail fibres (Fig. 9.14). Out lines of the genetic map are shown in Fig. 9.15. It is evident from the genetic map that bulk of the genetic information is involved in the synthesis of head and tail proteins, but important genes involved in replication and immunity are located at either end. The DNA molecule found within the virion is approximately 39 kilobase pairs long, but only 37.2 kilobase pairs makeup the actual Mu genome. This is because both ends of this DNA molecule contain host DNA. These host DNA sequences are not unique and represent DNA adjacent to the location where Mu was inserted into the genome of its previous host. Similarly, each virion arising from a single infected cell will have a different amount of host DNA, and the host DNA base sequence in each virion from the same cell will be different. A special segment of the Mu genome called G is invertible determines the kind of tail fibres and thus the host range of Mu is determined.

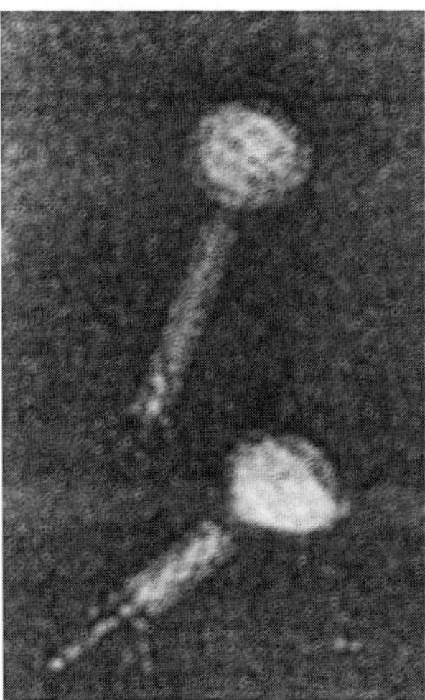

Fig. 9.14. Electron micrograph of phage Mu

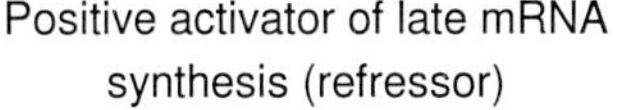

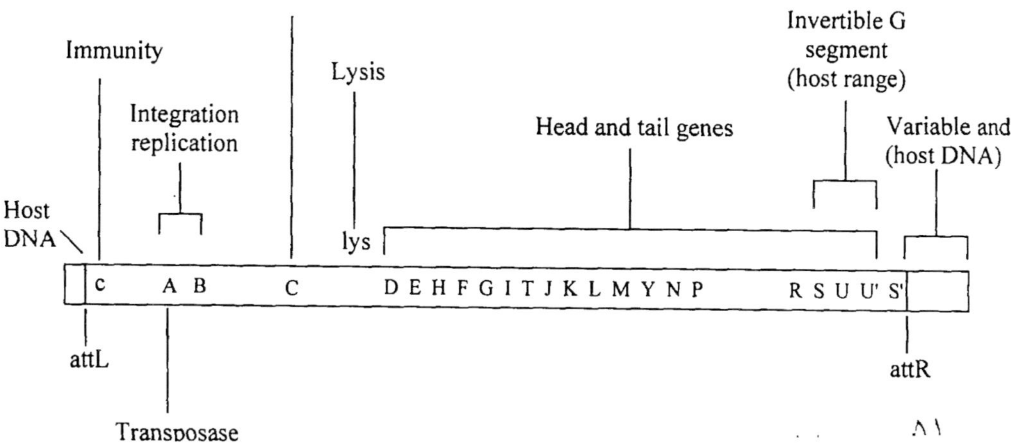

Fig. 9.15. Organization of the bacteriophage Mu genome

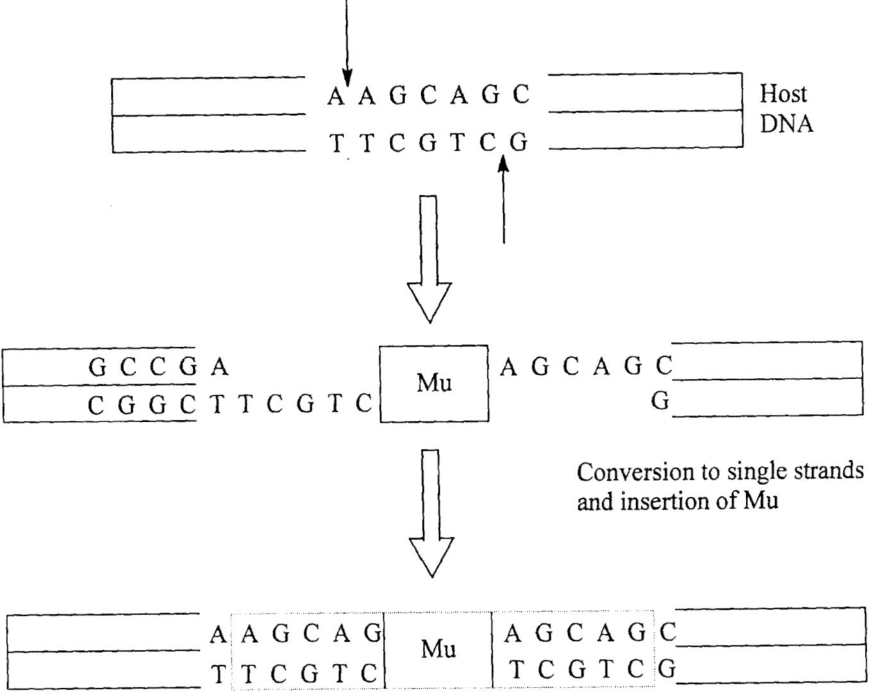

Fig. 9.16. Integration of Mu into host DNA, showing the generation of a five-base pair duplication of host DNA

Replication

Mu phage recognizes and attaches to different lipopolysaccaride components on the cell surface of its host. After phage adsorption, the DNA is injected into the cell along with the product of the *N* gene, which binds to the extremeties of the injected DNA, where it acts to noncovalently circularize and partially supercoil the infecting DNA. The injected DNA is protected from host restriction by a modification system in which about 15% of the adenine residues are acetoamidated. In contrast with lambda, integration of Mu DNA into the host genome is essential for both lytic and lysogenic growth. The DNA is inserted into a random location within the host genome (either in the main chromosome or plasmid) in a conservative manner with both infecting strands being integrated (Fig. 9.16). Integration requires the activity of the gene *A* product, which is a transposase enzyme. At the site where Mu DNA becomes integrated, a five-base-pair duplication of the host DNA arises at the target site. This host DNA duplication arises because staggered cuts are made in the host DNA at the point where Mu is inserted and the resulting single stranded segments are converted to the double-stranded form as a part of the integration gap filling processes. Duplication of short stretches of host DNA is typical of transposable element insertion.

Lytic/Lysogenic decision

At an indeterminate time after infection, the early operon is expressed and the phage transposes into host DNA. At this stage, it is not known whether the choice between lytic or lysogenic growth is made prior to, during, or after the first integration of the phage DNA. Phage Mu has a promoter, called Pc, which is a major promoter for expression of the repressor gene *C*. The repressor genes are transcribed and their protein products bind to three operator sites. These operators repress the expression of the early operons and thus inhibit the lytic cycle and pave the way for lysogenic cycle.

During lytic growth, including the induction of a lysogen, the early genes are expressed from promoter P_c. This results in the production of proteins from the *ner* gene and *A* (transposase) and *B* genes. The ner protein by binding to operator turns off expression of repressor operon and commit the phage to the lytic cycle. Replication through lytic cycle involves repeated transposition of Mu to multiple sites on the host genome. In the initial stages of transcription only the early genes of Mu are expressed. However, later the genes concerned with head and tail proteins are also expressed. Eventually, expression of the lytic function occurs and mature phage particles are released.

5. REPLICATION OF ssDNA PHAGES : COLIPHAGE ϕ x 174

ϕ x 174 is a member of the family *Microviridae*, genus *Microvirus*. It grows on various strains/species of *Enterobacteriaceae*, typically *E. coli*, *Salmonella* and *Shigella*. *E. coli* C is the best host for many isolates. The DNA of ϕ x 174 was the first DNA to be completely sequenced, a remarkable achievement when it was accomplished by **Frederick Sanger** and colleagues in 1977. It is also of special interest because it was the first genetic element shown to have **overlapping genes**.

Morphology

ϕ x 174 virion is an isometric phage with icosahedral symmetry (Fig. 9.17A). It has a particle weight of 6.2 x 10^6 daltons and sediments at 114s. The diameter of the particle, without the spikes, is roughly 26 nm. The protein capsid of the phage is assembled from 12 capsomers. Each capsomer is actually a cluster of five structural units. Thus, the virion is built up of 60 molecules each of F and J proteins of the main coat protein. There are 12 spikes projecting about 3 nm from the vertices of the icosahedron. Each spike is composed of five molecules of protein G and one molecule of protein H.

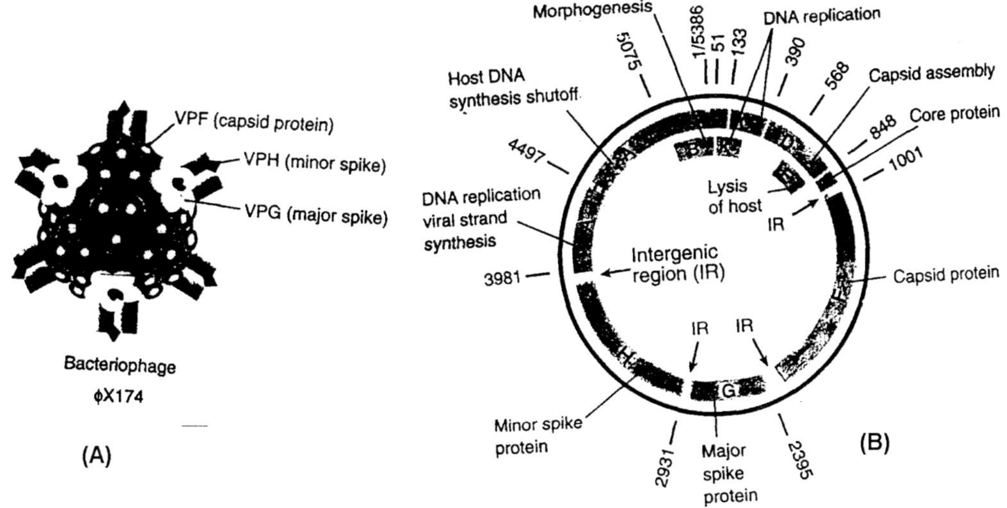

Fig. 9.17. **A.** Morphology of phage ϕ x 174. **B.** Genetic map.

Genome

The genome of ϕ x 174 consists of a single continuous circular molecule of ssDNA (Fig. 9.17B). The DNA has the same polarity as mRNA, and is thus a + strand. Sanger and his colleagues have mapped the entire nucleotide sequence of ϕ x 174, DNA and have shown that it contains 5386 nucleotides. A surprising revelation of the work of Sanger was that genes could overlap that the same DNA sequence could encode information for more than one protein using different reading frames. Together, 11 genes, in the order of A, A*, B, K, C, D, E, J, F, G, H, have been defined by virtue of their protein products. Eight of these (A, B, C, D, F, G, H and J) are essential for virus replication. Gene A* protein acts to suppress host DNA replication, gene E is responsible for cell lysis, and gene K appears somehow to enhance phage yields.

Replication of DNA

The mode of replication of ϕ x 174 is especially interesting since the phage has the problem of making an identical copy of single strand. The template cannot yield progeny molecules in a single step. Replication of DNA is completed through several successive steps (Fig. 9.18A).

i. Conversion of the parental ssDNA to a double stranded molecule called replicative form I (RFI) : As this occurs before transcription begins and hence it depends on host enzymes exclusively. The newly synthesized strand (-), now, functions as the template. This double stranded form is called replicative form I (RFI).

ii. Replication of RFI : Along with the RFI, a multifunctional protein, the gene A protein is also made. This A protein makes a single strand break (a nick) in the (+) strand at ori C and remains covatently linked to the 5'-P terminus. This nicked molecule is called RFII. *E. coli* proteins then cause the parental (+) strand to be displaced from RFII by the looped rolling-circle replication mode. When one round of replication is completed, the displaced (+) strand is cleaved from the looped rolling circle, recircularized, and used as a template for the synthesis of (-) strand. The result is a new RFI.

iii. Synthesis of (+) strand for encapsidation : Infact there is no switch over from synthesis of RFI to synthesize a free (+) strands before they can serve as templates for further synthesis of (-) strands. The capture is delayed until the newly assembled heads become available and begins when sufficient number of RFI copies are available. At this point, most of the DNA is engaged in looped rolling-circle replication, producing new RFI and some (+) strands for packing. Host DNA synthesis finally stops at this time. As more heads are made, all of the displaced (+) strands are packaged and synthesis of RFI stops for lack of a template.

Packaging of DNA

Packaging of ϕ x 174 DNA requires seven phage proteins, four of which are present in the finished particle. The steps of assembly are as follows (Fig. 9.18B).

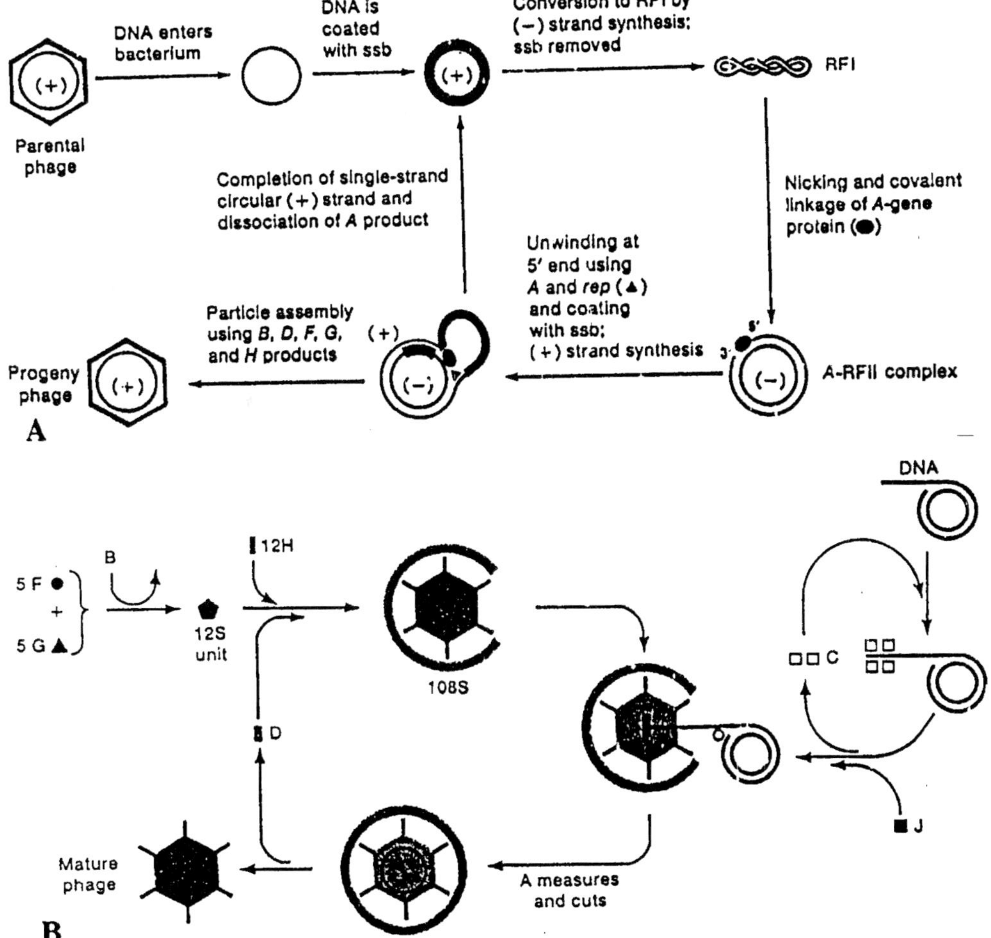

Fig. 9.18. **A**. replication of ϕ x 174. **B.** Assembly (From Freifelder. D. Molecular Biology)

i. The head protein (product of gene F) and the spike protein (gene G) combine and in presence of the gene B protein form an aggregate called a 125 particle.

ii. The gene H proteins are added to complete the spike.

iii. A prohead is formed from B, F, G and H proteins enclosed by frame like structure formed from D protein

iv. The gene-C protein directs the 5' end of the progeny DNA into the prohead. The gene-J protein organizes the DNA inside the head and stabilizes the structure.

v. The A protein cuts the displaced (+) strand from the looped rolling circle and circularizes the displaced strand.

vi. B and D proteins are then released and removed from the closing head leading to the formation of the finished phage particle.

The mechanism of lysis of phage filled cells is not well understood. The protein coded by gene E, presumably, plays a role in that function. Lysis, can be prevented by UV irradiation and mitomycin C. Active lysis is not an obligatory step in single stranded phage replication. Considerably more phage actually accumulates in the cell when the lysis is retarded. The entire life cycle takes more than 30 minutes time and the burst size is about 500.

6. REPLICATION OF ssDNA PHAGES : FILAMENTOUS PHAGE M13

When compared to the isometric phages like ϕ x 174, the filamentous phages appear and behave, in a paradoxical manner, both very differently and quite similarly. Three filamentous *E. coli* phages, M13 (Fig. 9.19), fd and f1 (*Inoviridae*) each of which contains a circular single-stranded DNA molecule have been carefully studied. The most studied member of this group is phage M13 which has found extensive use as a cloning vector and DNA sequencing vehicle in genetic engineering. This filamentous phage infects only male cells (containing F plasmid), entering after attachment to the male specific pilus. Even though this phage is linear helical in shape they possess circular ssDNA. The virion of M13 measures 6 x 860 nm. It has another interesting property of being released from the cell without killing the host cell. Thus a cell infected with phage M13 can continue to grow all the while releasing virus particles. Virus infection causes a slowing of cell growth, but otherwise a cell is able to coexist with its virus. Plaques are seen only as areas of reduced cell growth in the bacterial lawn.

Aspects of DNA replication in M13 phage are essentially similar to that of ϕ x 174 (Fig. 9.20). The progeny virions are released by a budding process. The virus particle is always released from the cell with the end containing the A-protein first (Fig. 9.21). As the assembly of the mature virions occurs on the inner surface of the cytoplasmic membrane just before budding process, there will be no accumulation of intracellular virus particles.

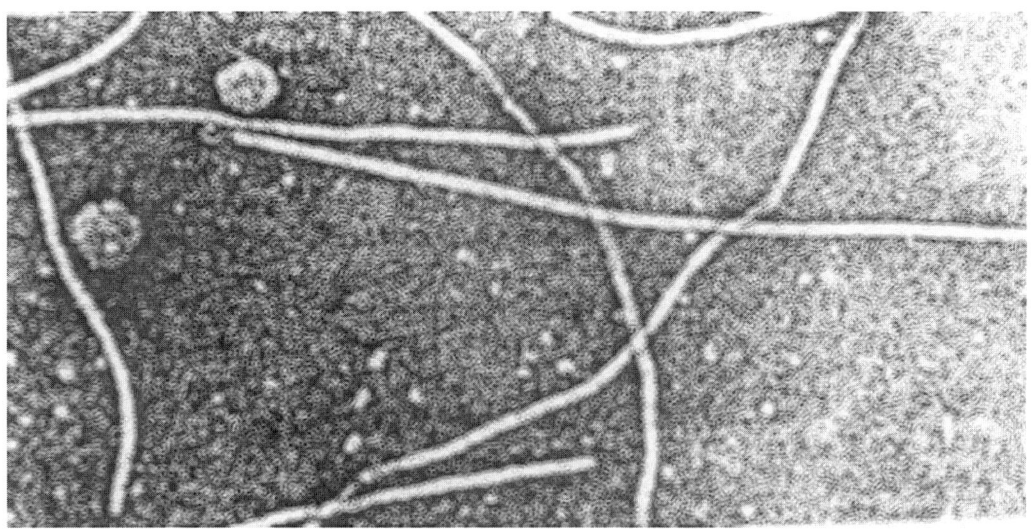

Fig. 9.19. Electron micrograph of the filamentous *E. coli* phage M13 [Courtesy of Robley Williams]

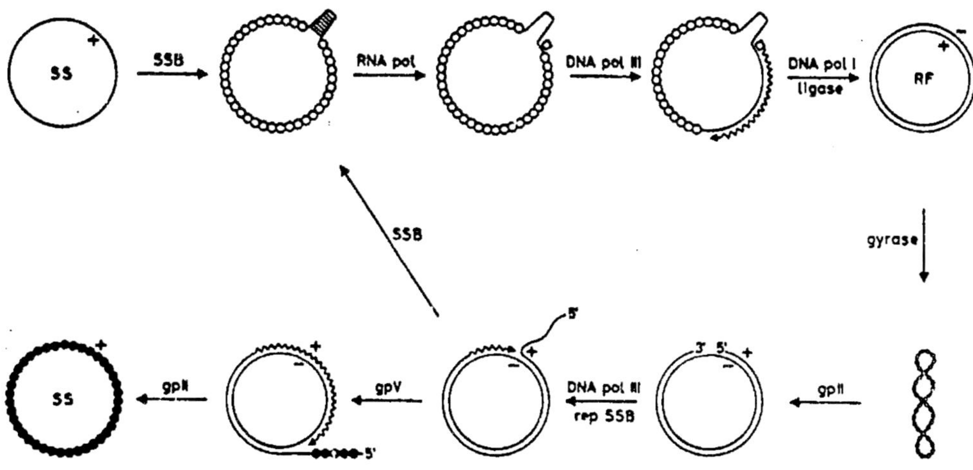

Fig. 9.20. The mechanism of DNA replication in phage M13. The upper part shows the discontinuous synthesis of the complementary (–) strand, and the lower part shows the continuous synthesis of the viral (+) strand. Small circles, SSB host protein small squares, RNA primers; dots, gpV (From Keepel, F., Fayer, O. and Georgopoulos, C., *in The Bacteriophages* Vol. 2)

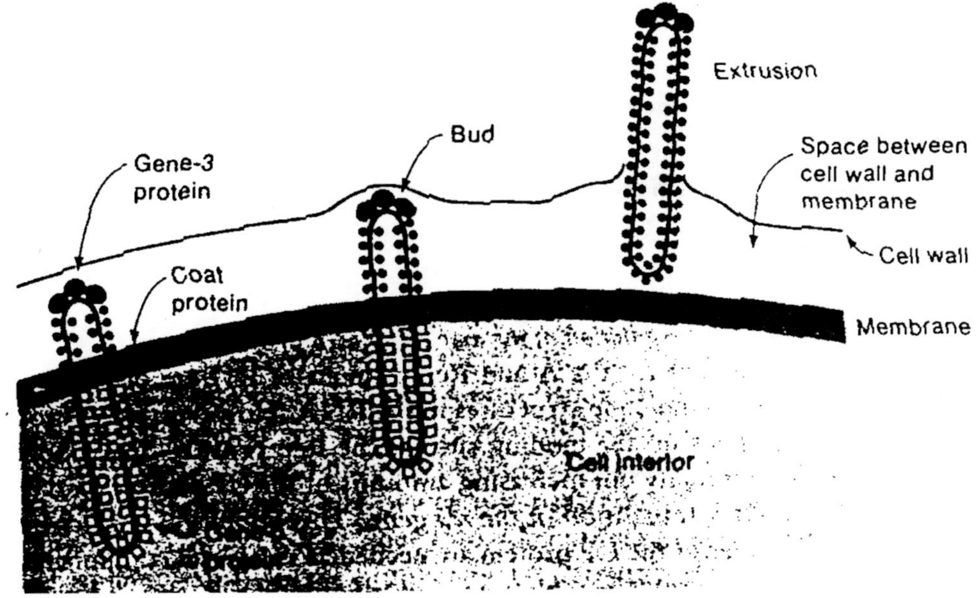

Fig. 9.21. Assmebly and extrusion of M13 phage.

7. REPLICATION OF ssRNA PHAGES : COLIPHAGES MS2, Qβ

The first single-stranded RNA phage was first discovered in 1961 by **Loeb and Zinder**. Since then several ssRNA phages have been isolated. The best known are *E. coli* phage f2, R17, MS2, M12 and Qβ which have served as model systems to explore a variety problems in molecular biology. The RNA coliphages form the family Leviviridae. Most research has centered around MS2 and Qβ, but it is assumed that features found in these two viruses hold good for other viruses of this family.

Morphology and genome

RNA phages are tail less icosahedrons (Fig. 9.22A). The diameter of the phage is 26 nm and the protein shell is about 2 nm wide. Each virion contains 180 copies of the coat protein and one copy of the maturation, or A, protein. Genetic map of MS2 is shown in Fig. 9.23. The genome is single positive single-stranded linear RNA, and consists of about 3600 nucleotides. The nucleotide sequences of MS2, R17 and Qβ phages are completely worked out. The genome contains three genes encoding a coat protein (CP), a maturation protein A, and an RNA replicase (Rep) (Fig. 9.22B, 9.23). The RNA molecule is both a replication template and mRNA. Hence, (i) neither DNA nor RNA polymerase is needed in the life cycle and (ii) regulation must occur at translational level.

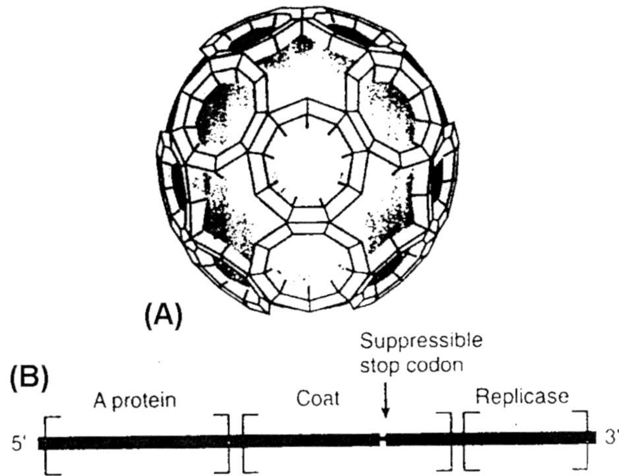

Fig. 9.22. Phage Qβ. **A.** Morphology, **B.** Genome

Replication (Fig. 9.24)

i. Attachment and entry: The RNA gains access to the interior of cell via pili. In *E. coli* the sex (or F) pili are employed as vehicle. Transfer of RNA via pili is not essential for infection, however, the cells that lack pili can be infected if they are converted to spheroplasts. The attachment of the phage to the sides of the pili proceeds via the maturation (A) protein. The coat protein is dispensable for this process. Contact of the phage with the pilus results in cleavage of the A protein which probably triggers the ejection of the tightly packed RNA from the virus shell. The 5' end of the RNA begins to move along the pilus towards the cell. Immediately after entry, a ribosome attaches to a binding site at the beginning of CP gene, which is the centermost gene in the RNA molecule.

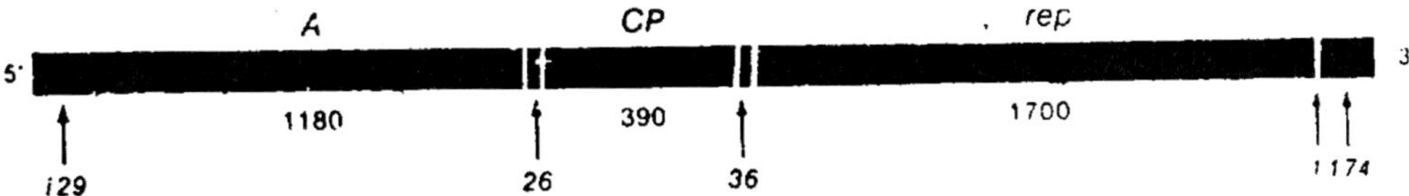

Fig. 9.23. The genetic map of phage MS2 showing the positions and number of bases in each gene. The italicized numbers are the lengths of leaders and spacers. **A**-Maturation protein; **CP**-Coat protein; rep-**RNA** replicase

ii. Translational regulation of three gene products: The ribosomal binding sites for the A and Rep proteins of the phage are blocked by the presence of stem-and-loop structures in the RNA. However, translation of the CP gene opens up the binding site for the Rep protein. Thus, both proteins are made initially but increasing amounts of the CP protein bind to rep site and partially black translation of the rep gene. Thus only a small and required level of Rep A protein (replicase) is synthesized.

iii. Replication of the phage RNA: Most of our knowledge on replication has been obtained in the Qβ system but it is assumed that the same principles apply to other phages. Qβ replicase is tetrameric holoenzyme and contains four different proteins called I to IV or subunits (α β γ and δ). Subunit-I was identified as ribosomal protein SI, and subunits III and IV are the translational elongation factors EF-Tu and EF-Ts. These three proteins, required for protein synthesis, are recruited by the phage to assist RNA synthesis.

Replication proceeds via the synthesis of a free negative strand. Annealed positive and negative strands are not a substrate for the replicase enzyme. The requirements for copying the positive and negative strand are different. To copy the negative strand, only subunits II, III and IV are required, whereas positive strand replication also needs subunit I. In addition, copying the positive strand is greately dependent on the product of the bacterial *hfq* gene called host factor. Replication of the negative strand depends on structural elements at both the 5' and 3' termini and is thus basically different from positive strand copying.

iv. Synthesis of A protein: The ribosomal binding site for the A protein is never available on a free (+) strand because of base pairing. However, replication of a (-) strand begins at the 3' – terminus adjacent to the A gene. Just after (+) strand synthesis begins there is a brief period during which the ribosomal binding site for the A protein is free, the complementary segment of the (+) strand having not yet been replicated during this time the A gene is translated. This is only for a brief period. Therefore, the number of A proteins is maintained roughly equal to the number of (+) strands. Each A protein later becomes bound to one RNA molecule at an unidentified binding site. It is presumed that the A protein facilitates the interaction of the RNA with the CP molecule.

v. Particle assembly : CP proteins, when synthesized in sufficient quantities, spontaneously aggregate around the newly synthesized (+) strand and form an icosahedral shell.

vi. Cell lysis and release : The assembled phage particles form huge crystalline arrays within the host. These crystals somehow damage the cell membrane, causing cell lysis in the absence of a lytic enzyme. Thus, the timing of lysis is not precise. Lysis occurs from 30-60 minutes after infection. The burst size is about 10,000.

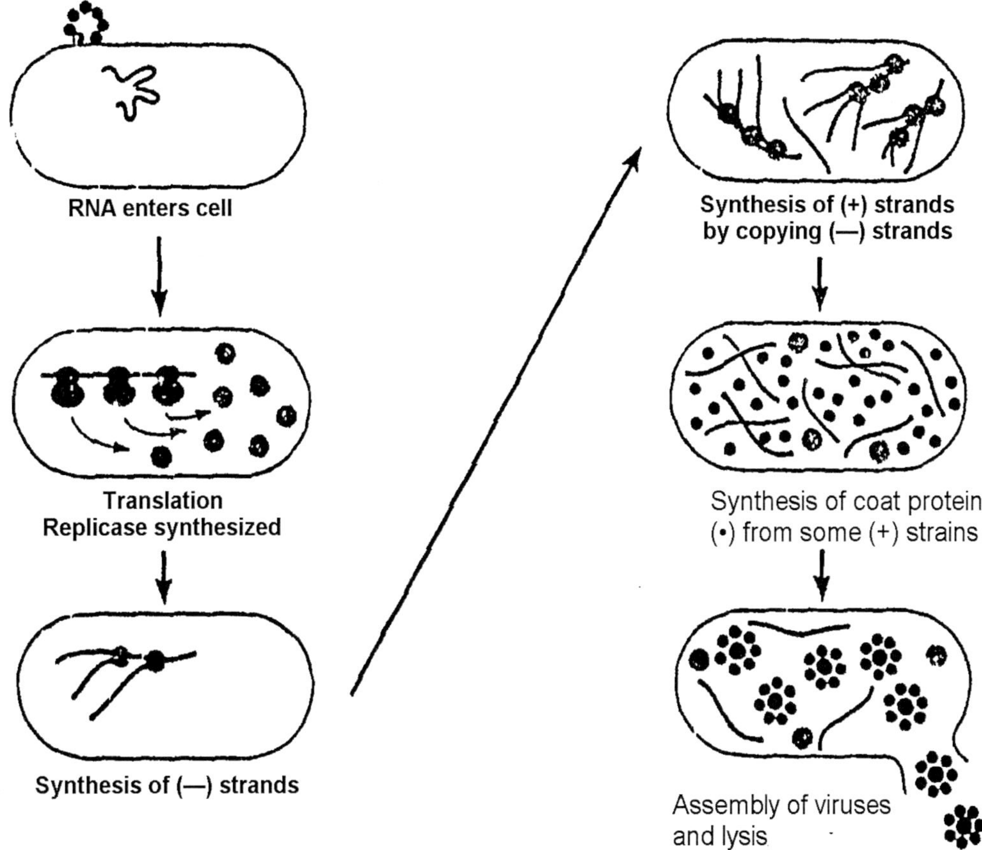

Fig. 9.24. The levivirus life cycle. (Adapted from Freifelder, D. *Molecular Biology*

8. REPLICATION OF dsRNA PHAGES : ϕ x 6

Phage ϕ x 6, until recently the only member of the family *Cystoviridae,* infects phytopathogenic pseudomonads. The first reports in 1973 described two unique features in this phage : (i) the special virus had a lipid envelope and (ii) the genome was composed of dsRNA. It is a virulent phage.

Morphology

The overall diameter of this enveloped phage is about 85 nm. The spike protein extends from the membrane surface and the membrane is approximately 6 nm thick. The nucleocapsid has a diameter of 58 nm and the polymerase complex about 45 nm (Fig. 9.25). The mass of the virion is about 99,000 kD and of nucleocapsid is 40,000 kD. The virion is composed of approximately 70% protein, 20% lipid and 10% RNA. The spike protein P3 is associated with the integral membrane protein P6 to form an adsorption fusion complex. In addition, there are three additional membrane associated proteins (P9, P10, P13). The lytic enzyme P5 is located between the membrane and nucleocapsid. Inside the shell of P8 is the icosahedral polymerase complex composed of four protein species P1, P2, P4, P7. P1 forms the structural dodecahedral framework upon which the rest of the polymerase complex proteins are attached. P2 contains an RNA – dependent RNA polymerase active site. P4, a potent unspecific nucleoside triphosphate phosphohydrolase, provides the energy for RNA packaging. P7 is necessary for stable encapsidation. The polymerase complex encloses three polycistronic genome segments, small (S, 2948 bp), medium (M, 4061) and large (L, 6374 bp). The membrane phospholipids are contributed by host.

Genome

The entire genome is cloned and sequenced and the genes are localized. Genes are well organized in three segments. The genes of the L segment code for the viral polymerase complex proteins P1, P2, P4 and P7 as well as a small protein P14 of unknown function. The M segment codes for the attachment fusion complex protein P3 and P6, and two small membrane – associated proteins P10 and P13. The genes of the S segment code for the nucleocapsid surface protein P8, the major membrane protein P9 and the lytic enzyme P5, as well as the nonstructural protein P12 needed in membrane assembly. Each genome segment has noncoding regions of a few hundred base pairs at each end. Similarity exists between the ends of the three segments. The 5' terminals contains the packaging (pac) site and the plus-strand synthesis start signal, whereas the 3' end encodes the functions needed for the minus strand synthesis initiation.

Replication

i. Adsorption and penetration : Protein P3 mediates the adsorption of the ϕ x 6 particle to the receptor, the host pilus. It is suggested that pilus retraction brings the virus particle into contact with the host outer membrane (Fig. 9.26). Protein P6 mediates the fusion of the virus membrane with the host membrane. This exposes the peptidoglycan layer to the virion associated lytic enzyme (P5). The peptidoglycan is locally digested, allowing the nucleocapsid to face the host cell membrane. The nucleocapsid penetrates the cell membrane via endocytosis and driven by the acidification of the nucleocapsid surface protein P8. Upon entry, the vesicle around the nucleocapsid gets dissociated and the genome is released to the cytosol of the host.

ii. Replication of RNA : Following the entry, the viral transcriptase starts the early transcription via a semiconservative mechanism (Fig. 9.27). Approximately, equal amounts of full length transcripts of all the genome segments are synthesized. However, only the L segment is actively translated. This leads to the assembly of polymerase complexes (procapsids) inside the cell. These particles enclose one of each type of messenger molecule. The packaging is followed by minus-strand synthesis (replication), leading to the dsRNA genomic segments inside the procapsid.

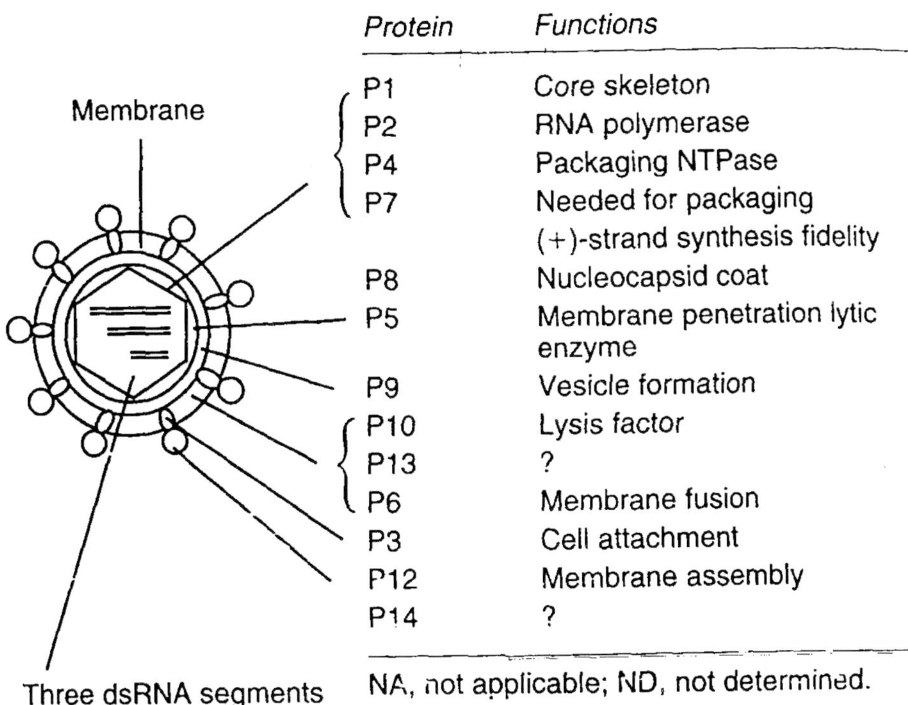

Protein	*Functions*
P1	Core skeleton
P2	RNA polymerase
P4	Packaging NTPase
P7	Needed for packaging (+)-strand synthesis fidelity
P8	Nucleocapsid coat
P5	Membrane penetration lytic enzyme
P9	Vesicle formation
P10	Lysis factor
P13	?
P6	Membrane fusion
P3	Cell attachment
P12	Membrane assembly
P14	?

NA, not applicable; ND, not determined.

Fig. 9.25. Structure, proteins and their functions of phage ϕ 6

iii. Assembly and release of virions : From the procapsid particles late genes are expressed. Majority of the transcripts originate from the M and S segments, which are then translated, leading to the synthesis of late proteins. The synthesis of nucleocapsid shell protein P8 leads to its assembly on to the polymerase complex. With the aid of the assembly factor P12, the nucleocapsid is enveloped inside the cell with a membrane derived from host cell membrane. Meanwhile, P3 spikes are assembled on particle. The mature virions are released by lysis. The cell lysis is brought about by the phage-specific lytic enzyme P5 accompanied by membrane protein P10.

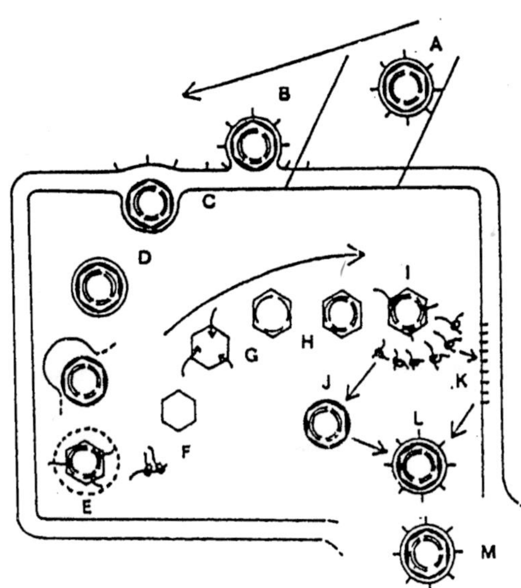

Fig. 9.26. Events of the ϕ 6 life cycles (A) Adsorption to host pilus, (B) Fusion of viral envelope with bacterial outer membrane, (C) Local digestion of cell wall peptidogl- ycam, (D) Penetration of nucleocapsid through cytoplasmic membrane, (E) Nucleo- capsid uncoating and early transcription, (F) Synthesis of early proteins and assembly of procapsids, (G) Packaging of viral ssRNAs, (H) dsDNA synthesis (I) Late transcription, (J) Synthesis and assembly of nucleocapsid shell protein, (K) Synthesis of viral envelope proteins on the cytoplasmic membrane, (L) Intrace-llular translocation of envelopes on the nucleocapsids, (M) Release of progeny phage by cell lysis. (From Olkkonen, V., Ph.D. thesis, Dept. of Genetics, Univ. of Helsinki, 1991).

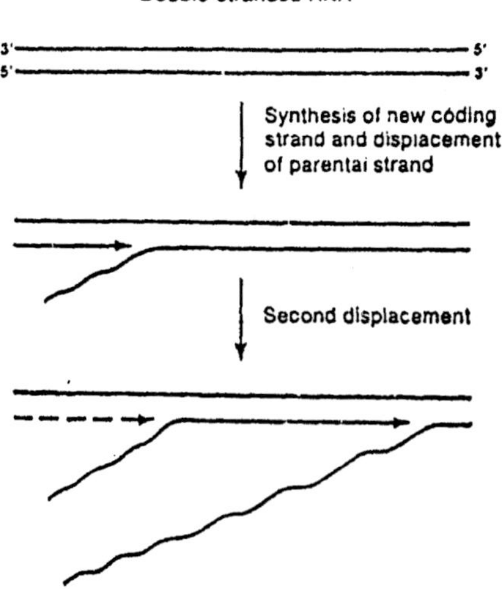

Fig. 9.27. Synthesis of mRNA in ϕ 6 by displacement of a parental RNA strand. (From Freifelder, D., *Molecular Biology, An Introduction to Prokaryotes and Eukaryotes)*

Further reading

Bamford, DH (1999). ϕ x 6 (*Cystoviridae*). In Encyclopedia Virology, Granoff, A. and Webster, R.G. eds. Academic Press, New York.

Cann, AJ (1997). Principles of Molecular Virology, Academic Press, New York.

Davies, HS *et al.* (1990). Microbiology, 4th ed. Philadelphia: JB Lippincot.

Freifelder, D (1987). Molecular Biology: A comprehensive introduction to prokaryotes and eukaryotes. 2nd ed. New York: Van Natrand Reinhold.

Hendrix, RW and Garcea, RL (1994). Capsid assembly of ds DNA viruses, Semin, Virol. 5, 15.

Kendrew, J (Ed. in Chief) (1994). Encyclopedia of Molecular Biology, Blackwell Science, Oxford.

Keppel, F, Fayet, O, Georgopoulos, C (1998). Strategies of bacteriophage DNA replication, in the Bacteriophages, Vol. 2 Calendar, R, Ed., Plenum Press, New York.

Model, P and Russel, M (1988). Filamentous bacteriophage. In the Bacteriophages, Vol. 2, Calender R (Ed.) Plenum Press, New York.

10

PHAGE-HOST INTERACTIONS

There are a number of reasons for interest in the interactions of phage-host interactions: (i) some phages have served as the models to guide studies on animal viruses that cause tumors, (ii) phages are important for medical reasons, either directly through involvement in pathogenic mechanisms or indirectly, through selective associations with pathogenic hosts which can be exploited for identification purposes, (iii) phages are also a major concern in industries, since they cause significant economic losses. Phages exhibit a great variation in morphology, nucleic acids, genetic information contents and modes of replication. As a consequence of these variations, different viruses show wide variations in biological effects on the host cell. Infections with some phages cause minimal effects, producing an infected cell with biological properties similar to uninfected cell. However, for many phage infections, the biological consequences for the cell are disastrous : the cell is destroyed in the process of producing virus progeny. Here we discuss these two types of effects under separate headings.

1. PRODUCTIVE INFECTIONS : LYSIS

In this type of infections, the infective phage jeopardises all the metabolic functions of the host leading to its death. Owing to the death and distruption of the cell, such infections are called **lytic** (or **cytocidal**). In persuance of this objective, phages employ different strategies to inhibit cellular processes. Host defense mechanisms against the virus are systematically incapacitated.

i. Inhibition of host cellular functions

Some of the simplest phages like phage fd establish a productive infection without killing or lysing the cells. The infected cells continue to replicate themselves as well as the virus, although the drain on cellular resources leads to poor bacterial growth. The virions are released by a nonlethal process of secretion. Most bacteriophages, on the otherhand, inhibit atleast some specific cell functions during infection and thus eventually kill the host cell during release of progeny viruses. Phage T4, one of the largest phages, has several of these weapons, some offensive and some defensive. For example, the product of phage gene 2 protects the end of the phage DNA strands from the attack of host exonuclease (exo V). T-even phages and also T_7 inhibit the transcription of host genes into mRNA. Host protein synthesis is also halted early during infection with T-even phages. T-even phages also encode nucleases that rapidly destroy host dsDNA. Some phages may inhibit nucleic acid synthesis of host even at precursor level. For example, T-even phages inhibit the synthesis of cytosine and indirectly contribute for the production of hydroxymethyl cytosine (HMC).

ii. Inhibition of host restriction enzymes

Many bacteria encode sequence specific DNA endonucleases or restriction enzymes that distinguish bacteria DNA from foreign DNA of viruses and degrade the latter. Thus, it is a antiviral defense mechanism. Phages like T_3 employ a strategy for neutralizing some host restriction enzyme systems that depend on methylation of the host DNA for protection. T_3 has the unusual ability to produce an enzyme, SAMase which hydrolyzes s-adenosyl methionine (SAM). This strange T_3 gene product has other effects on the host. Many DNA phages donot seem to possess the ability to inhibit restriction systems. Their DNA can therefore be degraded shortly after infection. However, if by chance the DNA is modified by the host system before it is restricted, the DNA can replicate and all progeny will also be modified as if they consisted of cellular DNA. The new phage will not be restricted upon infection of another cell with the same restriction system.

iii. Lysis of host cells

When a virulent phage replicates and produces progeny virions, there must be some mechanism for release of virus. Usually this involves complete disruption of the cell or lysis. Phages have developed different mechanisms for destablizing the cell membrane and cell wall. During normal infections late viral gene products disrupt the cell membrane. T4 gpt and phage lambda S both play this role. The second barrier, the peptidoglycan mesh of the wall may be attacked in one of two ways : T_4 encodes a **lysozyme** which cleaves the glycan component, and lambda produces an endolysin which breaks the peptide linkages. Interestingly, in many of the more complex phages, control genes have evolved to regulate the activities of enzymes involved in lysis.

Lysis is not mandatory for release of all phages. For example, phage fd particles are released without the lysis. As single stranded phage DNA passes through the membrane, the coat protein displaces another phage gene product (gpv) associated with the DNA inside the cell. The phage coat protein guides the release of progeny virions.

2. NONPRODUCTIVE INFECTIONS : LYSOGENY

Lysogeny involves extensive regulation of phage functions which allows the viral genome to remain for an indefinite time inside the infected cell without killing it or producing virions. Also, the replication of the phage genome becomes coordinated with replication of the host genome. Many a times, phage DNA integrates into host DNA molecule, thus its replication is linked to cell DNA replication. The phages that possess the strict control over their replicative urges can establish lysogeny, and these are called **temperate phages.** Temperate phages usually donot affect cellular replication functions, they may cause other cellular changes that exhibit profound biological effects, not only on the lysogens, but also on higher organisms serving as hosts to these bacteria.

i. Immunity

It is well known that a lysogenic cell is immue to superinfection by a closely related phage. Under ordinary conditions, the lysogenic cell is immune to further lysogenization and a superinfecting phage cannot even undergo a lytic, productive cycle. The *repressor* protein accounts for this phenomenon, because it immediately combines with the DNA of the superinfecting phage. The repressor renders the phage DNA unable to replicate. For instance, T-even phages prevent superinfection by causing the synthesis of a deoxyribonuclease that breaks down the DNA of a superinfecting phage. Temperate phages, like (λ), use this more subtle way but equally effective means of preventing a superinfection by simply ``turning off" the genes of the superinfecting phage.

ii. Integration

When there is a committment to lysogeny, phage DNA cannot replicate independently and further it gets integrated into host DNA that allows it to be replicated and distributed to daughter cells just like any other part of the host chromosome. Location of the integration alter the cellular functions of the host. Phage Lambda has a highly preferred location i.e. between *gal* and *bio* operons. If this site is missing, the phage DNA will sometimes be inserted at some other location in the host chromosome. Similarly, another temperate phage of *E. coli*, phage Mu, has the unusual ability to integrate with high efficiency virtually anywhere in the host DNA. Often this integration procedure results in the disruption of host gene, and this eliminates one or more functions. Thus, the temperate phages exert a powerful biological effects on the host.

iii. Induction

In order to serve as proper template for production of new (λ) DNA molecules, the integrated prophage must be excised and this excision process and commencement of lytic program is known as induction. Although the excision process is quite precise, occasionally the process goes awry and excision occurs somewhere else besides the attachment sites. The resultant phage circular DNA may contain cellular DNA from regions adjoining the integrated prophage. Such phages, known as **transducing phages**, transfer the cellular genes from one host to other. Further, the host cell is left with a deficient mutation and recipient new host becomes partially diploid.

3. CONVERSION OF HOST CELL PHENOTYPE

Apart from the above discussed effects of phages on host cell biology, phage infection alters some properties of the host. Changes in bacterial surface properties or pathogenic characters are generally referred to as **phage-induced phenotypic conversions**.

i. Changes in cell surface properties

The classical example of a prophage altering the surface of a lysogen involves sugar residues of polysaccharides of LPS in *Salmonella*. The changes brought about by phages can be due to phage products which interfere with antigen synthesis. These products may be inhibitors of specific cellular enzymes involved in polymerization of sugars forming the surface antigens. Also, bacterial enzymes which chemically modify surface sugars by addition of acetyl groups can be suppressed. Similarly, a number of other bacterial typing antigens have been shown to undergo similar alterations in response to lysogenic infections.

Phage conversion of surface antigens have another important consequence for the lysogen : altered susceptibility to infection with phages. The epsilon phages of *Salmonella* are the best examples to illustrate this point. In this case, the some sugar residues on the surface LPS that are altered by each phage are also part of the receptor for that same phage. By eliminating the phage receptors, conversion provides valuable added protection for the residence of the prophage. This mechanism of resistance to superinfection is often called **immunity**. It is also employed as a weapon of self defense by cells capable of producing bacteriocins.

ii. Pathogenic conversions

Phage infections also bring about the changes in the pathogenicity of the host bacterium. Two of the most remarkable and medically significant changes in bacterial host phenotype are brought about by prophages in *Corynebacterium diphtheriae* and *Clostridium botulinum.* These pathogens prod-uce two of the most potent toxins known and they are responsible for the severe pharyngitic disease diphtheria and food poisoning known as botulism respectively. A comparison with nontoxigenic variants of the bacteria have led to the discovery that these toxic proteins are actually products of proph-ages of the nontoxigenic strains. In the phages producing diphtheria toxin, it has been demonstrated that toxin production is remarkably free of regulatory influences of the prophage or cell. Repressed lysogenes produce toxin as efficiently as lytically infected cells and even when limitation of nutritional sources inhibits cell growth and reduces the overall rate of protein synthesis as much as 10-fold, the lysogen still produces toxin at the normal rate. Inhibitory effect of iron in the culture medium has been identified as a modulating influence on expression of the *tox* gene. Phage-induced conversion has now proved to be a fairly common phenomena of bacterial pathogenesis. In certain streptococci (Group A, beta hemolytic), conversion of nonlysogenic Group A streptococci to toxigenic strain is also thought to be due to phage conversion. Interestingly, certain strains of staphylococci produce hemolytic toxins, but in this case conversion by lysogenic infection with certain phage results in shut-off of toxin production. Since several toxins classified as bacterial in origin have by now been shown to be phage gene products, it will not be surpressing if others are added to this list in the near future.

4. MODULATION OF HOST-VIRULENCE BY PHAGE

Some bacteriophages encode factors that modulate the virulence of host bacterium against a third organism. This change in properties of the host can occur at different levels such as bacterial adhesion, spread through human tissues, exotoxin production or protection against immune defences.

The phage SM1 encodes PblA and PblB that promote binding of *Streptococcus mitis* to human platelets. The phage Gifsy-2 encodes a periplamic superoxide dismutase that protects the intracellular bacterium against oxidative stress. Some phages encode-invasins that allow the bacteria to penetrate the eukaryotic host tissue layers or cells.

For example, hyaluronidase breaks down hyaluronic, a constituent of the extracellular matrix to increase tissue permeability and allow bacterial spread. Some phage encoded virulence factors allow the bacteria to evade the eukaryotic host immune system. These factors can alter bacterial antigenicity and thus recognition by the host immune system or provide serum resistance.

Further reading

Auckermann HW and DuBow, MS (1987). Viruses of Prokaryotes, Vol I. Boca Raton : CRC Press.

Bickle TA and Kruger DH (1993). The biology of DNA restriction. *Microbiol. Rev.* 57: 434.

Knipe, DM (2002). Virus-host cell interactions. In Fields BN *et al.* (eds.) Virology, 3rd Edn. Raven Press, New York.

Kruger, DH and Reuter, M. (1999). Host controlled restriction and modification. In Encyclopedia of Virology Vol. II eds. Granoff, A and Webster, RG Academic Press, New York.

Norkin LC (1995). Virus receptors : Implications for pathogenesis and the design of antiviral agents *Clin. Microbiol rev.* 8(2) : 293.

Singer, RA (1976). Lysogeny and toxinogenicity in *Corynebacterium diphtheriae.* In Mechanisms in bacterial toxicology, ed. AW Bernheimer. John Wiley, New York, pp. 1-30.

Whitaker-Dowling, P and Youngner, JS (1999). Virus-host cell interactions. In Encyclopedia of Virology Vol III eds. Granoff, A and Webster, RG, Academic Press, New York.

11

IMPORTANCE AND APPLICATION OF PHAGES

Phages have been found in over 130 bacterial genera distributed all over the world. Most phages have been found in economically important bacterial groups like enterobacteria, bacilli, clostridia, lactobacilli, pseudomonads, staphylococci and streptococci. About half of the phages have been found in cultures of lysogenic bacteria. Phages by infecting the bacterial hosts alter their characteristics. Non pathogenic may become pathogenic or vice versa. Phages are also important in fermentation industry. Some times they cause colossal losses. Phages possess all types of nucleic acids and wide variation is found in the replication patterns. Because of their growth and short life cycles, many phages serve as good models for understanding the life at molecular level. Some phages are employed as cloning vectors. It is not exaggeration to say that many of the developments in molecular biology are due to the understanding of phages. Applications and economic importance of some phages are explained in this chapter.

1. PHAGES IN THE IDENTIFICATION AND CLASSIFICATION OF BACTERIA

Bacteriophages have been used in bacterial identification and classification since the pioneering work of Sonnenschein in 1925. The specificity of phage adsorption and lytic propagation on bacterial species and strains made them ideal agents for bacterial identification. We can use this to distinguish between strains of the same species. This is known as **phage typing**. Phage typing is relatively easy and lack necessity for sophisticated equipment. Many of the typing phages are extremely stable, readily available and can be stored for long periods. Additionally, automated phage typing allows rapid performance of multiple tests, a particular necessity in clinical situations. At one time phage typing was as popular as other means of bacterial identification including biochemical/physiological markers and serological tests. As a holistic approach, now-a-days, phage typing is used in the `battery of tests' for bacterial identification.

Phage typing is widely used in the identification of medically important bacteria. The best example of the use of phage typing is in distinguishing strains of *Staphylococcus aureus*. It has also industrial and environmental applications where it is desirable to keep track of a limited number of bacterial species and/or strains. Bacterial strains causing a particular industrial or epidemiological problem may differ from other strains in only a limited number of characteristics and phages can be used to track these. As the knowledge of bacterial characteristics improves, phage typing may also develop through the use of new and/or mutant phages for the characterization and identification of bacterial species and subspecies.

2. PHAGES TO SCREEN CARCINOGENS

Phages like lambda have also proved to be valuable tools for screening carcinogens. A test, similar to Ames test, known as ***Devoret test*** uses *E. coli* strains lysogenic for phage (λ). Many carcinogens, however, cause DNA damage and are inducing agents. The induction can be detected by a simple method. This test uses liver extracts in the agar to activate carcinogens. The Devoret test is more sensitive than Ames test. It requires lower concentration of carcinogens thus, it is capable of detecting weak carcinogens that are not detected by the Ames test.

3. PHAGES IN MOLECULAR BIOLOGY AND MICROBIAL GENETICS

Bacteriophages have played an important role in the development of molecular genetics. Studies on phages, in particular the *E. coli* phages T_4 and lambda and the *Salmonella typhimurium* phage P22, have led to the discovery of many basic phenomena concerning replication, transcription, regulation and recombination.

The application of restriction endonucleases has revolutionized molecular biological research. Bacterial viruses have led to the discovery of this class of enzymes. Restriction enzymes allow bacteria to monitor the origin of incoming DNA and to destroy if it is recognised as a foreign. Discovery of restriction enzymes in late 1960s by Arber and Smith was one of the first breakthroughs in the history of recombinant DNA technology. Using these restriction enzymes, restriction maps of different genomes are being prepared, which constitutes an important DNA marker.

The study of transcription of phage N4 has provided novel mechanisms for RNA polymerase promoter recognition, activation and the supercoil-dependant generation of nonocanonical DNA structures. N4 polymerase II is the smallest DNA dependant RNA polymerase described to date and an excellent candidate for structure-function studies.

4. PHAGES AS CLONING VECTORS

A number of phages, both dsDNA and ssDNA phages have been extensively used as cloning vectors.

Lambda phage : The unique sequence of virion DNA, the ability to package DNA *in vitro* into phage particles and the formation of stable lysogens have made lambda a convenient vector for DNA cloning. Many derivatives of lambda phage (e.g. EMBL3 and λ DASH) have been constructed with improved utility as cloning vectors. In the lambda genome, the genes necessary for lytic development comprise only about 60% of the 48 kb. Deletions of the most of the dispensible 40%, increase the possible insert length. Even longer inserts can be cloned into vectors containing the *cos* sites but lacking much of the functional phage DNA, which can be perpetuated in the presence of a helper lambda phage.

The filamentous phages, like M13, have been used extensively as cloning vectors. Since there is no constraint on the size of the DNA packaged, theoretically any segment of DNA can be inserted into a noncoding region with no deleterious effect. Other cloning vehicles, called phagemids, have been constructed; they combine the features of plasmid and the phage.

Other phages like P1 and SPOI can also be used as cloning vehicles. DNA fragments upto 100 kb can be cloned into P1 phage.

5. TRANSPOSABLE PHAGES AS GENETIC TOOLS

The properties of transposable phages like Mu that enable integration into random locations in host DNA, and that catalyze a wide variety of DNA rearrangements, have allowed molecular biologists to prepare a number of derivatives of these phages for *in vitro* genetic engineering. One important characteristics of these phages is to act as insertional mutagens and thus allow the creation of stable recessive mutations. The use of cloned antibiotic resistance markers within the phage genomes allows the facile cloning and characterization of the gene in which the prophage is inserted. The use of Mu and other transposable phages as *in vivo* gene manipulation vehicles has greately aided the studies of their host bacteria at molecular level. Transposable phages also provide model paradigms for the study of gene expression, DNA inversion and DNA transposition.

6. TRANSDUCTION AND LINKAGE MAPPING

Phage mediated genetic transfer in bacteria is known as transduction. Phage particles that contain bacterial DNA are called transducing particles. Transduction provides a valuable tool for linkage analysis of short regions of bacterial genome. Co-transduction can be used to determine a genetic map of the relative distances between genetic markers and the order of the markers. The use of co-transduction to determine gene order is based on the fact that the closer genes are to each other, the greater the probability they will be cotransduced. That is, the frequency of cotransduction is inversely proportional to the distance between two genes.

Another application of transduction in microbial genetics is localized random mutagenesis. The transducing phage lysate is subjected to mutagenesis before infection of the recipient strain. A marker linked to the gene to be mutagenized can then serve for selection of transductants. Screening easily reveals cotransductants carrying a mutation in the linked gene under study which had been the actual mutagenesis target.

The important application of transduction in the wide fields of fundamental and applied research is the general use of the concept of the host-vector system which has emerged from the understanding of structure and functions of specialized tranducing genomes. The obvious application of this is *in vitro* recombinant techniques. By this technology, molecular genetics, which was formerly mainly limited to microbial genes, has become accessible for application to the genes of any living organism.

P22 phage came to lime-light with the discovery of bacterial transduction by Zinder and Lederberg in 1952. It mediates generalized transduction of its host *Salmonella typhimurium.* P22 continues to be an indispensable tool of *Salmonella* genetics. The molecular genetics of P22 itself has been studied intensively. Research with P22 has contributed to our understanding of a number of biological processes, including replication, recombination and regulation, protein folding and assembly.

7. FINE STRUCTURE OF THE GENE

Elucidation of the fine structure of the gene became possible with bacteriophages. The most extensive fine structure map constructed to date is that of phage T_4 rII locus. Seymour Benzer (1955, 1956) isolated over 3000 independant mutant strains carrying mutations that mapped within the rII locus. His detailed fine structure analysis of these rII mutants showed that the mutations mapped at over 300 distinct sites, all located in the two genes that make up the rII locus. Benzer showed that genes were not divisible but could be subdivided by recombination. His work revealed that genes were composed of small units and that recombination could take place between but not within these units. He termed the unit of recombination as **recon.** Similarly terms **mutan** (unit of mutation) and **cistron** were also proposed.

8. PHAGE DISPLAY

The filamentous phages have been used as vehicles to enrich for peptides or proteins with specific properties, a process termed **phage display.** Sequences coding small peptides, or epitopes are commonly inserted into the N-terminal region of pIII. The resulting phage particles display these foreign peptides as part of the chimeric pIII at the tip of the bacteriophage. In this way, large `epitope libraries' are made which can be screened for a range of binding activities. Peptides have also been displayed as fusion proteins at the N-terminus of pVIII and reportedly with pVI. Any small peptides can be displayed fused to either pVI or pVIII. Large proteins can be displayed if enough wild type capsid proteins are available during assembly. To accomplish this, foreign proteins fused to the appropriate portion of PIII or PVIII are expressed from a plasmid in an infected bacterium.

The principle of the technique is that random fragments of DNA are incorporated into the genome of a bacteriophage (commonly the filamentous phage M13) so that the protein encoded by each DNA fragment is fused to one of the proteins that coats the phage particle. This produces a mixture of phages (known as a library) which display different proteins on their surface. Within this mixture, a small number of phages will display the protein that binds to the ligand of interest. If the surface of a well in a microtitre plate is coated with the chosen ligand, the required phage particles will bind to it and rest can be washed away. The bound phage, enriched for particles displaying the correct protein can then be eluted and amplified by infecting a suitable *E.coli* strain. This process, known as **panning,** can be repeated several times to obtain a pure phage preparation that provide a source of the gene required. In this way, genes of interest can be enriched from vast libraries simply as a function of the affinity of the corresponding polypeptide for a specific target molecule.

Phage display technology offers potential tools for development of therapeutic agents, vaccines, diagnostic reagents, as well as gene and drug delivery systems. Determination of interaction partners of organic (proteins, polysaccharides or DNAs) or inorganic compounds and also in vitro protein engineering are the major applications of phage display technology. More recently, a number of attempts have been made for using in vitro phage display technology in medical science by designing humanized antibodies or peptides and development of new pharmaceutics for various maladies such as cancer, autoimmune and inflammatory diseases, metabolic and allergic disorders. This methodology has usually been applied for the production and isolation of the antibodies, identification of peptide agonists and antagonists for receptors, characterization of novel protein/peptide-protein/DNA interactions, demonstration and mapping of highly carbohydrate-dependent epitopes and protein functional epitopes, recognition of specific antigens in or on the surface of the tumor cells as the therapeutic targets, in addition to molecular imaging with the use of fluorescently labeled phage. At present, > 20 phage display-derived antibody and peptides are in late-stage clinical trials or have been approved (Tables 11.1 and 11.2)

Table 11.1. Selected approved or under clinical development phage display-derived therapeutic antibodies.

Product	Manufacturer	Type	Target	Phase
Adalimumab	Abbott Laboratories	IgG1	TNFa	Approved
Belimumab (Benlysta)	GlaxoSmithKline	IgG1	BLyS	Approved
Raxibacumab (ABthrax)	GlaxoSmithKline	IgG1	*PA83 of Bacillus anthracis*	Approved
Ranibizumab (Lucentis)	Genentech	Fab fragment	VEGF-A	Approved
Ramucirumab (IMC-1121B, Cyramza)	Eli Lilly	IgG1	VEGFR2	Approved for treatment of stomach cancer/III
Necitumumab (IMC-11F8)	Eli Lilly	IgG1	EGFR	III

Table 11. 2. Selected approved or under clinical development phage display-derived peptides or peptide-based therapeutics.

Product	Manufacturer	Function	Indication(s)	Phase
Romiplostim (Nplate)	Amgen	Ligand for thrombop oietin receptor	Idiopathic (immune) thrombocytopenic purpura	Approved
Ecallantide (Kalbitor)	Dyax	Plasma kallikrein inhibitor	Hereditary angioedema	Approved
Peginesatide (OMONTYS)	Affymax, Takeda	Erythropo ietin mimetc	Chronic kidney disease associated anemia	Approved
AMG-386	Amgen	AMG-386	Inhibitor for binding of angiopoietin-1 and angiopoietin-2 to Tie2	III
CNTO530/CN TO 528	Ortho Biotech	Erythropo ietin mimetic	Chronic kidney disease associated anemia	I

Thus, in future, phage display technology will remain as a steady platform providing more and more useful therapeutic candidates for clinical development.

9. PHAGES AS THERAPEUTIC AGENTS

Early in 20th century, a key aim of bacteriophage researchers was to use these viruses as antibacterial therapeutic agents. Phage therapy was widely practiced in the former USSR until 1990 and surprisingly good results were reported in 1980s from Poland. This line of investigation was discontinued due to technical difficulties and the advent of antibiotics. However, the recent proliferation of antibiotic resistant strains has revived interest in this area.

The potential advantages and limitations of bacteriophages as therapeutic agents are presented in Table 11.3

Table 11.3: Potential advantages and limitations of bacteriophages in comparison to antibiotics

Advantages

- Specific of action
- Narrow spectrum of activity
- Higher safety
- Higher tolerability
- Easy administration
- Effect limited to site of infection
- Less expensive
- Possible additional benefits of after engineering

Limitations

- Absence of specific activity for a given bacterial strain
- Problem related to the formulation and stabilization of pharmaceutical preparations
- Possible emergence of bacterial resistance against bacteriophages
- Contribution of bateriophages in the development of antibiotic resistance
- Reduced activity due to immune system response to bacteriophages
- Genes for phage-encoded toxins, or genes for other bacterial virulence factors.

Recently, FDA approved phage therapy for patients who present a dangerous secondary bacterial infection due to SARS-CoV2 infection, responsible for the current COVID-19 pandemic.

Phages are also used to produce novel, nanostructured materials, i.e. wound dressings, packaging or a membrane for water treatment, reducing the risk of bacterial infections. There are recent examples of combining phages with synthetic and natural polymers (e.g. chitosan, alginates, collagen, fibres) eve in the form of electospun fibrous mats. As phages are not harmful to humans, it is possible to create edible and anti-bacterial materials.

10. PHAGES AS DRUG AND GENE DELIVERY VEHICLES AND BIOSENSING

Bacteriophages are also utilized as delivery carriers. They have been also applied as drug and gene delivery vehicles. In addition to therapeutic peptide display, they can serve as targeted nanocarriers, which encapsulate and deliver drug (chemical or biological) molecules to specific tissues. Such ideas have recently been applied in phage display vaccination by presenting appropriate antigen peptides on phage surface. Because of their immunostimulatory nature, these phage particles additionally show adjuvant activity, which is extremely valuable in designing vaccines. As an advanced application in effective gene therapy, the phage coat can truly protect the DNA from degradation after injection and display probe molecules on the phage coat for targeting specific cell types.

Lambda phage was used as early as 1971 to deliver the gene for galactose transferase into human fibroblast cells isolated from a patient with a deficiency in this enzyme. M13 was used for the targeted delivery of the GFP gene-expression cassette into breast cancer cells. Filamentous phage f1 with targeting antibodies displayed on the tip of the virion was used as a carrier of chemically conjugated antibiotics. Recently, Zhu et al. [2019] showed a prokaryotic–eukaryotic hybrid, composed of T4 phage and adeno-associated virus (AAV). The bacteriophage acted as the cargo, which might be reasonably easily modified. The eukaryotic virus acted as a "driver" that allowed for cargo delivery into mammalian cells.

In another remarkable aspect of application, filamentous phages have been applied in nanoscale biosensing. Their three-dimensional surface enables them to display various desirable peptides that result in providing the stronger multivalent bioreceptors for targeted molecules compared to conventional ones. There are numerous studies that have directly used whole engineered phage particles, reporter phages or phage-displayed peptides as the bioprobes in biosensors. As a successful example, Biotech Laboratories have marketed FASTPlaque-Response and FASTPlaqueTB for rapid detection of *Mycobacterium tuberculosis* in human sputum samples.

11. FOOD SAFETY-RELATED APPLICATIONS OF PHAGES

Bacteriophages have become of interest for other purposes such as alternative biocontrol strategies. They have been recommended as a feasible alternative for treatment and prophylaxis in cattle, where the most prevalent infectious diseases are bacterial agents. Furthermore, phages have been proposed as decontaminating agents of food products.

Direct application of lytic phages to ready-to-use food has shown to be significantly reduce contamination with various foodborne pathogens and most of them are employed as cocktail of phages. Given the available supporting data, some phage products have been approved by authorities for direct use on food and are commercially available. For example, the US Food and Drug Administration (FDA), approved the use of a mixture of six bacteriophages (ListShield, by Intralytix Inc., Columbia, USA), as a direct food additive for the control of *Listeria monocytogenes* in ready-to-eat (RTE) poultry and meat products. Similarly, another product called Listex P100 (manufactured by Micreos Food Safety, Wageningen, The Netherlands) and consisting of a single phage, was approved for use and it has received the GRAS status (generally recognized as safe).

Besides direct treatment of food, phage use is also projected for decontaminating inanimate surfaces of equipment, instruments and utensils used where formation of biofilm is very common, and thus helping to limit the risk of transmission of pathogens along the diary chain.

12. APPLICATION OF PHAGES IN AGRICULTURE AND AQUACULTURE

Phages can be suitable for the control of plant pathogens that cause major economic losses in agriculture by reducing the yield and the quality of products. Several studies showing phage effectiveness in crops have been published in recent years, based on this, a phage product (Agriphage, developed by company Omnilytics Inc., Sandy, Utah) containing specific bacteriophages against *Xanthomonas campestris* pv. *vesicatoria* and *Pseudomonas syringae* bacteria responsible for bacterial spot and speck in tomatoes and peppers, respectively, has been available since 2005. In addition, a new product (AgriPhage-Fire Blight by Omnylitics Inc, Sandy, Utah and Certis USA, Columbia, USA) has been approved for use against fire blight (caused by *Erwinia amylovara*) in apples and pears since 2019. These products were shown to effectively control bacterial spot, and significantly increase yields compared to the standard copper compounds used commercially.

Phages have also been considered for controlling infectious diseases in aquaculture. Several studies have been performed *in vitro* and *in vivo*, demonstrating the effect of phages on fish bacterial pathogens. Encouraging results have been obtained in the effective control of fish diseases caused by several species belonging to the genus *Vibrio*, or *Aeromonas* spp, highlighting the effectiveness of phage treatment as an excellent and feasible alternative to antibiotic treatment.

13. BACTERIOPHAGES FOR WASTEWATER PLANT TREATMENT

The contamination of wastewater plants by waterborne bacterial pathogens represents a global health concern, and also due to the high cost of common disinfecting methods in treatment plants. There are a number of potential waterborne bacterial pathogens, which are known to cause several diseases and there is an urgent need to find effective methods to counteract their growth and spread without impacting on pollution and/or AMR. In the search for ideal approaches to decrease waterborne pathogens, bacteriophages have been considered as indicators of bacterial contamination and as potential candidates for wastewater plant treatment. The rationale for the use of phages as indicators is based on their specificity, thus specific bacteriophages may be used as effective tracers of pathogens in order to monitor and improve disinfection methods. The direct use of bacteriophages for decontamination purposes has been proposed for the elimination of filamentous bacteria in ASP (Activated Sludge Process) systems and for the control of foam. However, effective application of phage biocontrol to wastewater treatment requires a complete and full understanding of the microbial communities dynamics, since the microbial population varies between different plants, and for this reason it is important to select and use specific phages able to target unwanted bacterial pathogens. One obstacle to this treatment is that a high amount of phages is needed to obtain a success.

14. BACTERIOPHAGES AS ENVIRONMENTAL SANITIZERS IN HOSPITALS

The use of bacteriophages has also been hypothesized as sanitizers in the hospital environment; as such an environment is colonized by bacteria that could potentially cause infections in the hospitalized patients. The potential use of phages as decontaminating agents in the hospital environment was also investigated and one such interesting paper was published in 2016. In studies evaluating the actual potential of phages to be used as decontaminants during routine hospital cleaning, their usability was investigated when additioned to ecofriendly detergents (Probiotic Cleaning Hygiene System (PCHS)) Based on the complementary characteristics of bacteriophages and probiotics, a potential synergistic effect of a combined system was therefore hypothesized. The results of the in vitro tests showed that phages were able to maintain their stability when diluted in PCHS, rendering them suitable for their use as a daily sanitation system. Importantly, phages could also target MDR bacteria, which are also often resistant to disinfectants. The in situ tests showed that the combined probiotic–phage application resulted in a stronger and faster decontamination activity compared to the individual probiotic and phage components alone, highlighting their possible synergistic effect. It seems to be a very promising strategy.

15. PHAGES IN CONTROL OF ALGAL BLOOMS

Cyanobacterial Harmful Algal blooms (CHABs) cause devastating impacts to fisheries, tourism, public health and ecosystem around the world. Cyanobacterial blooms occur in fresh water and marine environments, producing a variety of toxins, and poisoning risks to humans and animals. Chemicals can be used to kill cyanobacteria. Unfortunately, many of these chemicals are toxic to other forms of life, including fish and organisms they eat. The use of chemicals in natural lakes could create more problems than they solve, is not permitted.

Cyanophage is a double-stranded DNA virus that infects cyanobacteria and is detected in both freshwater and marine environments. Several studies have reported that cyanophage could be a major natural control mechanism of cyanobacteria population in aquatic ecosystems. Cyanophages with high abundance and genetic diversity have been isolated from lakes and ponds, which can infect cultures of known bloom-forming freshwater cyanobacteria. For instance, eleven lytic cyanophages were isolated on *Microcystis* while 12 species of lytic cyanophages each were found to infect *Anabaena* and *Planktothrix*. In a laboratory study, Tucker and Pollard (2005) found that isolated natural lake viral cocktail could decrease the abundance of *M. aeruginosa* culture by 95% within 6 days. The density of the cyanophage was positively correlated with the rate of *M. aeruginosa* cell lysis. The cyanophage replication time was 11.2 h, with an average burst size of 28 viral particles per host cell. The cell density of *M. aeruginosa* decreased 10-fold under the impact of phages with different genome sizes. Another study also demonstrated that infection by cyanophages may have a substantial effect on cyanobacterial succession in a hypereutrophic pond. The abundance of *M. aeruginosa* was negatively correlated with the cyanophage abundance. In addition to reducing cyanobacterial biomass, cyanophages also have a wider community-level impact due to the release of nutrients by cyanophage-induced cell lysis. Released intracellular nutrients could influence planktonic food web structure, community dynamics, and even biogeochemical cycles.

16. PHAGES IN INDUSTRIAL FERMENTATIONS

Industrial fermentations rely on growth and end products formation of microbes. These processes include the production of many types of food, aminoacids, organic acids and diverse foreign proteins from recombinant microbes. Any interference with these processes is an economic liability. Phages pose a major threat to industries which rely on bacterial growth. Due to their biological nature, bacteriophages are difficult to predict and to control. Thus phages are a significant concern to industrial microbiologists.

Phages have been responsible for disturbing industrial production of glutamic acid, butanol acetone, natto (fermented soy beans), antibiotics, enzymes, casein, sausage, wine, sourdrough bread, yogurt and cheese as well as sewage treatment. The introduction of continuous fermentation compounded the problem. Phage problems with industrial strains of *Clostridium, Bacillus, Lactobacillus, Lactococcus, Streptomyces, Escherichia, Streptococcus, Leuconostoc, Pseudomonas, Brevibacterium* and *Corynebac-terium* are identified. The segment of food industry that seems most seriously affected by phage is the fermented dairy foods industry. Phage infection of acetone and butanol fermentation by *Clostridium* species significantly decreases the process efficiency. As a result, scientists involved with the scale up of processes to production size should aware of the threat of phage infection and design equipment, processes and production facility layouts to provide a measure of defense against such problems.

17. BACTERIOPHAGE TOXINS AND DISEASES

Bacteriophages are associated with most bacteria, including pathogenic bacteria. Many bacterial pathogens produce toxic proteins that function as virulence factors and cause characteristic signs or symptoms of disease in humans and animals (Table 11.4). The ability of some pathogenic bacteria to produce toxins is determined by specific temperate phages. When a non toxinogenic bacterium becomes toxinogenic as a consequence of infection or lysozenization by a specific phage, the process is called **phage conversion** (or lysogenic conversion). Toxins that are produced as a result of phage conversion include diphtheria toxin in *Corynebacterium diphtheriae,* Shiga toxins in *E. coli,* pyrogenic exotoxins in *Streptococcus pyogenes,* enterotoxins in *Staphylococcus aureus,* botulinum toxins in *Clostridium botulinum* and cholera toxin in *Vibrio cholerae* (Table 11.4). The converting phages in these pathogenic bacteria contain the structural genes for corresponding toxins, but the various toxins differ greately in biochemical structure, immunogenicity, mode action and toxicity. The toxin genes have no essential role in the life cycle of the phage. The toxin genes can be expressed from prophage, from lytically replicating phage or from superinfecting, in replicating phage in bacteria with homologous lysogenic immunity.

Table 11.4 Some bacteriophage-encoded exotoxins

Virus	Host bacteria	Virulence factor	Gene	Type	Diseases (third eukaryotic organism)
β-phage	*Corynebactrium diphtheria*	Diphtheria toxin	tox	Exotoxin	Diphteria
Phage C1 Phage H-19B	Clostridium botulinum	Neurotoxin	C1	Exotoxin	Botulism
Enterobacteria phage 933W	*Escherichia coli*	Shiga toxin 1A, 1B, 2A, 2B	stx1AB stx2AB	Exotoxin	Dysentery
Phage CTXϕ	*Vibrio cholera*	Cholera toxin A, B	ctxAB	Exotoxin	Cholera
Phage ΦC3208	*Escherichia coli*	Hemolysin	hly2 stx2AB	Exotoxin	Hemolysis
Phage ϕCTX	*Pseudomonas aeruginosa*	Cytotoxin	Ctx	Exotoxin	Cytotoxicity (especially on leukocytes)
Phage N315	*Staphylococcus aureus*	Enterotoxin P	sep	Exotoxin	Vomiting
Phage Mu50A	*Staphylococcus aureus*	Enterotoxin A	sea	Exotoxin	Food poisoning syndrome
Phage ϕ13	*Staphylococcus aureus*	Enterotoxin A	entA	Exotoxin	Gastroenteritis
Phage ϕETA	*Staphylococcus aureus*	Exfoliative toxin A	eta	Exotoxin	Scaled-skin syndrome
Phage ϕ PVL	*Staphylococcus aureus*	Panton-Valentine Leukocidin S,F	lusK	Exotoxin	Leukocytes destruction
Phage T12	*Staphylococcus pyogenes*	Toxin A	speA	Exotoxin	Scarlet fever
Phage CS112	*Staphylococcus pyogenes*	Toxin C	spec	Exotoxin	Scarlet fever

Source: Common themes among bateriophage-encoded virulence factors and diversity among the bacteriophages involved.

Further reading

Auckerman HW (2004). Bacteriophages. In: The Desk Top Encyclopedia of Microbiology. Schaechter M (ed.) Elsevier, Academic Press.

Bertani LE and Six EW (1988). The P2-like bacteriophages and their parasite P4. In Calender R (ed.) The Bacteriophages, Vol. 2, P.73 New York, Plenum Press.

Bickle, TA and Kruger, DH (1993). The biology of DNA restriction. *Microbiol. Rev.* 57 : 434.

Harshey, RM (1998). Phage Mu. In Calender R (ed.) The Bacteriophages. P. 193. New York : Plenum Press.

Randall, KH (1999). Phage toxins and diseases. In Encyclopedia of Virology Vol. 2 2nd Edn. Academic Press, London.

Sanders, ME (1987). Bacteriophages of industrial importance. In : Goyal SM, Gerba CP and Bitton G (eds) Phage Ecology. p. 211 Chinchester Wiley.

Wilson, DR and Finley BB (1998). Phage display : applications, innovations, and issues in phage and host biology. *Can. J. Microbiol.* 44 : 313-329.

Principi, N., Silvestri, E and Sn Esposito, S (2019) Advantages and limitations of bacteriophages for the treatment of bacterial infections. Front. Pharmacol. 10: 513 – 522

Accolti, M et al (2021) Bacteriophages as a potential 360 – degree pathogen control strategy. Microorganisms 9:261 – 274

Paczency, J. Biele, K (2020) Application of bacteriophages in nanotechnology Nanomaterials 10: 1944 – 1969.

PART THREE

Plant Viruses

12

NOMENCLATURE AND CLASSIFICATION

Since the formation of ICTV (1973) renewed attempts were made to introduce Linnaen heirarchial system of classification and species concept to all viruses. The animal virologists were the first to agree to such a system. Later, phage virologists have also fallen in line with ICTV, however their species concept is different. For a long time, until 1990, the plant virologists have refused to classify their viruses in terms of families and genera and also to adapt species concept. Since long back, they have been classifying the plant viruses into `groups' (approximately 24). Though plant virologists have based the classification of these groups on the same characteristics like shape, composition and nature of nucleic acids, antigenic properties, they were categorical in that these `groups' donot correspond to any particular taxa in the heirarchial classification. They have putforward many valid arguments to vindicate their stand. Their arguments were : the binomial nomenclature can lead to undesirable characters (change of names several times) especially with organisms which are initially poorly characterized. Later their names are changed repeatedly and with time they are referred to in the literature by a number of names, resultng in confusion. The two major abstacles were identified in adapting species concept to viruses. First emphasis that is placed on phylogeny as an integral part of classification and second, the emphasis on the ability of two organisms to be capable of interbreeding as the ultimate test of conspecificity. Further, the establishment of taxa above the family level needs to be considered, however, there is no urgency for it.

As a part of the above objections, they have raised fundamental objections to the following rules of nomenclature of viruses promulgamated in 1966 (ICNV) and 1981 (ICTV).

Rule 4 : An effort will be made towards latinized nomenclature

Rule 12 : The genus and species epithet, together with the strain designation, must give an unambiguous identification of the virus and

Rule 13 : The species epithet must follow the genus name and be placed before the designation of the strain, variant or serotype.

With a lot persuation from ICTV, in the ICTV meeting in 1990 held at Berlin plant virologists have agreed for formal classification and nomenc-lature of plant virueses on par with other viruses. In turn ICTV has agreed to abolish rules 4 and 13 and to modify rules 5, 12 and 14 of the rules of the nomenclature. With these relaxations, plant virologists are now entitled legally to continue to name viruses just as they have done it in the past.

1. CURRENT NOMENCLATURE

i. Species concept : The delineation of the kinds of viruses that exist in nature, that is to say virus species is a practical necessity, especially for diagnostic purposes relating to the control of virus diseases. In day to day practice, virologists use the concept of a `virus' as being a group of fairly closely related strains, variants or pathovars. A virus defined in this way is essentially a species. In 1991, the ICTV accepted the concept that viruses exist as species, adopting the following definition (Von Rogenmortel, 1990). **A viral species is a polythetic class of viruses that constitutes a replicating lineage and occupies a particular ecological niche**. The species is the basis for any type of classification. The seventh report of the ICTV has provided a list of `species demarcating criteria' for each genus. This enabled the viruses to be differentiated as species and tentative species. Thus viruses are now recognised as `species' or tentative species. The viruses of tentative species have not been fully characterized to ensure that they are distinct and not strains of an existing virus or donot have the full characteristics of the genus to which they have been assigned.

ii. Naming of species : When a family or genus is approved by the ICTV, a type species or type virus is designated. Some virologists favoured using the English vernacular name as the official species name. Other virologists favoured serial numbering for viruses (species). However, both the approaches have serious drawbacks. The idea of Latinized binomial names for viruses was supported by the ICTV for many years but never implementated for any viruses. The proposal for latinization has now been withdrawn. Now, plant virus names are usually of the form `host name' plus `symptom name' plus `virus' e.g. Tomato streak virus, Cauliflower mosaic virus, Beet yellows virus. However, many hosts are shared by several plant viruses, and similarly many viruses have wide host range. The current rules forebid the form `host name' plus `virus' (rule 23).

iii. Acronyms : Abbreviations of virus names have been used for many years to make the literature more easy to read more succinct to present. The abbreviation is usually in the form of an acronym using the initial letters of each word in the virus name. ICTV has initiated a rationalization of plant virus acronyms and has subsequently updated the list regularly. The designation of the abbreviation is based on the following principles.

i. Abbreviations should be as simple as possible
ii. An abbreviation must not duplicate any other previously coined and still in current usage.
iii. The word `virus' in a name is abbreviated as `V'.
iv. The word `viroid' in a name is abbreviated as Vd. e.g. potato spindle tuber viroid (PstVd)

iv. Plant virus genera, families and orders: There is no formal definition for a genus, but it is usually considered as a `population' of virus species that share common characteristics and are different from other populations of a species. Currently, there are 72 genera of plant viruses. In some cases, there are numerous viruses which obviously belong to that family but for which there is insufficient information to place them either in existing genera or for creating new genera. These are listed as unassigned. Genera are named either after the type species (e.g. *Caulimovirus* after cauliflower mosaic virus) or are given a descriptive name, often from a Greek or Latin word for a major feature of the genus (e.g. *Geminivirus* from latin, *geminus* meaning twins) to describe the particles).

Similarly, genera are grouped together into families on common characteristics. Fifteen families are recognised for plant viruses. Twenty two of the genera have not yet been assigned to families and are termed **floating genera**. The acquisition of data on these floating genera together with changing attitudes on virus classification, will no doubt lead to the designation of further plant virus families. The family is either named after the type member of the genus (e.g. *Caulimoviridae)* named after the genus *Caulimovirus*) or given a descriptive name as with the genus, for a major feature of the family (e.g. *Geminiviridae,* a descriptive of the virus particles).

2. CLASSIFICATION

Two interconnected problems are usually encountered while making an attempt to classify viruses. First, related viruses must be placed in genera and higher taxa. For this purpose, the more stable properties of the virus such as amount and kind of nucleic acid, particle morphology and genome strategy must be used. The second problem is to be able to distinguish between related viruses and give some assessment of degrees of relationship within a group. Properties of virus for which there are many variants are more useful for this purpose. These include symptoms, host range, and amino acid composition of the coat protein. Certain properties such as serological specificity and aminoacid sequences are useful for defining groups and for distinguishing viruses within groups. To date, the taxonomic levels of order, family and genus have been defined by the ICTV. The criteria used for delineation of different taxa are presented in Table 12.1.

Table 12.1. Criteria for demarcating different virus taxa

I. Order
Common properties between several families including :
Biochemical composition
Virus replication strategy
Particle structure (to some extent)
General genome organization
II. Family
Common properties between several genera including :
Biochemical composition
Virus replication strategy
Nature of particle structure
Genome organization
III. Genus
Common properties within a genus including :
Virus replication strategy
Genome size, organization and/or number of segments
Sequence homologies (hybridization properties)
Vector transmission
IV. Species
Common properties within a species including :
Genome rearrangement
Sequence homologies (hybridization properties)
Serological relationships
Vector transmission
Host range
Pathogenicity
Tissue tropism
Geographical distribution

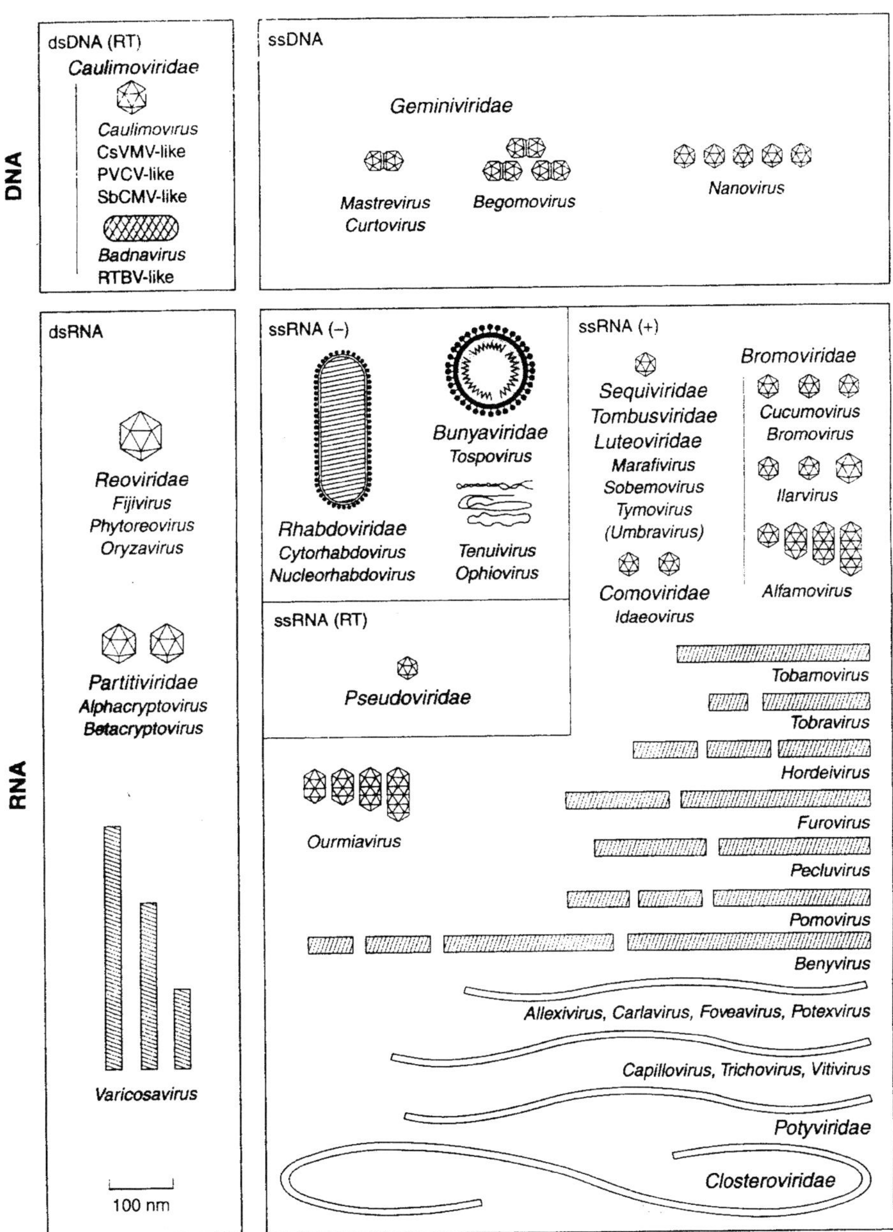

Fig. 12.1. Major plant virus families and genera

Classification of the plant viruses, as detailed in the 8th report of ICTV (2005) is presented in Table 12.2 and Fig. 12.1 This classification includes a single order *Mononegavirales;* 15 families and 72 genera with type species. In addition 21 are shown as floating genera. One species `tobacco necrosis virus satellite' demarcated as unassaigned species. Of

977 plant viruses listed in the ICTV seventh report, 701 are true species and 276 tentative species.

Table 12.2. Classification of plant viruses (ICTV, 2005).

Order	Family (Subfamily)	Genus	Type species
		Benyvirus	*Beet necrotic yellow vein virus*
	Bromoviridae	*Alfamovirus*	*Alfalfa mosaic virus*
		Ilarvirus	*Tobacco streak virus*
		Bromovirus	*Brome mosaic virus*
		Cucumovirus	*Cucumber mosaic virus*
		Oleavirus	*Olive latent virus 2*
	Bunyaviridae	*Tospovirus*	*Tomato spotted wilt virus*
	Caulimoviridae	*Badnavirus*	*Commelina yellow mottle virus*
		Caulimovirus	*Cauliflower mosaic virus*
		Tungrovirus (was ``Rice-Tungro bacilli-form like viruses")	*Rice tungro bacilliform virus*
		Soymovirus (was ``soybean chlorotic mottle like viruses")	*Soybean chlorotic mottle virus*
		Cavemovirus (was ``cassava vein mosaic-like viruses")	*Cassava vein mosaic virus*
		Petuvirus (was ``Petunia vein clea-ring-like viruses")	*Petunia vein clearing virus*
	Closteroviridae	*Closterovirus*	*Beet yellows virus*
		Crinivirus	*Lettuce infectious yellows virus*
		Ampelo virus	*Grapevine leafroll-associated virus 3*
	Comoviridae	*Comovirus*	*Cowpea mosaic virus*
		Fabavirus	*Broad bean wilt virus 1*
		Nepovirus	*Tobacco ringspot virus*
	Flexiviridae	*Allexivirus*	*Shallot virus X*
		Manadrivirus	*Indian citrus ringspot virus*
		Carlavirus	*Carnation latent virus*
		Capillovirus	*Apple stem grooving virus*

		Foveavirus	*Apple stem pitting virus*
		Potexvirus	*Potato virus* X
		Trichovirus	*Apple chlorotic leaf spot virus*
		Vitivirus	*Grape vine virus A*
		Furovirus	*Soil-borne wheat mosaic virus*
	Geminiviridae	*Mastrevirus*	*Maize streak virus*
		Curtovirus	*Beet curlytop virus*
		Begomovirus	*Bean golden mosaic virus-puerto Rico*
		Hordeivirus	*Barley stripe mosaic virus*
		Idaeovirus	*Raspberry bushy dwarf virus*
	Luteoviridae	*Luteovirus*	*Barley yellow dwarf virus - PAV*
		Polerovirus	*Potato Leafroll virus*
		Enamovirus	*Pea enation mosaic virus-1*
	Nanoviridae	*Nanovirus*	*Subterranean clover stunt virus*
		Babuvirus	*Babana bunchy top virus*
		Ophiovirus	*Citrus prorosis virus*
		Ourmiavirus	*Ourmia melon virus*
	Partitiviridae	*Alphacryptovirus*	*White clover cryptic virus 1*
		Beta cryptovirus	*White clover cryptic virus-2*
		Pecluvirus	*Peanut clump virus*
		Pomovirus	*Potato mop-top virus*
	Potyviridae	*Potyvirus*	*Potato virus Y*
		Rymovirus	*Rye grass mosaic virus*
		Bymovirus	*Barley yellow mosaic virus*
		Macluravirus	*Maclura mosaic virus*
		Ipomavirus	*Sweet potato mild mottle virus*
		Tritimovirus	*Wheat streak mosaic virus*
	Reoviridae	*Fijivirus*	*Fiji disease virus*
		Phytoreovirus	*Rice dwarf virus*
		Oryzavirus	*Rice ragged stunt virus*
Mononegavirales	*Rhabdoviridae*	*Cytorhabdovirus*	*Lettuce necrotic yellows virus*
		Nucleorhabdovirus	*Potato yellow dwarf virus*
			Tobacco necrosis virus satellite
	Sequiviridae	*Sequivirus*	*Parsnip yellow fleck virus*
		Waikavirus	*Rice tungro spherical virus*

		Sobemovirus	*Southern bean mosaic virus*
		Tenuivirus	*Rice stripe virus*
		Tobamovirus	*Tobacco mosaic virus*
		Tobravirus	*Tobacco rattle virus*
	Tombusviridae	*Tombusvirus*	*Tomato bushy stunt virus*
		Carnovirus	*Carnation mottle virus*
		Necrovirus	*Tobacco necrosis virus A*
		Dianthovirus	*Carnation ringspot virus*
		Machlomovirus	*Maize chlorotic mottle virus*
		Avenavirus	*Oat chlorotic stunt virus*
		Aureusvirus	*Pathos latent virus*
		Panicovirus	*Panicum mosaic virus*
	Tymoviridae	*Tymovirus*	*Turnip yellow mosaic virus*
		Marotivirus	*Maize rayadotine virus*
		Maculavirus	*Grapevine fleck virus*
		Umbravirus	*Carrot mottle virus*
		Varicosavirus	*Lettuce big-vein associated virus*
Viroid	*Pospiviroidae*	*Pospiviroid*	*Potato spindle tuber viroid*
		Hostuviroid	*Hop stunt viroid*
		Cocadviroid	*Coconut cadang cadang viroid*
		Apscaviroid	*Apple scar skin viroid*
		Coleviroid	*Coleus blumei viroid 1*
	Avsunviroidae	*Avsunviroid*	*Avocado sunblotch viroid*
		Pelamoviroid	*Peach latent mosaic viroid*

Further reading

Martelli, GP (1992). Classification and nomenclature of plant viruses : State of the art. Plant Diseases 76 (5) : 436-442.

Matthews, REF (ed.) (1983). A critical appraisal of viral taxonomy. Bcoa Raton, Fla, CRC Press PP 256.

Mayo, MA (1996). Recent revisions of the rules of virus classification and nomenclature. *Archives of Virology* 141 : 2479-2484.

Mayo MA and Pringle CR (1998). Virus Taxonomy. *J. Gen. Virol.* 79 : 649-657.

Van Regenmortel MHV (1989). Applying the species concept to plant viruses. *Arch. Virol* 104 : 1-17.

Van Regenmortel MHV (1990). Virus species, a much overlooked but essential concept in virus classification. *Intervirology* 31, 241-254.

Van Regenmortel, MHV (1999). How to write the names of virus species. *Arch Virol* 144 : 1041-1042.

Van Regenmortel *et al.* (1999). Virus taxonomy, classification and nomenclature of viruses. Seventh Report of the ICTV. Academic Press, New York.

Gorbalenya et al. (2020) The new scope of Virus Taxonomy: partitioning of virosphere into 15 heirarchial ranks, Nature Microbiology 5: 668 – 674

13

REPLICATION STRATEGIES

There are more than 957 viruses which can infect plants. The genomes which are encapsidated have representatives of single-and double-stranded DNA and RNA. The vast majority have single stranded RNA of (+) type. For some viruses the genome needed for infection is divided between two or more segments which may be encapsidated in the same particle or in separate particles (multicomponent). Morphologically, also there is a lot of variation in plant viruses. Majority are naked without the envelope and few are enveloped. Plant viruses vary in their breadth of their host ranges. There are some viruses which can infect many host (e.g. TSWV, CMV). On the other hand, there are other viruses, which exhibit very limited host range (e.g. OYDV). For all plant viruses the successful replication in a host depends upon the interactions of two genomes, the virus and the host thus while discussing the replication of plant viruses, one has to consider all these aspects. Our understanding of the ways in which plant viruses replicate has increased remarkably over the past few years. This is, inpart, because the complete nucleotide sequence of many plant viral genomes have been established, allowing the number, size and amino acid sequence of putative gene products to be determined. We now have this information for representatives of most plant virus genera. Determination of the gene products has been greately assisted by the use of well established *in vitro* translation systems.

For many aspects of replication, plant viruses are intimately dependent on the activities of the host. Prominent among them are : (i) viruses use amino acids and nucleotides synthesized by host cell metabolism to build viral proteins and nucleic acids, (ii) the energy required for the viral protein and nucleic acids synthesis is provided by the host cell, mainly in the form of nucleoside triphosphates, (iii) viruses use the ribsomes, tRNAs and associated enzymes and factors of the host cell's protein synthesizing system for the synthesis of viral proteins, (iv) many viruses code for an enzyme or enzymes involved in the synthesis of their viral nucleic acids, but they may not contribute all polypeptides involved. Hence, for these polypeptides they are dependant host, and (v) structural components of host cell, particularly membranes are involved in virus replication.

1. GENERAL ASPECTS OF REPLICATION

Like phages, in plant virus replication several stages can be differentiated. The stages in the initially infected cells are (Fig. 13.1):

i. Virus entry

ii. Uncoating

iii. Preliminary translation

iv. Replication

v. Secondary translation

vi. Assembly.

i. Virus entry : A mechanical damage is required for a virus to enter a plant cell. The machanical damage is necessary for the virus particles to cross the cuticle and cellulose cell wall and finally to be placed on the plasma membrane. The proposals for possibilities for the existence of cell surface receptors for viruses of plants have been putforward. However, the available evidence points against cell surface receptors for the entry of viruses into plant cells and the current thinking is that virus particles are introduced directly into the cell, be it of host or non-host. This is the principal difference between plant viruses and other two main groups of viruses.

It has been assumed that differences between viruses of different genome types with respect to their entry are likely to be, if any. However, after this stage the events surrounding their replication and expression differ greately.

ii. Uncoating : Until recently, the event of uncoating of plant viruses was a `black hole' of knowledge. A major breakthrough came from experiments by Wilson and co-workers. They showed that the treatment of the usually very stable TMV particles with rabbit reticutocyte lysate or wheat germ extract relaxed the structure at the 5'-end in such a way as to allow the attachment to ribosomes. The ribosomes then translocated along the RNA dislodging coat protein subunits as the RNA was translated. These workers characterized intermediates of this contranslational disassembly (termed striposomes) which comprised partially uncoated TMV rods complexed with polysomes. These striposome complexes were then isolated from recently inoculated epidermal tissue. But in some viruses like cowpea chlorotic mottle virus (CCMV), southern bean mosaic virus (SBMV) and cowpea mosaic virus (CPMV), it is not known how the uncoating process occurs. However, it has been hypothesized that within the cell there exist some conditions or mechanisms which destabilize particles structure leading to cotranslational disassembly. For instance, one mechanism could be phosphorylation of capsid proteins which for poliovirus results in destabilization of the viral capsid.

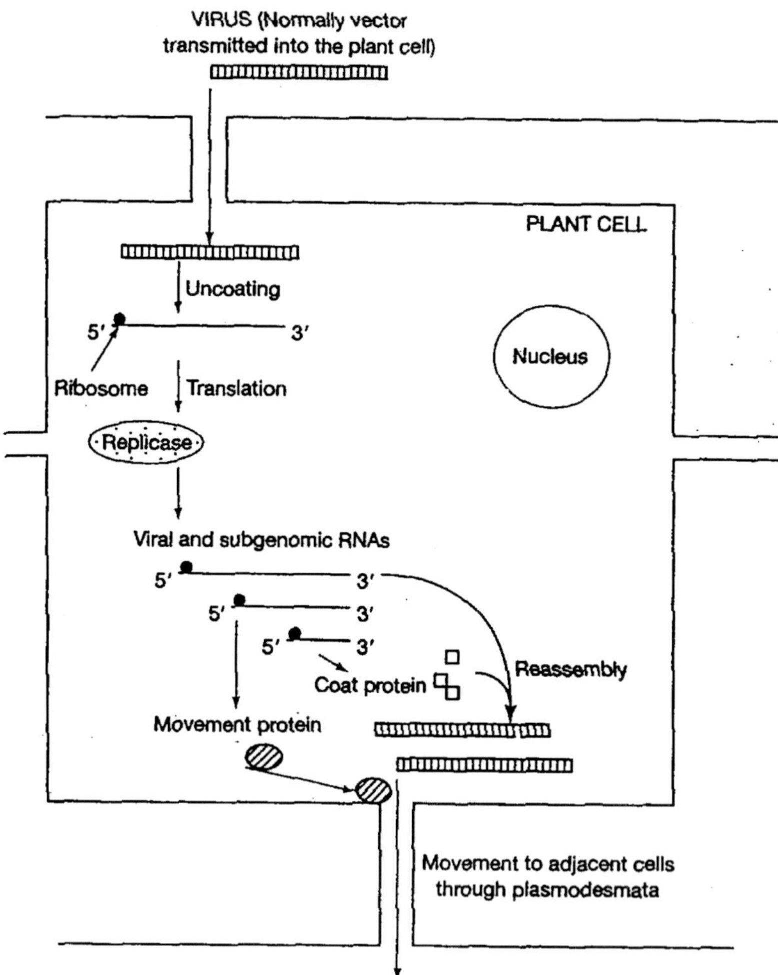

Fig. 13.1. General scheme of virus spread within plants. The virus enters the plant cell, and following replication and production of more coat protein, the virus reassembles and moves between adjacent cells through the plasmodesmata, facilitated by the movement protein (From Dickinson, M. Molecular Plant Pathology, 2003)

iii. Preliminary translation : Plant viruses face a critical problem i.e. the eukaryotic ribosomes usually translate the 5'-cistron in a polycistronic mRNA and cistron downstream of the 5' cistron is effectively closed. Plant viruses employ two basic strategies to overcome this problem. They either have all the genetic information in one open reading frame which is translated to give a polyprotein, which is later cleaved to functional proteins. Alternatively, the genome is divided into monocistronic segments which can be translated individually. This division occurs either during transcription to give a collection of 3'-coterminal subgenomic mRNAs (e.g, TMV) or in multic-omponent viruses, several of the RNA species are monocistronic. Some viruses employ a combination of both strategies.

The preliminary translation of genomes of many, it not all, plant viruses occurs simultaneously with uncoating. The gene products required early in infection are those involved in nucleic acid replication. All the viruses which have undivided genomes with a subgenomic strategy have the RNA polymer-ase genes at their 5'-ends. Thus, these genes are open for translation and have been shown to be expressed during contranslational disassembly. For multicomponent viruses several of the genes can be translated during disassembly.

iv. Replication : All viruses, irrespective of the composition of their genome, have to produce mRNA for the expression of their genome. Furthermore, all viruses encode proteins which are involved in the replication of their genomes. Therefore, mRNA production is an important early event in the infection of cells. Replication of viruses containing different types of genomes is described in detail in the next section of this chapter. However, for comprehension purpose, out lines of different viruses are given below.

All single stranded (+) RNA viruses encode one or more proteins which are involved in the RNA dependent RNA polymerase complexes responsible for viral replication. The composition of the RNA replication complexes has not fully been determined. The (-) ssRNA and dsRNA viruses both encapsidate enzymes which are capable of transcribing mRNA from the viral genome. The transcription of mRNAs from DNA viruses involves the host DNA dependent RNA polymerase (Pol II) operating on a double stranded covalently closed form of the DNA. Therefore, in ssDNA containing viruses (e.g. Geminiviruses) the DNA must be converted to double stranded DNA. In pararetroviruses, like caulimoviruses, encode a reverse transcriptase which uses the RNA transcripts, for DNA synthesis thus removing the formation of viral DNA completely from the host DNA replication system.

v. Secondary translation : The same considerations concerning the limitations of eukaryotic ribosomes, as were discussed for primary translation, apply to secondary translation. It is often the proteins required late in replication that are produced at this stage. Coat proteins and cell-to-cell movement proteins are the best example for late proteins. However, data from protoplast experiments only partially supports the concept of early and late expression of viral genes.

vi. Assembly : Experimental evidence suggests that many viruses, especially simple RNA viruses can be disassembled into their constituent coat protein and nucleic acid and again can be reassembled *in vitro*. It is thought most likely that virus assembly *in vivo* follows the same lines as that *in vitro*. The coat proteins of several viruses have been shown to form virus-like particles around a range of polycations. However, except in the case of TMV which forms short rod-shaped particles containing chloroplast RNA, there is no evidence for non-viral nucleic acid being encapsidated. This would suggest that viral replication and encapsidation are coordinated.

2. REPLICATION OF SINGLE STRANDED POSITIVE RNA VIRUSES – TOBACCO MOSAIC VIRUS

Virion morphology and genome

The genus *Tobamovirus* comprises 15 members of which TMV is the type member. TMV and the closely related tomato strain of TMV (now called tomato mosaic virus, ToMV) are the best characterized viruses on the molecular level.

The TMV is a rigid rod of approximately 18 nm diameter and is extra ordinarily stable (Fig. 13.4A). The structures of the disc shaped aggregate of the coat protein and of the virion have been determined by x-ray diffraction analysis. In the virion, the 2130 coat protein subunits are arranged in a right handed helix with 16 1/3 subunits per turn. The RNA is located between successive turns of the helix with the phosphates 4 nm from the axis of the helix. The genome of TMV is monopartite, positive sense ssRNA. It is 6.4 kb and contains 4 ORFs. The 5'-terminis is capped (m7 GPPPG) followed by an AU-rich 5'- UTR of bout 70 nt. The 3'- UTR consists of a sequence that folds into five pseudoknots two of which are transfer RNA (tRNA)-like structures at the terminus, which can accept histidine *in vitro*. The significance of pseudoknots is that they introduce structure into RNA molecules, which can potentially provide specificity in terms of interaction with proteins or catalytic reactions.

The first ORF encodes a 126 kD protein but read through the UAG termination codon generates a 183 kD protein (Fig. 13.2). These two proteins function as a methyl transferase helicase and RNA dependent RNA polymerase (RdR) respectively. The last five codons of the 183 kD reading frame overlap the 30-kD reading frame, the latter inturn terminates in two nucleotides prior to the initiation codon of the ORF that encodes the 17.5-kD coat protein. The 30 kD protein is known as movement protein that facilitates the movement of virions from infected cell to adjacent non-infected cell through plasmodesmata. The 17.5 kD protein functions as coat protein.

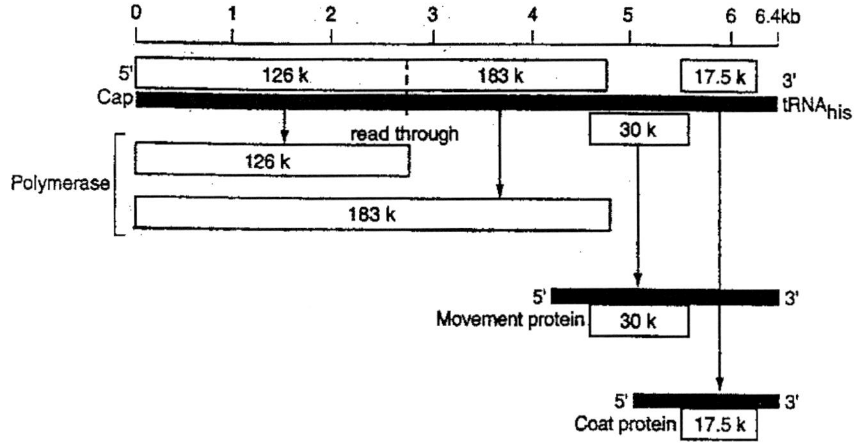

Fig. 13.2. Organization of the TMV genome

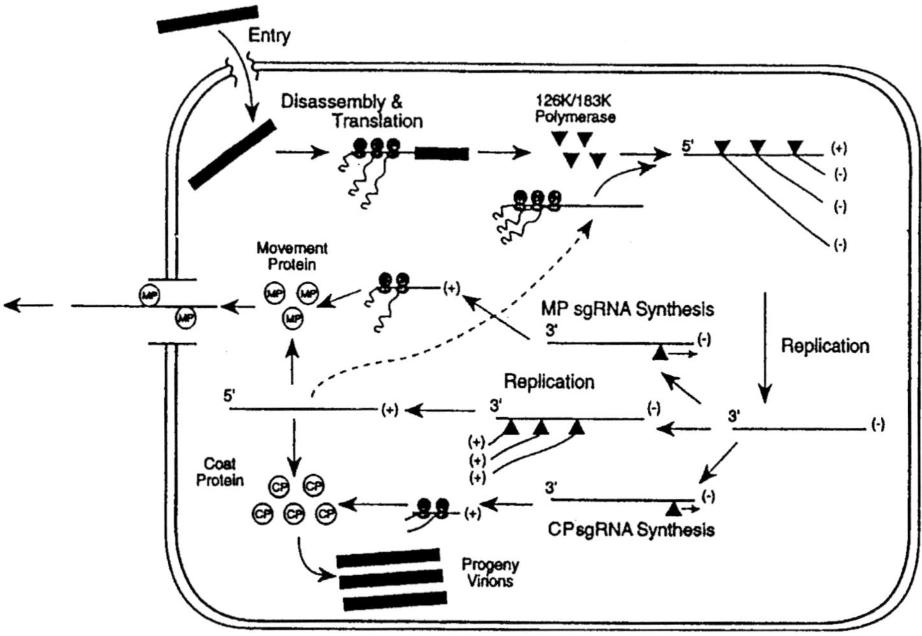

Fig. 13.3. Diagram showing the stages in TMV replication. All the events shown are presumed to occur in the cytoplasm of infected cells (From Fields Virology, 3rd ed.)

Replication cycle

i. Entry and uncoating : TMV is mechanically transmitted and is readily transmitted to healthy plants during contact with infected plants or virus contaminated implements. After entry the uncoating begins by removal of a few coat protein subunits from the end of the virion bearing the 5-terminus of the RNA (Fig. 13.3). The ribosomes then drive a co-translational disassembly process in which protein subunits are very rapidly dislodged. Simultaneously the 126 kD and 183 kD ORFS are translated. However, it is not known how the regions near the 3'-end of the viral RNA become uncoated.

ii. Translation and transcription : Like many cytoplasmic positive sense RNA viruses, TMV replication occurs in association with cytoplasmic membranes. Following initial 5'-to –3' disassembly, the 126kD and 183 kD proteins are translated from the genomic RNA. Readthrough (also called termination suppression) is context dependent : in TMV the two codons just downstream of the suppressed UAG are critical for efficient readthrough. As a result of this readthrough translation, the two proteins are read in the same frame. The efficiency of readthrough regulates the relative amounts of the 183 kD and 126 kD proteins. The 126 kD protein contains conserved motifs typical of methyl transferases and helicases and the 183 kD protein domain contains the GDD motif typical of viral RdRp. The 183 kD protein together with the 126 kD protein, make up the subunits of the replicase. Replicase acts to replicate full-length complementary negative strand RNA. This then serves as the template for the synthesis of additional copies of full-length genomic RNA or if it reconizes internal subgenomic promoters, the three subgenomic RNAs each of which is functionally monocistronic. The smallest sgRNA (subgenomic) is the most abundant and is translated into the viral coat protein (CP). The middle sized sgRNA, which accumulates to much lower levels than the CP sgRNA, is translated into the viral movement protein (MP). The longest sgRNA found in infected cells accumulates to low levels and is predicted to encode a 54 kD protein. However, the functions of this protein are not clearly known.

iii. Assembly : The detailed model for the *in vitro* assembly of virus particles was proposed many years ago, and it is assumed that the same mechanism operates in infected cells. Assembly begins by the association of an aggregate of coat protein molecules with a specific origin-of-assembly sequence that includes nucleotides 5444-5518 in the viral RNA. The protein molecules have been assumed to be present as two layer, ``20S" disks, each containing 34 protein molecules (Fig. 13.4). Part of the origin-of-assembly sequence is inserted via the central hole of the disc-between two layers of protein molecules. Later the disc switches to a protohelical ``lockwasher-like" configuration. Discs are then successively and rapidly added to the growing helical rod as segments of the RNA bearing the 5'- terminus are drawn through the central cavity to become inserted between newly-formed turns of the helix. Encapsidation of the 3'-terminal part of the RNA occurs more slowly and probably involves smaller aggregates of the coat protein.

The life cycle of TMV is completed in about 40 minutes and the burst size is 10^6 to 10^7 per cell.

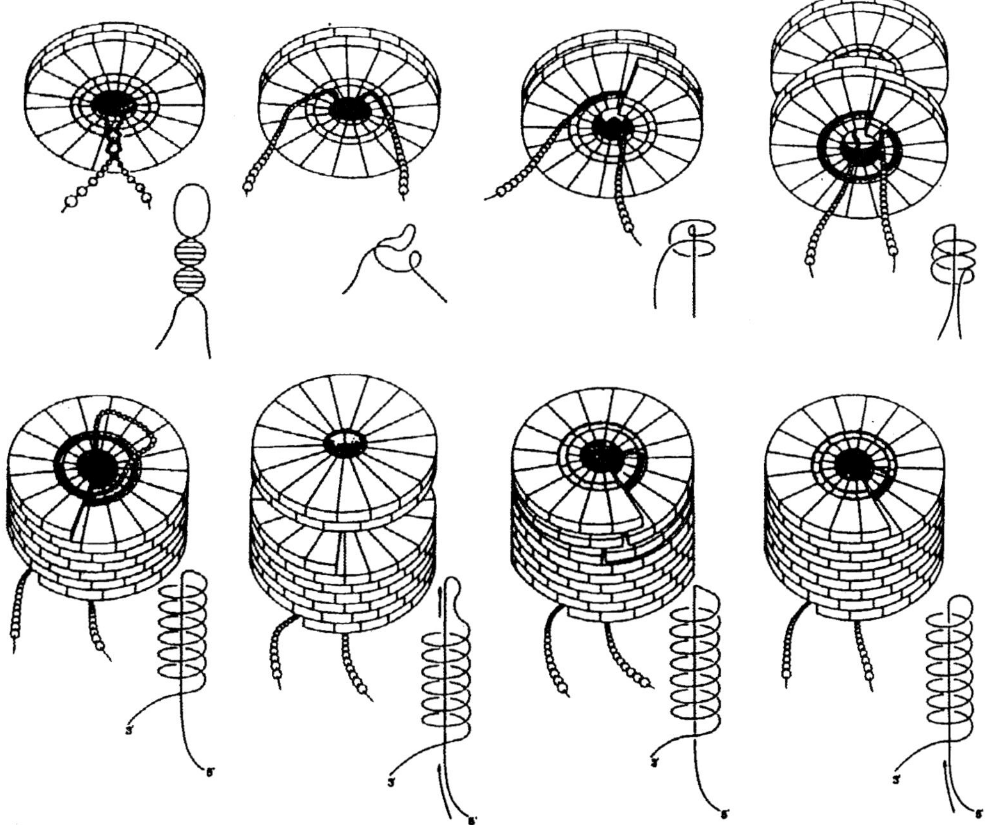

Fig. 13.4. Proposed mechanism of assembly of TMV. **Upper row:** Initial interaction of origin - of - assembly sequence in viral RNA with 20 S disk and switch to helical configuration. **Lower row:** Elongation of helical rod by addition of disks and insertion of RNA via the central hole (From Butter)

3. REPLICATION OF SINGLE STRANDED NEGATIVE RNA VIRUSES – PHYTORHABDOVIRUS

The *Rhabdoviridae* members infect both animals and plants. Plant viruses have been categorized as rabdoviruses based on ultrastructural exam-ination of infected plant cells or laboratory virus preparations.

They identify themselves as enveloped virus particles which are bullet shaped in unfixed preparations or bacilliform when fixed. Approximately, 75 plant viruses are currently listed as possible rhabdoviruses. About 15 of these viruses have been sufficiently characterized. Ultrastructural variations induced in the host cell favoured the division of this family into two genera : ***Cytorhabdovirus*** and ***Nucleorhabdovirus***. Cytorhabdoviruses are observed to form at cyto-plasmic membranes and accumulate in the cytoplasm (e.g. lettuce necrotic yellow virus-LNYV). On the other hand, nucleorhabdoviruses form nuclear inclusions and appear to bud from the inner nuclear membrane to accumulate in perinuclear spaces (e.g. sonchus yellow net virus-SYNV).

Virion and genome structure

Virions of rhabdovirus are 45 to 100 nm wide and 130 to 250 nm long. Virions are composed of a lipid envelope from which project spikes made of the viral encoded G glycoprotein (Fig. 13.5 A,B). Within the envelope, the virion consists of a helical RNP that consists of the viral genomic unsegmented negative-sense ssRNA, surrounded by N protein to form nucleocapsid. Nucleocapsid is also associated with viral P and L proteins that make up the RdRp. The location of the matrix protein M has not been clearly demonstrated but it is assumed to be involved in coiling the nucleocapsid and interacting with the internal domains of G.

The genome is about 11 to 14 kb. The complete sequence of the SYNV genome shows an organization that is same as VSV, with an additional gene sc4. The 3' leader sequence is followed by the genes encoding N, P, M, G and L in the same order (Fig. 13.5C). The sc4 gene is located between the P and M genes. The sc4 protein is found in virions and appears to be phosphorylated *in vivo*. It has been suggested that sc4 might be viral movement protein. It could also play a role in aphid transmission. SYNV induces the formation of viroplasms in the nuclei of infected plant cells. The viral N, L and P proteins have been immunolocalized to these nuclear viroplasms, which appear to be the sites of viral replication. It is also the source of RdRp activity.

Replication

Rhabdoviruses are transmitted by aphids, leaf hoppers or plant hoppers depending on the virus. Upon entering a cell, the viral nucleocapsid cores are released into the cytoplasm possibly in association with the ER (Fig. 13.6). Later these are transported into the nucleus where SYNV RdRp is activated and primary transcription occurs. The viral RNAs are transported to the cytoplasm, where the translation occurs and the newly synthesized N, L and P are transported into the nucleus. These proteins continue transcription of viral mRNAs and replicate the viral genome. Transcription of entire (-)ssRNA results in the formation + ssRNA. This (+) ssRNA is again transcribed into (-) ssRNAs that serve as the genomes of progeny virions. Assembly of the nucleocapsids takes place inside the nucleus.

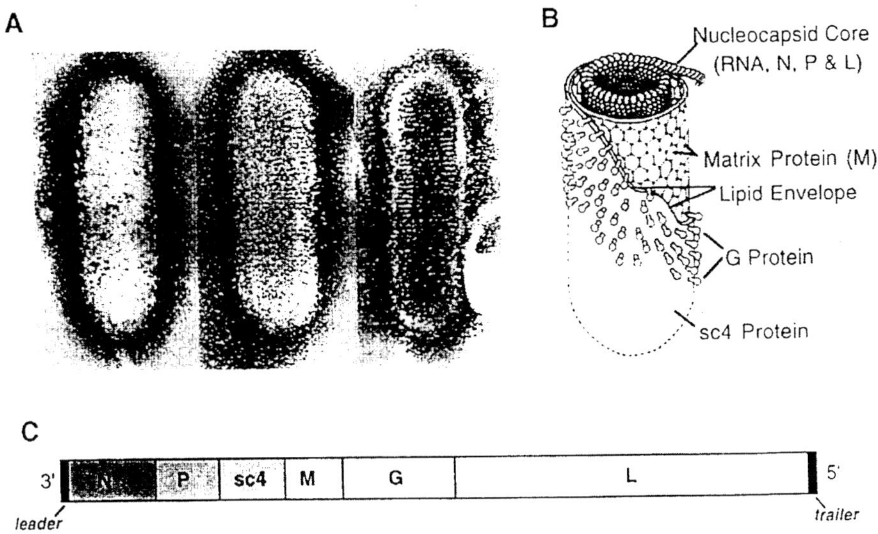

Fig. 13.5. Structure and genome organization of sonchus yellow net virus (SYNV), a nucleorhabdovirus (Courtesy Jackson *et al.,* 1999)

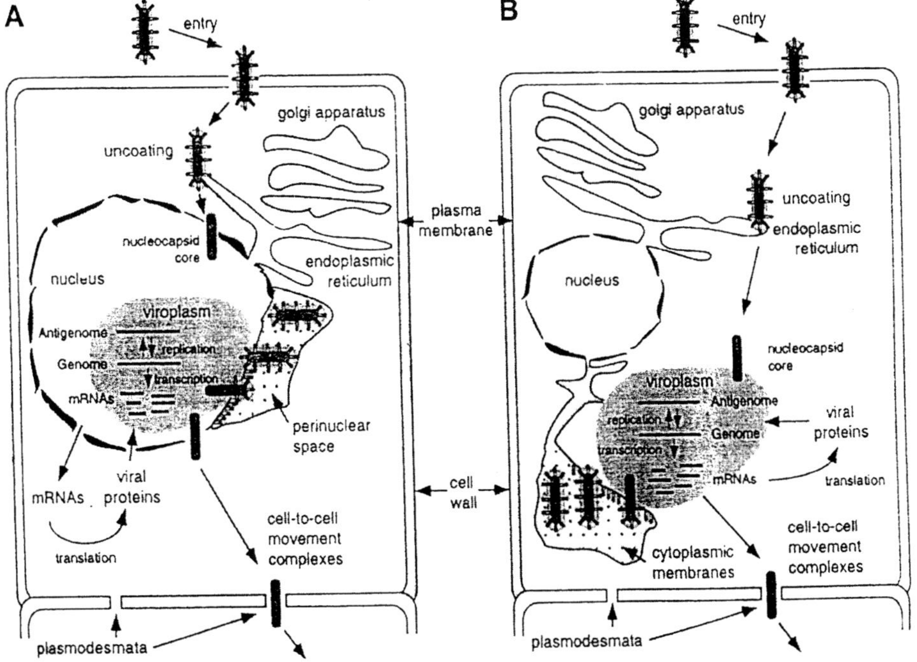

Fig. 13.6. Multiplication of (A) nucleorhabdoviruses and (B) cytorhaboviruses. (Courtesy Jackson *et al.,* 1999).

As tunicomycin interrupts the SYNV budding at the nuclear envelope and causes nucleocapsids to accumulate at the nuclear periphery (Fig. 13.6A), hence it is thought that newly synthesized G is transported to the inner envelope, and virus buds at these sites into the perinuclear space between the inner and outer nuclear membranes. This model proposes that viral cores lacking an envelope are exported from nucleus and move cell to cell.

LNYV has not been completely sequenced, but genome mapping studies suggest that it has a genome organization similar to that of SYNV. Replication of cytorhabdoviruses occur inside the cytoplasm in association with ER membranes. The virions appear to assemble by budding into vesicles that are derived from ER (Fig. 13.6B).

4. REPLICATION OF DOUBLE STRANDED RNA VIRUSES : PHYTOREOVIRUS

The double stranded RNA (dsRNA) viruses are grouped into two families: the *Reoviridae*, which have genomes of 26-28 kb in 10-12 segments, and the *Partitiviridae* that have 3-4 kb genomes in 2-3 segments. Furthermore, the *Reoviridae* utilize a conservative mechanism of replication whilst the *Partitiviridae* replicate semiconservatively. Like the (-) strand RNA viruses, these viruses contain RdRp protein within their protein coats.

Plant members of the *Reoviridae* family are placed in three genera : *Phytoreovirus* with 12 dsRNA genome segments, the type member being Rice dwarf virus (RDV); *Fijivirus,* with 10 dsRNA genome segments, the type member being Fiji disease virus (FDV); and *Oryzavirus* with 10 dsRNA genome segments, the type member being Rice ragged stunt virus (RRSV). These viruses replicate in their insect vectors as well as plants.

Virions morphology and genome

Reovirus (e.g. *Fijivirus*) particles are double-shelled, icosohedral structure with `A' type spikes on the 12 vertices of the outer shell and `B' type spikes on the inner shell or core. The outer shell is fragile and easily breaks down, leaving the inner shell bearing 12B spikes. The genome of *Fijivirus* is divided between 10 RNA segments. The coding assignments for those that are known are summarized in Table 13.1. The gene product from segment 3 is the major core protein and those from segments 1 and 7 contribute both enzymatic activity and core. The product from segment 2 has been attributed to the `B' spike on the inner core. Three of the products, those from segments 8, 9 and 10 give proteins described as non-structural, and no function has yet been assigned to the products of segments 4, 5 and 6.

Table 13.1. Genome properties of members of the genus *Fijivirus*

Segment no.	Size (bp)	Protein size (kD)	Protein function
1	ND	-	-
2	ND	-	-
3	ND	-	-
4	ND	-	-
5	ND	-	-
6	ND	-	-
7	2193	41.2	NS/TUP
		36.4	NS
8	1927	68.1	Core/PNTP
9	1900	39.9	NS/VP
		24.2	NS
10	1801	63.3	MOS

ND – Not determined; NS – Non structural protein; PNTP – NTP binding protein; TUP – Protein forms tubular structure; MOS – Major outer shell protein.

Replication

Plant reoviruses replicate in the cytoplasm as do those of infecting mammals. Infection is followed by the formation of densely staining viroplasms in the cytoplasm. Viral RNA appears to be synthesized in the viroplasm, where the mature particles are formed. The mature particles then migrate into the cytoplasm.

In reoviruses, the (-) sense strands are synthesized by the viral replicase on a (+) sense template that is associated with a particulate fraction (Fig. 13.7). DsRNA is formed within the nascent cores of developing virus particles and remains within these particles. A problem faced by reoviruses is the selection of progeny for packaging. Each of the particles contains the genomic complement of one copy of each of 10 or 12 dsRNA species. The selection of a correct set of 12 genomic RNAs could involve the (+) ss strand. The base paired inverted repeats could be the recognition signals. It may be that other virus encoded `scaffold' proteins transiently present in the developing core are involved in RNA recognition. The packaging of the 12 segments presumably involves 12 different and specific protein-RNA and/or RNA-RNA interactions.

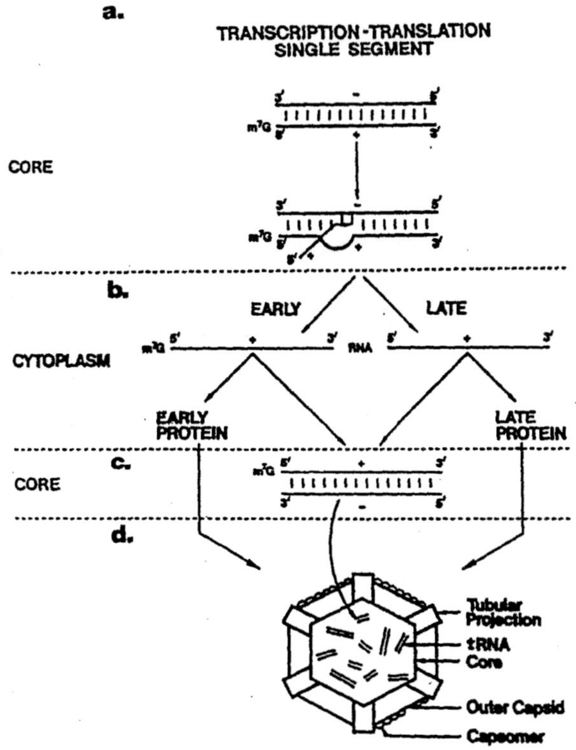

Fig. 13.7. Reovirus replication. RNA transcription from each of the ten segments occurs within viral cores. (a) Transcription from a single segment. (b) Single-stranded (+) RNAs leave viral cores and are translated within the cytoplasm into early viral proteins. (c) Single-stranded (+) RNAs are also partially encapsidated to form cores in which complementary strands are synthetized. (d) Progeny cores mature to virus particles or (a) serve as sites for late mRNA synthesis. Late mRNA, unlike early mRNA, is not capped (b). (From Fenger, T.W., in *Textbook of Human Virology*, 2nd ed., Belshe, R.B., Ed., Mosby-Year Book, St. Louis, 1991, 74).

5. REPLICATION OF dsDNA VIRUSES : PARARETROVIRUSES – CAULIFLOWER MOSAIC VIRUS (CaMV)

Cauliflower mosaic viruses (*Caulimoviruses*) are a group of plant viruses composed of double stranded (ds) DNA. These viruses have been subject of study because they cause economic losses in crop plants. In addition, they have made significant contributions to our knowledge of the molecular biology of plant viruses. CaMV was the first plant virus demonstrated to contain DNA and also the first virus genome to be complety sequenced. The CaMV genome has served as a source for the most widely used promoter in plant genetic engineering. The 35S promoter of CaMV is capable of directing a high level of constitutive transcription in most types of plant tissues.

The family *Caulimoviridae* comprises the plant infecting pararetro-viruses. The viruses of this family have circular dsDNA. The two genera of *Caulimoviridae-Caulimovirus,* with more than 12 viruses and *Badnavirus* with about 13 viruses are distinguished based on virion morphology. CaMV, the type member of the caulimoviruses has been a model for investigation of replication and gene expression.

Morphology

Caulimoviruses are icosahedral (T = 7), 45-50 nm in diameter and sediment between 200 and 250S. The main capsid protein with electrophoretic mobilities of 37 and 44 kD is derived from ORF IV (Fig. 13.8). In addition, CaMV particles contain minor amounts of other polypeptides, a 15 kD nucleic acid binding protein derived from ORF III, protease and reverse transcriptase derived from ORF V, and host casein kinase II. In cells infected with CaMV large numbers of virus particles accumulate in typical stable inclusion bodies, the matrix of which is the virus-encoded multifunctional inclusion body protein of 62 kD, derived from ORF VI.

Genome

The genome of CaMV consists of circular dsDNA of 8 kb. It is organized into six closely spaced or overlapping major open reading frames (ORF). One or two additional ORFs (VII, VIII), and a large intergenic regions are also found in the genome. The DNA contains a single gap in the negative (transcribed) strand and two gaps (with overlapping residues) in the positive strand. The proteins encoded by different ORFs and their functions are precised in Table 13.2.

Table 13.2. ORFs and the proteins encoded in CaMV

ORF	Size of the encoded protein (in kD)	Function
I	41-45	Movement protein
II	18-19	Vector transmission mediation
III	15	DNA binding protein
IV	57	Polyprotein precursor of coat protein
V	37-42	Components of mature virion
	80	Reverse transcriptase
VI	62	Development of symptoms

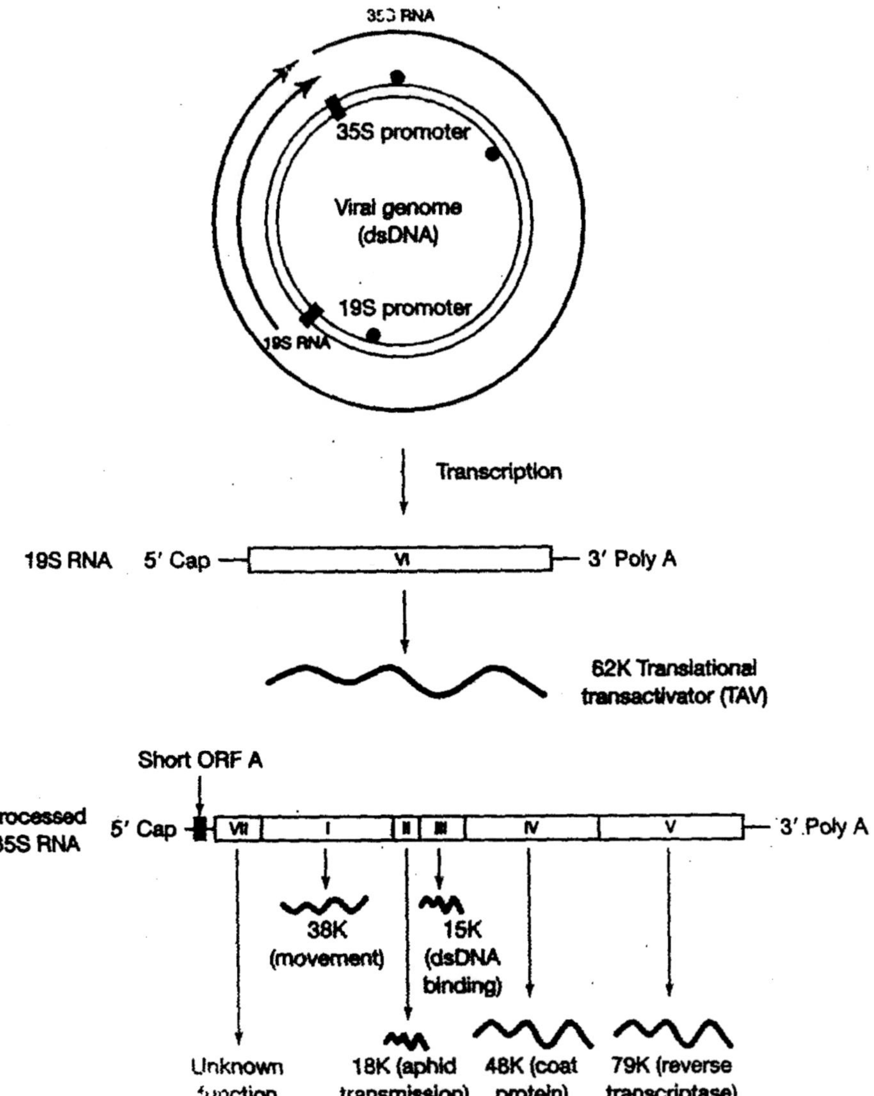

Fig. 13.8. The genome of *Cauliflower mosiac virus*. The dsDNA genome is closed at the 3 discontinuities (marked as black dots) and two transcripts, the 19S and 35S RNAs are produced. (From Dickinson, M. Molecular Plant Pathology)

Replication : Cauliflower mosaic virus particles are transmitted from plant to plant by aphids in a nonpersistent manner. Infections can also be established by manual inoculation. The details of initial entry and establish- ment are not clearly known. After disassembly, the viral DNA is transported to the cell nucleus and there after removal of overlapping nucleotides and closing of the gaps, it becomes associated with histones in a minichromosome configuration. Two polyadenylated RNAs are transcribed from one strand of the viral DNA. These are the 35S RNA and 19S RNA (Fig. 13.9).

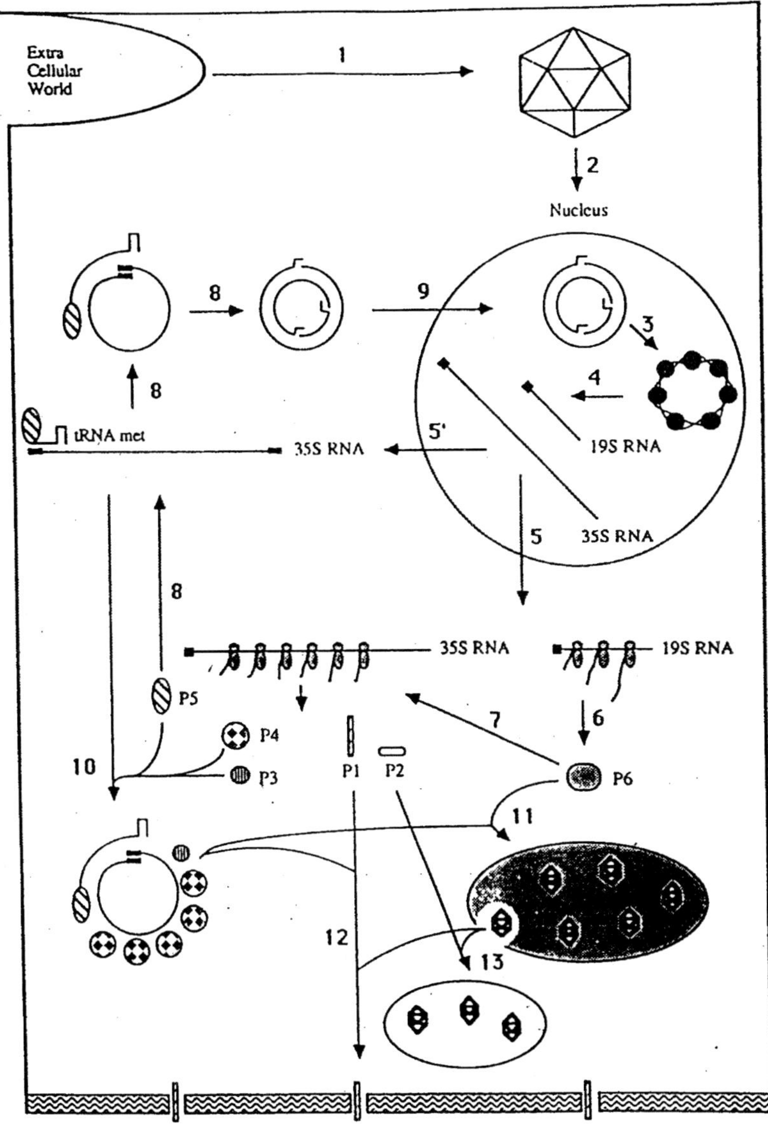

Fig. 13.9. Replication of cauliflower mosaic virus 1. Entry, 2. Uncoating and entering into nucleus, 3. minichromosome formation, 4. Transcription of 19s and 35s RNA 5. migration of transcripts into cytosol, 6. Translation of 19s RNA into P_6 protein (post transcriptional activation factor), 7. Activation of 35s RNA translation, 8. reverse transcription, 9. entry of DNA into nucleus, 10. provirion formation, 11. viroplasm formation, 12 movement of virion into adjacent cell through plasmodesmata, 13. formation of electron-lucent inclusions for aphids transmission; P1, P2, P3, P4, P5, P6 are different protein factors.

The 35S RNA promoter of CaMV has been well characterized and has played a central role in the development and use of transient and constitutive transformation systems in plants. These two RNAs are transported to the cytoplasm where they serve as mRNAs. The ORF IV protein is produced by translation of the 19S RNA. The polycistronic 35S RNA serves as the messenger RNA for the other proteins. Two mechanisms have been proposed to regulate the translation of the 35S RNA. Cis-acting elements in the leader sequence permit ribosomes to be shunted from the 5' end to downstream position and to effect translation of the internal genes of the 35S RNA. This requires a trans-acting mechanisms involving ORF VI protein.

The synthesis of progeny viral DNA occurs in the cytoplasm in viral inclusion bodies composed of the ORF VI protein. A host tRNA serves as a primer to initiate reverse transcription of the RNA. A 180 nucleotide direct repeat at each end of the RNA permits strand-switching at the five end of the template and the completion of synthesis of negative strand DNA. Partial digestion of the used template, probably by the viral RNase H, leaves remn-ants of the RNA that may serve as primers for synthesis of plus strand DNA.

Reverse transcription of 35S RNA is thought to occur in the previrion complex with coat protein, and therefore, to be closely associated with the assembly of progeny virions. However, little is known about the mechanisms that are involved in assembly of caulimoviruses. A P6 protein encoded from 19S RNA forms a matrix of the viroplasm in which virions finally accumulate. P1 protein, a movement protein facilitates the movement of viroplasm to adjacent cells through cytoplasm. Still, another protein P2 functions as aphid transmitting factor. It converts the viroplasm so that it is easily acquisitioned by aphids.

6. REPLICATION OF SINGLE STRANDED DNA VIRUSES : GEMINIVIRUSES

Genimi viruses, first named by Harrison (1977), are a group of plant viruses characterized by their circular ssDNA genome and a unique geminate (gemini-twin moons) particle morphology. *Geminiviridae* to which gemini-viruses belong contains approximately 60 members. *Geminiviridae* comprise fourteen genera and 520 species based on genome organization, vector specificity and host range properties (ICTV, 2020 release). The geminiviruses contain either one (monopartite) or two (bipartite) circular DNA molecules and the viruses are transmitted by either leaf hoppers or white flies. In addition to their importance as destructive pathogens of important crops, the geminiviruses are of considerable interest as potential vectors for the expression of genes in plants.

Virion morphology

Purified preparations of geminiviruses predominantly contain geminate particles of 18 x 30 nm in size (Fig. 13.10). In electron micrographs each of the two segments appears to be pentagonal with the faces in contact longer than others. Each incomplete icosahedron (T=1) contains 110 protein subunit (29 kD) arranged in 22 capsomers. The physical deposition of the DNA within the particle is not known. Particles of the subgroup I viruses contain small DNA molecules that are thought to be primers for second strand synthesis.

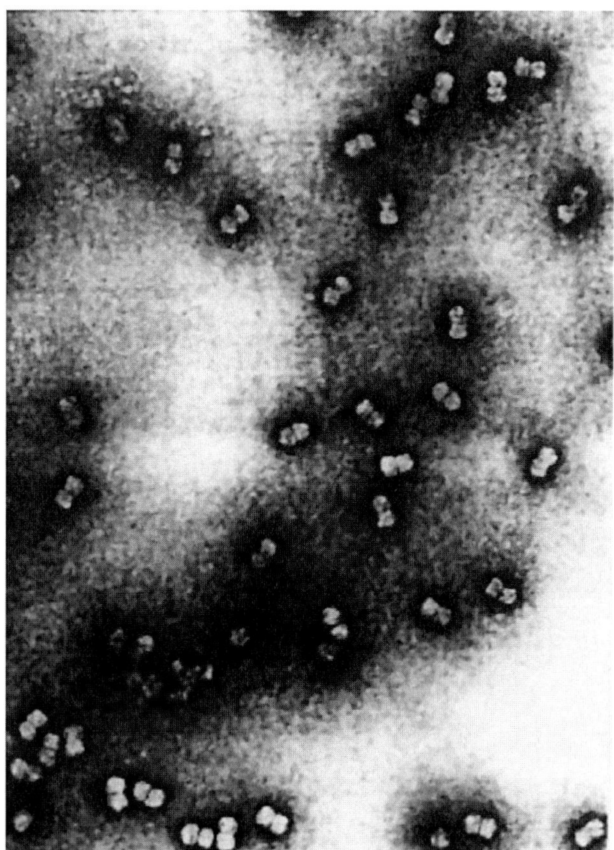

Fig. 13.10. Electron micrograph of Geminivirus (Courtsey R. Lastre)

Genome

Genomes of geminiviruses consist of one or two small (2.7 – 3.0 kb), covalently closed, circular ssDNA molecules (Fig. 13.11). The DNAs are characterized by a bidirectional arrangement of conserved ORF in which coding regions are positioned in both the virion (v) and complementary (c) senses of the DNA. The DNA components of the viruses with bipartite genomes are referred to as A and B. DNA A of African cassava mosaic virus (ACMV) contains six ORFS, two in the virion and four in the opposite sense. DNA B consists of two ORFs, one in each sense. DNA A of another subgroup III virus tomato golden mosaic virus, contains fewer ORFs. Each component contains an intergenic region of variable size (e.g., 300 b in DNA A and 600 b in DNA B of ACMV). For any given geminivirus, a sequence of 200 b (the common region) in the intergenic region of component A is the same as that in component B but, except for a highly conserved sequence of some 30 b, is different from the common region in other bipartite geminiviral genomes. The 30 b sequence has the potential to assume a stem-loop structure and contains an absolutely conserved nonanucleotide motif (TAATATTCA) in the loop sequence.

Replication

Many events of the geminivirus replication cycle have recently been elucidated. However, still there are certain points that are poorly understood. Interestingly, the elucidation of the replication cycle has revealed several aspects of the normal cell cycle as geminivirus replication depends on upon many host functions. Geminiviruses replicate in differentiated cells that are in the G_1 phase and have closed most of their DNA replication machinery. Thus geminiviruses reactivate the replication activities and make the host cell to revert back to S phase.

Two initial observations pointed to geminiviruses using the rolling-circle mechanism of replication. One observation was provided by two dimensional electrophoresis of extracts of ACMV - infected plants that revealed five putative replication intermediates.

i. Subgenomic (-) strand DNA associated with genomic (+) strand DNA,

ii. Unit length (-) strand DNA

iii. Virion DNA ranging from one of two genome lengths – concatemeric virion as DNA, ds DNA, and partially ssDNA.

Rolling circle replication is a two-step process, in the first phase of which the ss(+) strand is the template for the synthesis of (-) strand to generate a dsDNA, a replicative form (RF). This RF has two functions. It is the template for transcription of mRNAs and it is also the template for (+) strand synthesis generating free ssDNA (Fig. 13.12).

i. Minus strand synthesis : A small oligonucleotide complementary to the 3' intergenic region functions as a primer. ACMV replication intermediates indicated that (-) strand synthesis is primed within the 5' intergenic region. Little is known about the proteins and mechanisms of (-) strand synthesis. It is presumed that the virus particles are targeted to the nucleus, where this stage of the replication occurs. Generally, it is believed that (-) strand synthesis is affected by host factors.

ii. Plus-strand synthesis : The priming of geminivirus (+) strand synthesis is through a DNA cleavage at a specific site. The sequences of the progeny were observed to be consistent with (+) strand DNA synthesis by rolling circle initiating or terminating within the conserved hairpin sequence in the 5' intergenic region. The initiation site was mapped to the conserved non a nucleotide sequence, TAATATTAC, in the loop of the hairpin. The geminivirus Rep protein is site-specific endonuclease that nicks and ligates (+) strand viral DNA at the same position.

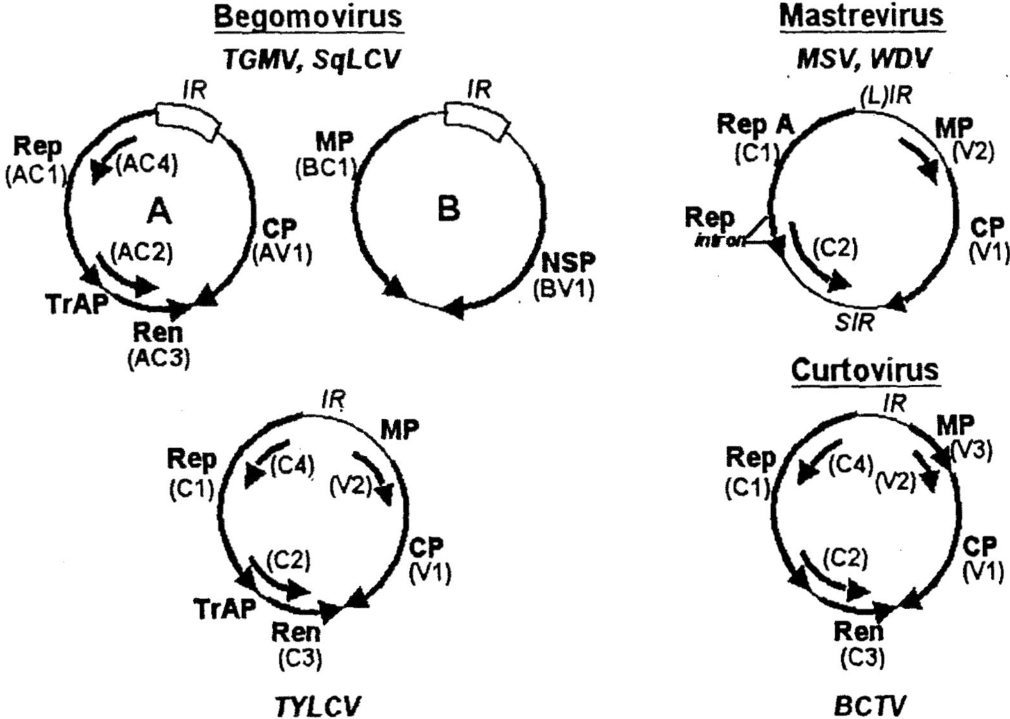

Fig. 13.11. Genome organization of different genera in the *Geminiviridae. IR – Intergenic region, MP – Movement protein, CP = Coat protein, REN – Replication enhancer protein, NSP – Nuclear shuttle protein TrAP – Transcription activator protein* (From Fields virology)

There are two major types of molecular organization of the (+) strand DNA replication origin. The replication origin for mastreviruses consists of a large cis-acting region where the Rep protein forms multiple complexes. On the other hand begomoviruses contain one binding site.

iii. Plus strand origin : In order to produce virus sense DNA i.e. (+) strand DNA from ds template, a nick is introduced into the virus sense strand. The 3' – end of the nicked DNA then acts as a primer for DNA synthesis displacing original virus strand as the template strand is copied and thereby producing concatermeric DNA. After the full round of replication, the nascent DNA is cleaved and regulated to produce a circular ssDNA. The circular ssDNA might enter into DNA replication to serve as a template for the generation of additional dsDNA or it might be sequestered for the purpose of encapsidation. The regulation of this process may be under the control of coat protein production. Absence of coat protein in the early phase of replication may lead to the chanelling of DNA back into replication whereas its presence in the latter stages of the infection cycle leads to encapsidation.

Defective interfering (DI) DNAs have been detected in geminvirus-infected plants indicating the fact that the replication sometimes may go wrong skipping certain areas of template.

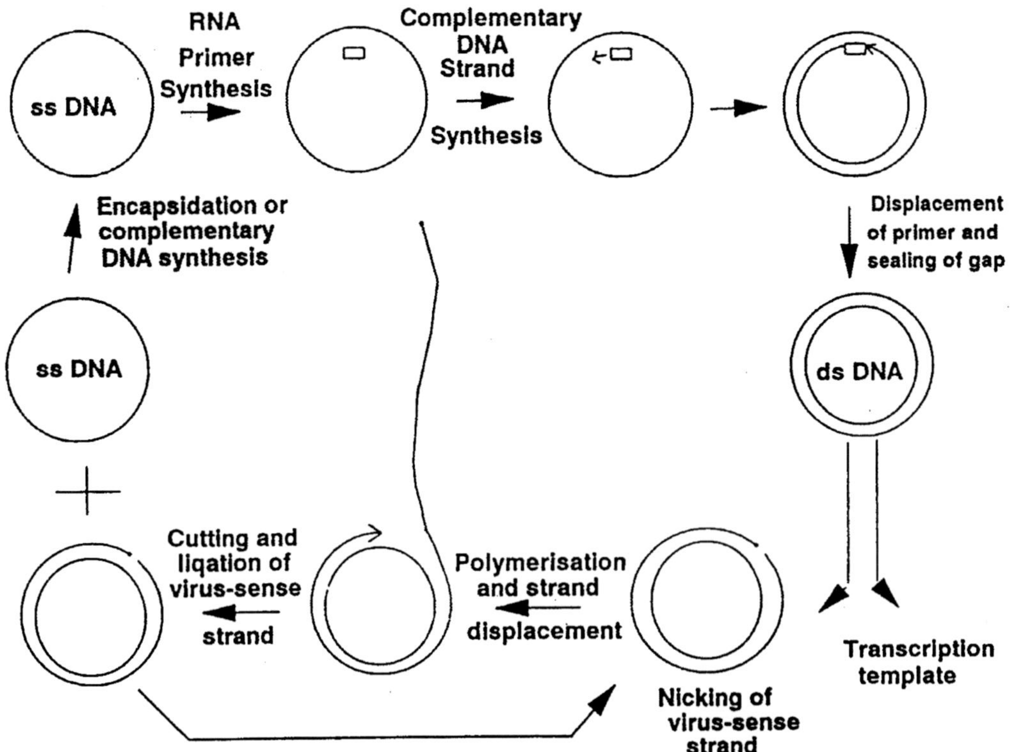

Fig. 13.12. Replication model for Geminivirus

Further reading

Agrios, GN (1997). Plant Pathology, San Diego Academic Press.

Cann, AJ (1993). Virus Replication. In Principles of Molecular Virology Cann AJ (ed.) Academic Press London.

Granoff, A and Webster RG (1999). Encyclopedia of Virology Vol. I, II and III. San Diego, Academic Press.

Lazarnwitz, SG (2001). Plant Viruses. In Fields Virology, Knife DM *et al.,* (eds.) Lippincott Williams & Wilkins, Philadephia.

Hull, R (2002). Matthew's Plant Virology 4th eds. San Diego Academic Press.

Shaw, JG (1996). Plant Viruses. In Fundamental Virology, Fields BN *et al.,* (eds.) Philadelphia, Lippincott-Raven.

Zerbini FM et al. (2017) ICTV virus taxonomy profile: Geminiviridae. *J. Gen. Virol.* 98(2): 131 – 133

14

VIRUS-HOST INTERACTIONS

Viruses are introduced into one or few of cells of healthy host. It is not end of the story instead it is only the beginning of the story. To establish itself and induce the disease, the virus has to spread through and replicate in almost all the plant parts. At this stage viral genome and host genome confront one another, with the virus attempting to establish infection and the host attempting to resist it. In most cases, it is a do-or-die situation for both. It is interesting to note that the virus wages this fight exploiting the host machinery. In a successful infection the host looses the ground to the virus. Nevertheless, in majority of the cases, virus does not allow the death of cells as it is detrimental to its own survival. While allowing the survival of host cell, the virus wants to take the advantage. In order to resist the virus onslaught the host cell deputes its entire metabolism. Protein and nucleic acid metabolisms in particularly are drastically affected. All the various microscopic or macroscopic symptoms of diseases originate in biochemical aberrations induced directly or indirectly by the virus. Many workers described differences in compositon or in rate of some processes between healthy and infected tissues. Aspects commonly investigated are : changes in enzymes, hormones, amino acids, nucleic acids, proteins, lipids, carbohyd-rates and rate of photosynthesis, respiration and transpiration. In viral disease changes are associated with the following :

1. BIOCHEMICAL CHANGES

i. Enzymes

Following the infection, changes in specific enzymes have been observed. These changes may be direct or indirect. Infection may affect enzyme activities through change in the amount of enzyme inhibitors or activators released. Increase in the activity of enzymes like catalase, peroxidase and polyphenol oxidase has been reported for many host-virus combinations. Significant increase in nuclease activity has also been reported. The increase may be because of their direct involvement with virus synthesis. In different host-virus combinations increase in the activity of acid phosphatase, aconitase and starch synthetase was also observed. In contrast, in certain cases decrease in the activity was also reported for example, pentose phosphate isomerase and succinyl CoA synthetase activity was reported to decrease following the infection with TMV in tobacco plants.

ii. Hormones

Virus infections influence hormone activities in infected plants and thus hormones play a significant role in the induction of a disease. Quantitative effects of infection on concentration have been shown for all the major groups of plant hormones. Systemic virus infections impair the normal levels of auxin with concomittant morphological alterations. Both increased and decreased levels of auxins are reported in different plant-viral combinations. Starting of virus infected plants has been associated with a decrease in gibberellins. Similarly, changes have also been observed in cytokinins and abscisic acid. Cytokinin contents may be either decreased or increased. The most affected plant growth regulator in viral infections is ethylene which is associated with senescence in plants. Ethylene production generally increases with virus infections.

iii. Low molecular-weight compounds

There are numerous reports on the effects of virus-infection on concen-tration of low-molecular weight compounds in various parts of infected plants. The large amount of data, however, is difficult to interpret in relation to virus replication, because the information varies with different host-virus combinations.

Accumulation of free amino acids in virus infected plants has been reported by several workers. Generally, an increase in amides such as glutamine and aspargine has been observed. In contrast, in many host-virus combinations deficiency of both soluble and insoluble nitrogen compounds may occur during periods when the virus is rapidly being synthesized. Disturbances in phosphorus contents have also been observed in virus infected plants. These variations in phosphorus contents can be attributed to the use of phosphorus by virus at the expense of some insoluble source of phosphate in the leaf. However, no definite conclusion could be drawn with regard to the effect of virus infection on host phosphorus metabolism. Variations in organic acid synthesis during virus replication has also been reported. Some essential elements like manganese, copper and zinc are signi-ficantly reduced in amount by virus infection. On the other hand, infection may have slight effect on potassium, calcium, magnesium, iron, boron and aluminium. Reduction in specific mineral elements may be associated with certain virus disease symptoms.

A majority of viral diseases exhibit yellow mosaic mottling or generalized yellowing of leaves. Such changes are due to change in the contents of different leaf pigments. Frequently, there is loss of chlorophyll with concomi-ttant increase in carotenoid pigments. The reduction in amount of leaf pigments may be due to either to inhibition of chlorophyll and xanthophyll synthesis or to the destruction of pigments in mature chloroplasts. In few cases, the virus infection may result in the increase of chlorophyll content.

iv. Proteins

Virus infection may lead to a significant decrease in host protein synthesis. During the period of active replication, host protein synthesizing machinery may be diverted for viral protein synthesis. It is surprising to note that the coat protein of a virus such as TMV can represent about half of the total protein in the diseased leaf. Effects on host protein synthesis are not necessarily correlated with amounts of virus produced. In mosaic and yellow type of symptoms, a substantial decrease in protein-ribulose biphosphate carboxylase-oxygenase (rubisco) has been reported by several workers.

v. Lipids

Not much information is available on the changes of lipids in viral diseased plants. However, the available information reports both increase and decrease in lipid contents in diseased plants than the healthy ones.

vi. Carbohydrates

Some viruses appear to have little effect on carbohydrates in the leaves, while others may alter both their rate of synthesis and rate of translocation. The rate of accumulation of starch is differently influenced under dark and light conditions. Accumulation of carbohydrates and sugars in virus infected leaves has been reported. Infections with some viruses such as BYDV, can induce damage to the phloem that restricts translocation of photo assimi-lates. These accumulate in the leaf lamina and possibly induce symptoms such as thicknened leaves and leaf yellowing and rolling. Generally, a rise in glucose, fructose and sucrose has been observed in virus infected leaves.

vii. Photosynthesis

Virus infections usually affect the process of photosynthesis. Reduction in carbon fixation is the most commonly reported effect in leaves showing mosaic or yellows diseases. This reduction usually becomes apparent some days after infection. Photosynthetic activity may be reduced by changes in chloroplast structure, by reduced content of photosynthetic pigments or rubisco, or by reduction in specific proteins associated with the particles of photosystem II. However, such changes appear to be secondary. Chloroplast colour may change from normal green to colourless. There may be reduction in number of grana and chlorophyll content. The biochemistry of photosynth-esis and related pathways may be affected as a result of virus infection. Carbon assimilation may also be affected. Diversion of the photosynthetic products from sugars to organic acids has also been observed in some cases. For example, instead of sugar formation, there may be production of organic acid and amino acids. Such changes, however, are usually transient. During the rapid replication of virus there may be diversion of early products of carbon fixation away from sugars and into pathways that lead more rapidly to the production of building blocks for the synthesis of nucleic acids and proteins. The most general result of virus infection is a reduction in overall photosynthetic activity and may arise from a variety of biochemical and physiologiocal changes.

viii. Respiration

Many studies have been made on the effects of virus infection on rates and pathways of respiration. However, it has been found not possible to correlate the results to the process involved in viral replication. Nevertheless, for many virus-host combinations, where necrosis occurs there is a rise in respiration rate, which may begin before symptoms appear and continue for a time as disease develops. In chronically infected plants, respiration is often lower than normal. In systemically infected plants there is no detectable change in the pathway of respiration. In host-virus combinations, where local necrotic lesions develop, there is an increase in respiration as necrosis develops. This increase was attributed to the activation of hexose monophos-phate shunt pathway. Increased activity of the enzymes of glycolytic pathway and TCA cycle has also been recorded.

ix. Transpiration

In chronically virus-infected leaves, transpiration rate and water content have been found to be generally lower than in corresponding healthy tissues. The reported effects over the first 1-2 weeks after inoculation vary. The results obtained by various workers are difficult to generalize because different viruses and host species have been used under different conditions.

x. Nucleic acids

It is widely assumed that the small RNA viruses have little effect on host cell DNA. Virus infection may have some effect on host cell DNA synthesis, but such effects are likely to be fairly small. Virus infection, however, may affect RNA synthesis in different host-virus combinations. Chloroplasts as well as cytoplasmic ribosomal RNA synthesis may be affected as a result of virus infection. The effect of infection may vary with age. The changes observed in young and old leaves may significantly vary. Effects of virus infection on ribosomal RNA synthesis and the concentration of riboso-mes may differ with the virus, time after infection, and the host and tissue concerned. A reduction in chloroplast ribosomes without a marked effect on cytoplasmic ribosomes is a fairly common feature for mosaic diseases.

2. CYTOLOGICAL, HISTOLOGICAL AND ANATOMICAL CHANGES

Virus infections generally change the overall morphology, arrangement and appearance of cells. Changes may occur even at cell organelle level. Important cell organelles like nucleus, chloroplasts and mitochondria are drastically affected. In some cases, disintegration of organelles like ribosomes and ER may occur.

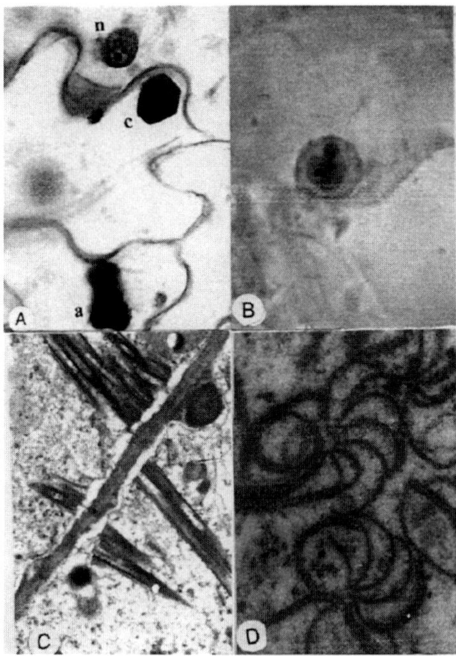

Fig. 14.1. Viral cell inclusion bodies. A. Crystalline (c) and amorphous (a) bodies ind-uced by TMV. B. Amorphous nuclear inclusion body induced by squash leaf curl geminivirus. C. Cylindrical and irregular inclusion bodies induced by potyviruses. D. Typical pinwheel like inclusion bodies induced by potyviruses (From Agrios, 1997).

i. Nuclei : Some viral infections induce intranuclear inclusions of various types and may affect the nucleolus or the size and shape of the nucleus. Nuclei may become filled with virus particles and become greatly enlarged (e.g. geminiviruses). Consequently, the nuclei may stain less readily, become granular, chromatin may be reduced. Nucleoli may be swollen or hypertrophied and disintegrated. Some times peri-nuclear space may be filled with viral particles (e.g. nucleorhabdoviruses). Crystalline plate-like inclusions and other types of intranuclear inclusions have been reported for some viruses (e.g. Bean yellow mosaic virus).

ii. Chloroplasts and mitochondria : Viral induced structural and biochemical degeneration of chloroplasts have been observed in many cases Chloroplasts may become swollen, rounded and clumped together in the cell (e.g., TYMV). There may be accumulation of starch grains and their size and number may be abnormal. They may become colourless and there may be fragmentation of chloroplasts. Changes may also be induced in mitochondria. There may be regeneration or aggregation of mitochondria in virus infected cells (e.g. TRV). Sometimes mitochondria may be filled with vesicles also.

iii. Cell wall : The cell wall in virus infected cells may be abnormally thickened. There may be deposition of electron-dense material between cell wall and plasma membrane. Protrusions may also develop on the cell wall. Some virus infections induce callus deposition, which probably may help in restricting virus movement. There may be the proliferation of the plasmalemma to produce vesicles.

iv. Inclusion bodies : One of the major effects of virus infection on host is the development of inclusion bodies. Inclusion bodies are the intracellular structures produced *de novo* as a result of virus infection (Fig. 14.1 CD). Some inclusion bodies are large enough to be seen in the light microscope, whereas others can be detected with the electron microscope. The inclusion bodies may be nuclear or cytoplasmic. They may contain virus particles, virus related materials or host cell constituents in a normal or degenerate condition. They are of various forms and size.

Inclusion bodies may be amorphous or crystalline or pin-wheel. Amorphous inclusion bodies also called `X-bodies' are amoeboid virus induced structures in the cell where components of viruses are synthesized and assembled. Crystalline bodies are nearly pure aggregates of virus which mostly accumulate in cell cytoplasm producing three dimensional structures (herring bone effect). Normally in these inclusions virus particles aggregate in an orderly fashion. Some times, cell ribosomes may be trapped within these banded bodies. Crystalline inclusions are commonly seen in epidermal and hair cells. In tobacco leaves infected with TMV, leaf and epidermal cells in yellow green areas contain crystalline inclusions (Fig. 14.1A).

Some viruses induce complex inclusions. These inclusions possess a central tubular core and from this central core radiate out curved arms to give a pinwheel effect. Pin-wheel inclusions are characteristic in the cytoplasm of infected cells of potyviruses (Fig. 14.1D). Virus particles may be intimately associated with the pin-wheel arms and are antigenic. Virus induced inclusions may be sometimes of help in classification and diagnosis of viruses.

3. VIRUS-PLANT HOST INTERACTIONS AT SUBCELLULAR AND MOLECULAR LEVELS: GEMINIVIRUSES AND HOST PLANT AS A MODEL

Geminiviruses are plant-infecting viruses characterized by a single-stranded DNA (ssDNA) genome. The defense and counter-defense strategies between geminiviruses and plants have become an interesting area of research and serve as a model to understand virus-host interaction at cellular and molecular level. Geminiviruses have genome and can code for only 5–7 proteins. Neverthless, viruses are capable of overcoming the defense strategies of plants, due to the complex nature of viral proteins, which are designed to interact dynamically with diverse corresponding plant proteins. Virus-host interactions result in changes in host gene expression patterns, reprogram plant signaling controls, disrupt central cellular metabolic pathways, impair plant's defense system, and effectively evade RNA silencing response leading to host susceptibility.

The geminivirus genome encodes replication-associated protein (Rep/AC1/C1) only, and thus is dependent on the host replication machinery to amplify its genome. In order to overcome the inadequate host DNA replication factors in the differentiated cells, the virus redirects host gene expression to accumulate host DNA polymerases to assist the viral genome amplification. Geminivirus proteins interact with host transcription machinery, DNA replication and cell division-related proteins, plant metabolic process proteins, defense-related proteins, and stress-related proteins. Rep plays a pivotal role in stimulating the replication of viral and plant DNA. It can bind to the subunits of host DNA polymerase complex, ssDNA binding proteins, and recombination/repair proteins to interfere in host DNA replication process. Geminivirus encoded proteins have been demonstrated to interact with host protein kinases implying their role in altering signal transduction pathways. It is also evident that diverse host gene expression pathways including genes involved in systemic acquired resistance (SAR), WRKY, TFs, cell cycle and PCNA are also altered following virus infection. Upregulation of defensin-like proteins, PR1 proteins and other LRR domain containing proteins, along with some transcription factors such as NAC or WRKY, have been associated with the induction of defense-related genes and innate immunity. Interactions of viral and cellular factors, but may also indirectly affect host physiological processes. Viral infection also affects hormone biosynthesis. Several host genes associated with auxins/IAA, ethylene-responsive factors, cytokinin and gibberellins were found to be differentially regulated during ToLCNDV infection.

Geminiviral proteins have also been reported to impede host protein degradation pathways to redirect protein metabolism in the plant cell. The majority of viral proteins were shown to interact with plant ubiquitination, SUMOylation and protease-mediated degradation machinery to alter plant developmental and defense processes. It has been suggested that geminivirus proteins have functionally evolved the capability to the modulate the protein degradation system to its advantage. Recently, it has been demonstrated that the beta-satellite DNA molecule associated with geminiviruses induces host calmodulin-like protein (CAM), which is an endogenous RNA silencing suppressor. Studies on tomato leaf curl infection in tomato suggest that viral infection results in the alteration of metabolism of carbohydrates, starch, sugars and amino acids. Up-regulation of phosphoenol pyruvate phosphate and acetyl Coenzyme A transcripts during infection suggests changes in this important biochemical network, central to cellular metabolism, which includes glycolysis and TCA cycle. Capsid protein (CP) has also been shown to hinder glyoxylase and shikimate pathways. The role of stress-responsive proteins in the development of virus infection was also reported in various studies.

4. MACROSCOPIC SYMPTOMS

Symptoms may be defined as visible abnormalities in plants that make us to identify a diseased plant from the healthy. Any attribute of the plant may be changed as a result of virus infection. Symptoms are reflections of host response to virus infection. Virus-like symptoms may also be caused by other plant pathogens such as viroids, phytoplasmas, nematodes or mineral deficiencies. Symptoms are of particular importance since viruses are ordinarily invisible and can be recognised only by their effect on their hosts. However, field diagnosis based on symptoms alone should only serve as a guide. Symptoms can provide only a partial diagnosis because : (i) similar symptoms can be produced by different viruses, (ii) symptoms may be extremely variable; the same virus can produce a range of symptoms, depending on environment and host genotype, (iii) a lack of symptom does not necessarily mean that the infection is latent or masked and (iv) mixed infection with several viruses may have an additive effect on the host, resulting in more severe marked symptoms. For the sake of comprehensive understanding, the viral disease symptoms can be categorized into the following :

External symptoms

i. General appearance of the plant : Deviations resulting from viral infections in the general appearance of the plant include : abnormal colour either in part or whole plant, stunting rosetting, witches broom, decline and necrosis. Shortening of nodes and internodes which produces a bunched appearance is called rosetting. Witches broom type of symptoms result from excessive budding and branching, stunting and shortening of nodes. Loss of vigour of the whole plant or parts of the plants is referred to as as decline. Necrosis is death and browning of tissues.

ii. Colour deviations : Colour deviations are the predominant type of symptoms produced in viral diseases. Colour deviations may occur in leaves, flowers, fruits and even in roots.

Leaves : Colour deviation type of symptoms on leaves include : chlorosis, bleaching, yellowing, reddening, browning, bronzing, mosaic, mottle, local lesions, ringspots, streaking, vein yellowing, vein clearing, vein banding and vein necrosis (Fig. 14.3). Chlorosis is the weakening of the green colour. Bleaching is the disappearance of all colour usually giving white appearance. Chlorosis associated with yellow pigments is known as yellowing. Abnormal synthesis and accumulation of anthocyanin pigments results in reddening of the leaves. Production of dark melanin-like substances cause browning and blackening of leaves. Bronzing is the result of necrosis and collapse of epidermal cells covering the still green and apparently healthy mesophyll. Pale green, yellow or chlorotic areas, sharply bordered by small veins are called mosaic symptoms (Fig. 14.2). Mottling is the discolouration of areas which are rounded shape and often diffusely bordered. Local lesions range from small pin-point size chlorotic or necrotic areas to large irregular patches. If chlorotic or necrotic areas are separated by normal green tissue by single or concentric rings, they are called ring spots. Elongated, sharply defined chlorotic patches are called streakings. Yellow colouration of the veins due to lack of chlorophyll is called vein yellowing. When veins appear translucent rather than chlorotic or yellow, it is called vein clearing. Discoloured areas along the veins is called vein banding. Vein necrosis is the death of vascular tissues resulting in browning.

Fig. 14.2. Typical mosaic symptoms of tobacco mosaic disease (From Agrios, 1997).

Flowers : In flowers, colour deviation may be in the form of intensification, weakening or change of pigments in the epidermal layer of the petals. Flower breaking usually consists of flecks, streaks, or sectors of abnormally coloured tissue.

Fig. 14.3. Symptoms caused by plant virus diseases. 1. Potato leaf showing venal necrosis, 2. yellow vein mosaic of bhendi, 3. Golden yellow mosaic of bean, 4. Stripe mosaic in grass, 5. Leaf curl and formation of enations in tomato, 6. Mosaic in *Abutilon striatum.*

Fruits : In fruits colour deviation may occur on the whole fruit or parts of the fruit (e.g., marbling, mottling and spotting).

Roots : Viral infections may lead to the development of lesions on root tissues.

iii. *Malformation* : Malformations or irregular abnormal growth may occur on any part of the plant. However, usually the leaves are the targets.

Leaves : Malformations on leaves include distortion, epinasty, narrowing, reduction, thickening and enations. Distortions of leaves are due to crinkling, curling and twisting. Downward curling of the leaves is known as epinasty, whereas upward curling is known as hyponasty. Reduction in laminar tissue results in narrowing of leaves. Outgrowths of the leaf blade, is referred to as enations. (Fig. 14.3)

Flowers : Malformations on flowers include various kinds of distortions, production of abnormal flowers and reduced flower size.

Fruits : Malformed fruits reveal abnormalities like reduction in size, defor-mation and irregular shapes, production of tumorous swellings and abortive seeds.

Stem : Distortion and shortening of internodes are common malformations on stems e.g. apple viruses, citrus tristeza streaks.

Root : Malformation type of symptoms on roots include decay and dieback, tumors and proliferation of side roots. Soil borne virus diseases induce such symptoms.

iv. Miscellaneous symptoms: Viral infections cause a variety of symptoms other than described above. They include : wilting, defoliation, premature leaf drop, premature or delayed flowering, abnormal fruit-flower, woodiness of the fruit, gummosis, bark scaling, wood pitting, shoot swelling, graft incompatibility.

v. Inapparent symptoms: Under certain environmental conditions, no visible symptoms are produced even though a virus is present in the plant. This temporarily latent infection is generally due to environmental factors such as temperature, light and nutrient excesses or deficiencies. Certain viruses never induce symptoms in certain hosts and are, therefore, permanently latent.

Internal symptoms

The external symptoms induced by viruses are frequently the reflections of anatomical and histological changes within the plant. These changes may be in the form of necrosis, hypoplasia, hyperplasia. These may occur singly or together in any particular disease.

Further reading

Bos, L (1978). Symptoms of virus diseases in plants 3rd edn. Pudoc, Wageningen.

Christie, RG and Edwardson, JR (1977). Light and electron microscopy of virus inclusions. Florida Agric. Exptl. Stn. Monograph No. 9.150 pp.

Edwardson, JR and Christie, RG (1978). Use of virus induced inclusions in classification and diagnosis. *Ann. Rev. Phytopathology,* **16** : 31-55.

Esau, E (1956). An anatomists view of virus disease. *Am. J. Bot.,* 43 : 739-748.

Fraser, RSS (ed) (1987). Biochemistry of virus infected plants. Research studies Press. Letchworth, U.K.

Gibbs, AJ and Harrison, BD (1976). Plant Virology. The Principles. Edward Arnold : London.

Hull, R (2002). Matthew's Plant Virology, IV edition, Academic Press, New York.

Martelli, GP. and Russo, M (1977). Plant virus inclusion bodies. *Adv. Virus. Res* **21** : 175-266.

Stutic, D and Sinclair, JB (1991). Anatomy and Physiology of diseased plants, CRC Press, Boca Raton, FL.

Verma, HN (2003). Basics of Plant Virology. Oxford and IBH publishing Co. Pvt. Ltd. New Delhi.

Ramesh SV *et al* (2017) Geminiviruses and plant hosts : A close examination of the molecular arms race. *Viruses* 9: 256–275

15

TRANSMISSION

Plant viruses being obligate parasites in order to survive must spread from diseased plant to another susceptible but healthy plant. They are not transmitted by wind or air. They need a biological agent that can carry the virus from diseased plant and/or field to healthy plant and/or field or they require atleast a physical contact between diseased and healthy plants. Further, plant viruses cannot enter their host by themselves. They can enter host tissue through wounds or with the help of other organisms which can acquire them from an infected plant then introduce them into a healthy plant. Because of the static position the hosts, plant viruses evolved a device to recruit vectors for transmission, or are carried in seed and pollen. The knowledge of the ways of virus transmission is important for various reasons : (i) A disease is identified as a virus disease only if it can be transmitted to healthy individuals by some means, (ii) the mode of virus spread in the field is essential for the development of effective control measures, (iii) the relationship between the virus and its vector is of considerable biological interest, and (iv) mode of transmission is also very important for laboratory investigations.

Any mobile biological agent (other than host parts like pollen, seed) that is capable of transmitting the virus from infected plant to healthy susceptible plant is called a **vector**. Plant virus vectors include insects, mites, nematodes fungi. Other modes of transmission include seeds, pollen grains, dodder and different versions of mechanical transmissions.

1. INSECT TRANSMISSION

Of the vector transmitted viruses, 70% are transmitted by arthropods mainly phloem feeding homoptera. Nearly 99% of these arthropod vectors are insects. Different groups of insects such as aphids, whiteflies, leafhoppers, plant hoppers, beetles, mealy bugs and thrips are able to transmit different viruses (Fig. 15.2).

Insect transmission of plant viruses has traditionally been categorized according to the nature of the association with the vector (Table 15.1). Stylet borne viruses that are acquired and carried on the mouth parts are termed **non-persistent** because they are lost quickly (minutes) from the insect once it feeds (fig. 15.1). Foregut borne viruses are known as **semi-persistent**, whilst those that pass through the gut into the haemolymph and then to the salivary glands are known as **persistent** or **circulative**, because they can often be transmitted for as long as the insect lives. Persistent viruses inturn are sub-divided into the **propagative** viruses, which are able to replicate in both plant and the insect vector and **non-propagative**, which can only replicate in plants.

Insect transmission of the viruses may be horizontal or vertical. Transmission of viruses between the individual of the same generation is known as **horizontal transmission**. However, certain viruses are transmitted from parents to offsprints through eggs (transovarial). This type of transmission is known as **vertical or hereditary transmission**.

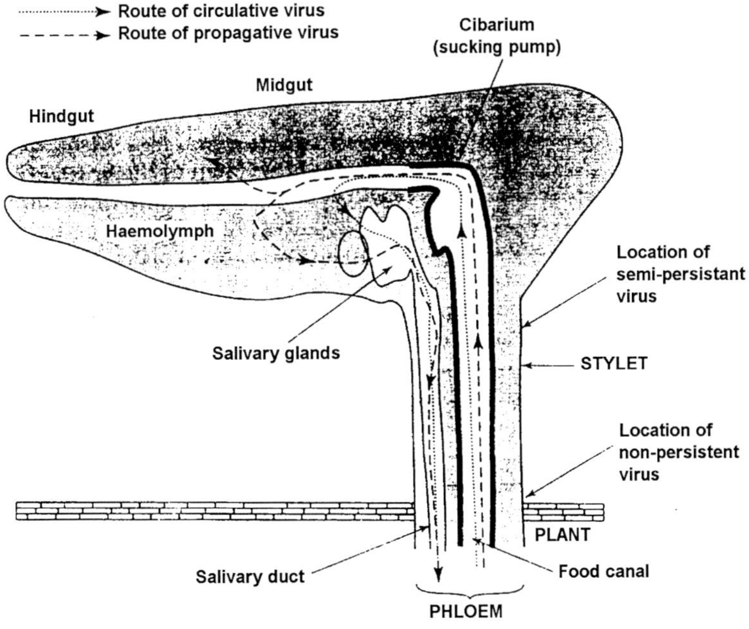

Fig. 15.1. Stylised diagram of an insect feeding system showing the extent of virus uptake. Non-persistent viruses are carried on the stylet, semi-persistent viruses on the foregut and persistent viruses are taken up into the haemolymph and salivary glands and can be divided into the circulative viruses and the propagative viruses (From Dickinson, M. Molecular Plant Pathology).

i. Aphids : More than 190 aphid species are known to transmit more than 160 virus diseases. The most common virus transmitting genera are *Aphis, Myzus, Brevicorne, Rophalosiphum, Macrosiphum* and *Toxoptera.* Most aphid tranmitted viruses induce mosaic diseases. Some also produce a yellow-type disease. Aphids transmit the viruses through all the three ways viz, non-persistent, semi-persistent and persistent-circulative. Persistent-propagative mode by tranmission is not recorded. Viruses transmitted by persistent mode include different morphological forms and different taxonomic groups e.g. rigid rods, flexuous rods, spherical and bacilliform. Non-persistently transmitted viruses are acquired rapidly from plants, usually in a matter of seconds and similarly they become non-infective in a matter of seconds. Circulative aphid-borne viruses are less common than stylet-borne diseases.

Table 15.1. Mode of insect transmission of plant viruses

Stylet-borne Non-persistent Non-circulative	**Foregut-borne semi-persistent Non-circulative**	**Persistent circulative**	**Persistent propagative**
Alfamovirus	*Badna virus*	*Begomovirus*	*Torpovirus*
Carlavirus	*Caulimovirus*	*Curtovirus*	*Marafivirus*
Cucumovirus	*Closterovirus*	*Mastrevirus*	*Phytoreovirus*
Fabavirus	*Sequivirus*	*Enamovirus*	*Fijivirus*
Potyvirus	*Trichovirus*	*Luteovirus*	*Oryzavirus*
	Waikavirus	*Polerovirus*	*Cytorhabdovirus*
		Nanovirus	*Nucleorhabdovirus*
		Umbravirus	*Tenuivirus*
		Bromovirus	
		Carmovirus	
		Comovirus	
		Sobemovirus	
		Tymovirus	

ii. White flies : *Bemisia tabaci* is the most important and widespread vector in tropics and subtropics. The whiteflies which are present on the underside of leaves become viruliferous after a short feeding on the diseased plants. The virus-vector relationship is mostly of circulative type. The virus doesnot multiply in the insect vector. Latent period may be a few hours to a day. The vector may retain the virus from a few days to twenty five days. Whitefly transmitted viruses are generally not mechanically transmitted and produce yellow, mosaic, leaf curl and rugose mosaic type of symptoms.

iii. Leaf hoppers : More than 30 species of leafhoppers have been reported to transmit 30 different viruses. The most common genera of virus-transmitting leafhoppers are : *Aceratagallia, Agallia, Cicadulina, Circulifer, Dalbulus, Empoasca, Eutattix, Graminella, Javesella, Macrosteles, Nephotettix.* Leaf hoppers are phloem feeders. They mainly transmit phloem associated viruses. The leafhopper transmitted viruses are generally persistent circulative type. Acquisition period varies from 30 minutes to several hours. The virus is not lost by the vector after moulting. Leaf hoppers have also developed propagative type of transmission. For example, rice dwarf virus is carried through the eggs (transovarial passage) of female parents, *Nephotettix cincticeps.*

iv. Mealy bugs : The main genera of mealy bugs known to transmit viruses are *Planococcus, Pseudococcus* and *Dysmicoccus.* Mealy bugs are phloem feeders and feed by sucking. Interestingly mealy bug transmitted viruses are also sap transmitted. Viruses are generally nonpersistent or semipersistent. Acquisition feeding time is few hours and the virus is retained for a few days (3-4 days). Crawling nymphs are more effective vectors than adults. Cocoa swollen shoot virus is transmitted through mealy bugs.

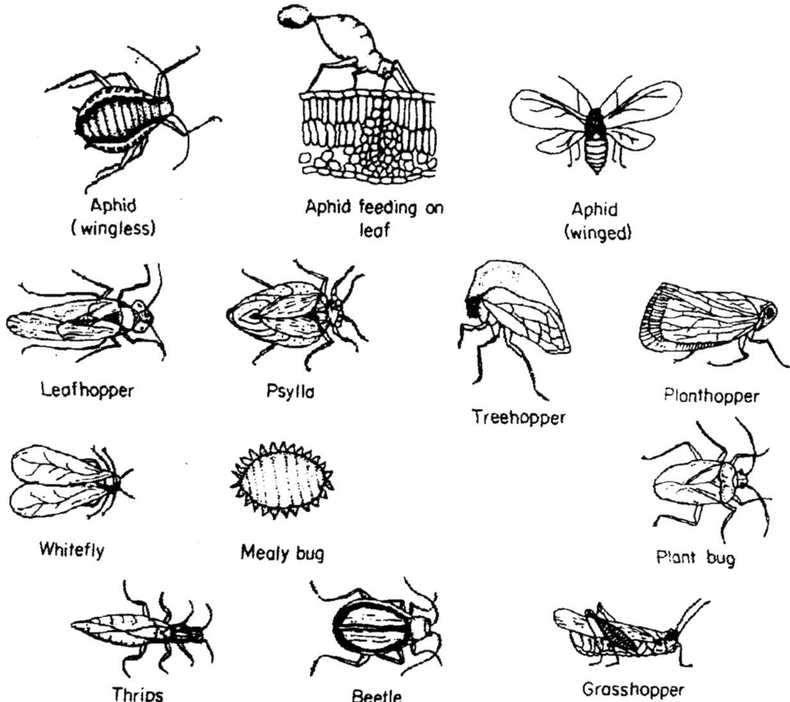

Fig. 15.2. Insect vectors of plant viruses. (From Agrios, 1997)

v. Thrips : Thrips belonging to three genera are known to transmit viruses. They are *Thrips, Frankliniella* and *Scirtothrips*. Thrips usually feed on young tissues. They feed by sucking contents of sub-epidermal cells of the host. They transmit the virus in a circulative manner and may retain infectivity for life. Several days elapse between the vector acquiring the virus and becoming infective. Tomato spotted wilt virus is reported to be transmitted by thrips.

vi. Beetles : About 24 beetle species have been reported to be vectors. The most common virus transmitting beetles are flea beetles (*Phellotrata* spp) mustard beetles (*Phaedon* spp) and Cucumber beetles (*Acalymma* spp.). These are known to transmit about 45 viruses. The virus vector relationship is of circulative type but there is no evidence of multiplication of virus in the vector. Increased acquisition feeding by the vector, increases transmission rate. Vectors remain viruliferous for a few days. Important viruses like cow pea mosaic, turnip yellow mosaic, squash mosaic and southern bean mosaic are transmitted by beetles.

vii. Mite transmission : The most common virus transmitting mite genera are : *Aceria, Brevipalpus, Eryophyes*. The virsues are carried in the alimentary canal of the insects. The relationship is circulative but not propagative type. The virus is not transmitted to offsprings. The virus is acquired within minutes of feeding on diseased plants and may persist in the vector for a few days. Mites are easily dispersed by wind because of their small size and delicate nature. Four confirmed examples of mite transmitted viruses are: ryegrass mosaic, wheat spot mosaic, pigeon pea sterility mosaic and wheat streak mosaic.

2. TRANSMISSION BY NEMATODES

Soil-inhabitory nematodes of the genera *Longidorus, Paralongidorus* and *Xiphinema* are able to transmit the nepoviruses and dianthoviruses. Nematodes of the genera *Trichodorus* and *Paratrichodorus* transmit members of the *Tobravirus* genus. They acquire the virus by feeding on infected roots, and the viruses have been shown to attach to the surface of the food canal, although there is no evidence that they replicate in the nematodes. Viruses are retained in nematodes from weeks to several months. They persist for about two weeks in *Trichodorus* and *Longidorus* and for about 8 months in *Xiphinema*. However, they are lost during molting.

3. TRANSMISSION BY FUNGI

A number of viruses such as the *Furovirus, Pocluvirus, Benyvirus* and some genera of the *Potyviridae* are transmitted by soil borne zoosporic protozoa *Polymyxa* and *Spongospora* species. These organisms possess many fungal like characteristics, but only the necroviruses are transmitted by true fungi of the *Olpidium* species. These fungal and fungal-like organisms transmit plant viruses either non-persistently, when the virus is absorbed to the surface of motile zoospores, or persistently, in which case the virus can become internalised into both zoospores and the resting spore stage. These fungi are not themselves pathogens of plants, but when they become associated with roots, the virus is transmitted into the plant. Different fungi show specificity in virus transmission e.g. *Olpidium brassicae* transmits tobacco necrosis virus but not cucumber necrosis virus, which is transmitted by *O. cucurbitacearum.*

4. TRANSMISSION BY SEED AND POLLEN

Around 20% of all plant viruses can be transmitted through seed of one or more plant species, and are mostly contained within the embryo. They can invade the developing embryo in one of two ways, either indirectly by moving into the gametes prior to fertilization, or directly by infecting the embryo after fertilization. This capacity for seed transmission is determined by both host and viral genes. Seed transmitted viruses are also spread by nematodes fungi, aphids and beetles. Seed transmission provides an effective means of spread of the virus over long distances. Survival of virus between growing seasons and during times adverse to plant growth is also facilitated when the virus is seed borne. When the virus-bearing pollen from diseased individuals brings about the infection of the ovule bearing plants, the virus is said to be pollen transmitted and pollen borne.

Some of the commonly occuring virus like barley stripe mosaic, tobacco ring spot, bean common mosaic are pollen and seed borne, whereas viruses like cowpea mosaic, cucumber mosaic, mung and urd bean mosaic, tomato mosaic etc. are seed borne.

5. TRANSMISSION BY DODDER

Dodder plants are vines and are parasitic members of the family. Convolvulaceae. The leafless stems of this plant wind around suitable plants, sending out haustoria which connect with the vascular tissues of host. Branches from the initial dodder stem entagle other plants and parasitize them. Virus can pass along the connecting dodder stems between plants. The virus is acquired from the vascular bundles of the systemically infected plant by the haustoria of dodder and after translocation through the dodder phloem, it is introduced in the next plant by the new dodder haustoria produced in contact with the vascular bundles by the inoculated plant. Viruses such as CMV and tobacco rattle replicate in the dodder and are more efficiently transferred than TMV which does not multiply in the dodder and is merely carried out in the dodder which acts as a passive biological bridge connecting two plants. Dodder sometimes may harbour a dodder latent mosaic virus and produces severe symptoms in plants parasitized by the infected dodder.

6. CONTACT TRANSMISSION

Contact transmission occurs (i) when infected and healthy plants come in contact and then leaves rub together, (ii) by the action of animals, (iii) by action of humans, either occidentally, or deliberately for experimental purposes. During rubbing of plants, hair epidermal cells containing virus may be broken and virus may be liberated into similarly damaged cells on healthy plants, thereby bringing about transmission. This type of transmission results in intensification of disease within closely planted crops. Animals (other than insects) may transfer virus by first rubbing against infected plants and then entering healthy crops. Virus attached to their body surface enters mainly through minor damage or injury to plants.

Contact transfer of virus between plants can be brought about by man, tractor wheels, agricultural implements, e.g. PVX can survive on clothing, tools and mechanical devices used in agriculture and horticulture. Clothing contaminated with TMV may remain infective for several months. Viruses may also be transferred from infected to healthy plants by cutting knives when pruning. Viruses such as TMV and tomato mosaic may spread when pricking out seedlings or harvesting fruits and flowers by hand.

7. VEGETATIVE AND GRAFT TRANSMISSION

In systemically infected plants virus is present in nearly all organs and tissues of the plants, therefore, the parts from mother plants used for vegetative propagation e.g., cuttings, tubers, bulbs, corms, bulbs or rhizomes will always contain virus and so it will be transmitted to the progeny. Care should be taken not to propagate ornamentals like carnations, chrysanthemums and dahlia and field crops like potato, banana from diseased or suspected plants. Apical meristems of systemically infected plants are, however, virus-free and by tissue culture virus-free plants may be grown from tips of virus infected plants.

Graft transmission of viruses has been known for a long time. Graft transmission is being used as a diagnostic test and interpreted as indicating viral etiology for disease. Mycoplasma like organisms (MLO), now called phytoplasma, are also transmitted by grafting.

Further reading

Bennett, CW (1969). Seed transmission of plant viruses. *Adv. Vir. Res.,* **14** : 221-261.

Costa, AS (1976). Whitefly-transmitted plant diseases. *Ann. Rev. Phytopath.* **14** : 429-449.

Green, SK (1984). Guidelines for diagnostic work in plant virology. Technical Bulletin 15. (2nd Ed.). Asian Vegetable Research and Development Center, Taipei.

Harris, KF (1982). Arthropod and nematode vectors of plant viruses. *Ann. Rev. Phytopathol.,* **19** : 391-426.

Hall, R (2002). Mathew's Plant Virology. IV edition, Academic Press, New York.

Raski, DJ and Hewitt, WB (1963). Plant parasitic nematodes as vectors of plant viruses. *Phytopathology;* **53** : 39-47.

Slack, SA and Foulton, JP (1971). Some plant virus-beetle vector relations. Virology, **43** : 728-729.

Verma, HN. (2003). Basics of Plant Virology. Oxford and IBH Publishing Co. Pvt. Ltd. New Delhi.

Watson, MA and Plumb, RT (1972). Transmission of plant pathogenic viruses by aphids. *Ann. Rev. Entomol,* **17** : 425-452.

16

DIAGNOSIS OF PLANT VIRAL DISEASES

Till now, several hundred virus diseases of plants have been described and new ones continue to be reported. Very often the causal agents are described as new viruses on totally inadequate grounds. The literature contains too many examples of known viruses re-described under new names, or of viruses so inadequately described that it is doubtful whether they really are new. Although some viral diseases can be diagnosed quickly by visual examination of symptoms, many require thorough investigations for accurate diagnosis. The `diagnostic phase' in some instances may not need to go beyond determination of particles morphology to determine to which group the virus belongs, followed by an appropriate serological and molecular tests to identify the virus. If the virus under investigation is not easily assigned to any known taxon, the diagnostic phase passess into the full `descriptive phase' in which every property possible is investigated.

Further, during the last few years there has been an explosion in plant virus outbreaks, because of either direct or indirect activities of humans. The ease of transporting plant materials and virus vectors, and collecting relatives of crop plants from wild, have spread viruses widely throughout the world. Thus, most important step in managing a virus disease is correct identification. Tests for diagnosis should be fast, accurate, and inexpensive. Apart from conventional methods, several new techniques, based on serology and genome characteristic, are now available for plant viral diagnosis. The prudency of the investigators lies in the selection of an appropriate method.

1. DIAGNOSIS USING CRUDE VIRUS PREPARATIONS

It often happens that plants are infected by one or more viruses that may or may not cause overt disease, and isolation of a virus from a diseased plant is no guarantee that the agent of the disease has been found. It is important, therefore, first to be sure that the causal virus has been isolated, i.e. that there is not a mixture of viruses present. Two methods may be required to identify whether the infection is due to a single virus or a mixture of viruses : (i) testing for possible mixtures by inoculation of a series of specific indicator plants, and by serology, light or electron microscope or a combination of these, and (ii) separation of mixtures by using differential vectors, test plants or single local lesion passages. Finally, the isolate should be reinoculated to the original host which is then observed for typical symptoms of the disease. The spirit behind this whole exercise is to test the Koch postulates.

1.1 Host range

This is generally not a very precise or reliable guide to virus identification and exhaustive host tests are usually no longer in practice. However, the host plant in which the virus occurs may point to the literature on candidate viruses and in certain cases narrows the scope of identification. In any case, a preliminary host range study is useful in discovering convenient laboratory tests for subsequent multiplication, purification and tests of infectivity.

1.2 Diagnostic hosts

Plant virologists have identified certain species or varieties of host plants that will give clear, characteristic and consistent symptoms for a particular virus under green house conditions. Such hosts, called indicator or diagnostic hosts, provide one of the most basic tools for routine dianosis. For example, groundnut bud necrosis virus can be identified with inoculation on *N. glutinosa, P. hybrida, Vinca rosea;* Peanut mottle virus with *P. vulgaris* cv topcrop; CMV with cowpea, *Datura metal* and TSV with cowpea topcrop bean.

1.3 Symptom expression

Symptom expression guides the diagnosis but not usually a precise indication of virus identity. In few cases the type of symptom may indicate a distinctive virus or virus group. However, environmental factors light, temperature can often have over riding influences on the symptoms expressed. Similary, minor strain differences between virus isolates may also responsible for large symptom differences, while widely differing viruses may produce similar effects. A short description of symptoms, therefore, may have its place but need not to be treated as a substitute for more reliable and informative data.

1.4 Mode of transmission

The natural means of transmission is highly diagnostic as many viruses are transmitted by specific vectors. However, there are also virus groups without specific vectors, or for which the vectors are not known, or for which the main route of transmission is by pollen, seed or vegetative propagation. Though the search for a vector is a long and painstaking process, the mode of transmission is indispensable to the adequate description of a new virus.

1.5 Electron microscopy

Electron microscopy of particles is often very useful at the crude extract stage and sufficient diagnostic information can often be obtained, especially with the aid of immunoelectron microscopy. However, better particle detail is usually seen in purified preparations.

1.6 Cytopathology

Both light microscopy and electron microscopy are useful to characterize the cytopathic effects (CPE). Many cytochemical and enzyme digestion tests, more difficult at the electron microscope levels, are available for the light microscope. Electron microscopic examination of inclusions can assist the rapid identification and characterization of many plant viruses. Some individual viruses can be distinguished by the inclusions they induce. Viruses within some groups induce distinctive inclusions which are diagnostic for infections at the groups level and individual viruses within a group may some times produce distinctive effects.

1.7 *In vitro* properties

Before the advent of serological and molecular based diagnostic tests, *in vitro* properties of the viruses were widely used as criteria for diagnosis of plant viruses. Even today also, some virologists consider them as `gold stand- ards' for diagnosis and taxonomy. *In vitro* properties include : thermal inacti- vation point (TIP), longevity *in vitro* (LIV) and dilution end point (DEP). However, none of these is a good taxonomic indicator and this is especially true for viruses that are neither particularly stable nor unstable. Although these *in vitro* properties are of limited value in diagnosis, they may be useful in separating mixtures of viruses or for designing purification schemes.

2. DIAGNOSIS USING PURIFIED PREPARATIONS

Purification, with its attendant aims of high quality and quantity, is often the major stumbling block in virus characterization. Precipitation of viruses from clarified extracts with polyethylene glycol (PEG) can often furnish virus of sufficient quantity and purity for limited characterization by electron microscopy, sedimentation, and serology. However, highly purified preparations should be used for determination of the physicochemical properties of the virus and its properties of nucleic acid component and for the production of high quality sera. The purification details are described in chapter 5.

2.1 Fine structure

One of the fastest ways to assign a particular virus to a taxonomic group is to examine preparations by electron microscopy using negative stain procedure. For comparison, excellent photographs of viruses in negative stain are found (Maromorosah, 1977; Williams & Fisher, 1974). Based on particle morphology and size, a virus can be tentatively assigned to a taxonomic group. Viruses with elongated particles are especially easy to assign in this manner because the size range of most groups of such viruses is sufficiently distinct from that of the other groups. Similarly, geminate, bullet shaped, and enveloped viruses (e.g. TSWV) can be identified.

2.2 Serology

Plant viruses like other foreign proteins are antigenic. Upon infection into experimental animals like rabbits, guinea pigs, mice etc., provoke the production of specific antibodies in the blood stream. Antigenically active part of a plant virus is its protein coat and serological tests depend on the surface properties of viral coat protein.

Serological reactions are extremely helpful in rapid diagnosis of virus diseases in seed material, in strain differentiation and identification of a virus causing a plant disease. These reactions also permit recognition and measurement of virus material when the material otherwise could not be detected by infectivity tests. A spectrum of serological procedures have been developed.

i. Tube precipitation test : This test is very commonly used. Various dilutions of clarified virus suspension and antiserum are mixed in narrow thin walled glass tubes and incubated at 37°C in a water bath. In a positive reaction precipitation is observed in the mixture (Fig. 16.1). The type of precipitate formed depends on the shape of the antigen (virus particle). Elongated virus particles form floccular precipitate whereas spherical particles form dense granular precipitate.

ii. Ring interface test : In this test, a virus containing suspension is carefully layered on top of the viscous, undiluted antiserum suspension in a small glass tube. In case of positive reaction, a ring or band rapidly becomes visible at the interface. It is a useful quick test, but somewhat insensitive.

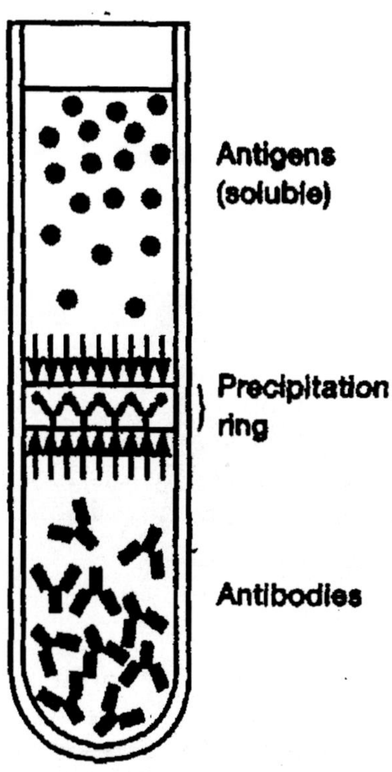

Fig. 16.1. Tube immunoprecipitation test

iii. Micro-precipitin test : This test, a miniature version of the precipitation test, is done to economize antiserum. Drops of series of dilution mixtures (antiserum and viral suspension) are spotted at the bottom of a petri-dish. The precipitates produced are observed with a microscope against dark background. The precipitation varies, depending on the ratio of concentrations of antigen and antibodies.

iv. Agglutination tests : In the simple chloroplast agglutination tests, a few drops of crude freshly expressed diseased plant leaf sap is mixed with double the amount of diluted antiserum on a microscope slide. Due to the interaction of antigen and antiserum, chloroplasts and chloroplast fragments along with small particles of host material clump or co-precipitate together, which can be observed with the naked eye or with a microscope. With crude sap from healthy plants, there is no clustering of chloroplasts and they remain uniformly distributed in the sap.

v. Gel-diffusion test (Ouchterlony test): This is basically a precipitation test but conducted on a agar gel medium either on a slide or petriplate. The reactants, antiserum and virus solutions are placed in wells cut in the agar on the bottom of a petridish. The antibodies and virus diffuse into the agar towards each other from the adjacent wells and meet at particular locations in the form of white bands (Fig. 16.2). Bands can be recorded by direct observation with appropriate lighting. Since both the reactants diffuse in the test, it is also known as double-diffusion test. Antigenic analysis of isometric viruses is perhaps best done by gel diffusion test. The elongated viruses, and those which have tendency to aggregate or diffuse poorly into the agar are not suitable for this method. However, they can be compared by precipitation tests in liquid medium. This test is ideal for distinguishing the viral strains.

Fig. 16.2. Double immunodiffusion test (From Agrios, 1997)

vi. Immunoelectrophoresis : In this test components of the antigen mixtures are first separated in an electric field in an agar gel containing an appropriate buffer. Thereafter, antiserum is put into a long trough at one side of the well parallel to the direction of the electrophoretic migration and an immunodiffusion test is conducted. Antigenic components and antibodies then diffuse together and produce bands which can be visually observed. Closely related strains may then be distinguished.

vii. Immunoelectron microscopy : Immuno microscopical methods have the great advantage of being able to be applied to tissue homogenates and of requiring very low quantities of virus and antiserum. Relationship can be studied in different ways : (i) by the differential trapping of virus particles on electron microscope grids coated with antisera to different viruses; (ii) by endpoint dilution of an antiserum which effectively coats (decorates) the virus particle, (iii) by the observation of a clumping reaction at different serum concentrations. Pretreatment of grids with protein A of *Staphylococcus aureus* may enhance virus adsorption but meaningful results occur only under certain conditions.

viii. Enzyme linked immunosorbent assay (ELISA) : ELISA, widely used throughout the world, is an extremely sensitive, simple and accurate method. Clark and Adams (1977) were the first to utilize the ELISA method for diagnosing plant viral diseases. It is used for both detection and assay of viruses that occur in low concentration. It can be applied to viruses of various morphological types in both purified preparations and crude extracts. It is also very convenient method for epidemiological investigations where a large number of samples are required to be screened.

ELISA technique utilizes the ability of antibodies raised in animals to recognize proteins, usually the coat protein of the virus interest. Antibodies are fixed to the surface of a well within a microtitre plate, and a sap extract from the plant is added to the well (Fig. 16.3). If the virus of interest is present in the plant, it will bind to the antibodies fixed on the surface. Any unbound extract is washed-off before a second antibody that recognizes the first antibody is added. The second antibody allows for indirect detection of the virus because it has reporter molecule attached to it, usually an enzyme that acts on a substrate that changes colour, which is detected visually by a calibrated microtitre plate spectrophotometer (ELISA reader). With careful observation, ELISA can be quantitative as well as qualitative. A major const-raint of the method is the requirement for polyclonal or monoclonal antibody sera specific for each virus of interest that doesnot cross react with plant pro-teins, but cross absorption with plant sap avoids this problem substantially.

ELISA – based methods, Suchas direct tissue blot immunoassay (DTBIA), double antibody sandwich (DAS) ELISA, and tissue – print (TP) ELISA are the most popular in viral identification. A modification of ELISA named voltametric enzyme immunoassay, detects the change in electrical conductivity of the substrate, rather than a colour change, when acted upon by an enzyme attached to a secondary antibody. This method is claimed to be an order of magnitude more sensitive than ELISA. It was used to detect cucumber mosaic virus. Further, from time to time, various modification of ELISA were developed and used for the rapid detection of various plant viral diseases in various host plants. (Table 16.1)

Table 16.1: Serological tests based on ELISA and its modified methods increase font size employed in plant viral disease diagnosis for various important crops

Virus Name	Host Plant	ELISA based methods
Rice stripe mosaic virus (RSMV) Rice	Rice	ACP-ELISA Dot-ELISA Tissue print-ELISA
Rice dwarf virus (RDV)	Rice	PTA-ELISA, Dot-ELISA
Squash mosaic virus (SQMV), Cowpea severe mosaic virus (CPSMV), Cucumber mosaic virus (CMV), Cowpea aphid borne mosaic virus (CABMV), Papaya lethal yellowing virus (PLYV)	Squash Cowpea Papaya	PTA-ELISA
Tomato yellow leaf curl Thailand virus (TYLCTHV)	Tomato,Pepper, Eggplant, Okra, Cucurbit	TAS-ELISA
Beet soil-borne virus (BSBV)	Sugar beet	TAS-ELISA
Tomato yellow leaf curl Thailand virus, (TYLCTHV)	Tomato	TAS-ELISA
Beet necrotic yellow vein virus (BNYVV)	Sugar beet	TAS-ELISA
Potato virus Y (PVY)	Potato	TAS-ELISA DAS-ELISA
Potato virus S (PVS)	Potato	Tissue print-ELISA Dot-ELISA
Rice-infecting viruses	Rice	DAS-ELISA
Badnaviruses	West African Yam	PAS-ELISA
Comovirus, Cowpea severe mosaic virus (CPSMV), Squash mosaic virus (SQMV)	Cowpea, Squash	IP-ELISA

ix. Tissue blot immuno assay (TIBA) : Tissue blotting, like ELISA, utilizes antibodies raised against viruses. Sap from the plant tissue is expressed onto blotting paper; nitrocellulose or nylon membranes and the virus is detected by labelled probes, often chemiluminiscent. The procedure is less labour-intensive than ELISA, rapid, sensitive, simple (no virus extraction is required), inexpensive, suitable for surveys of 1000-2000 samples per day, and the samples can be collected in the field and processed some time later. Kits are available for a number of viruses.

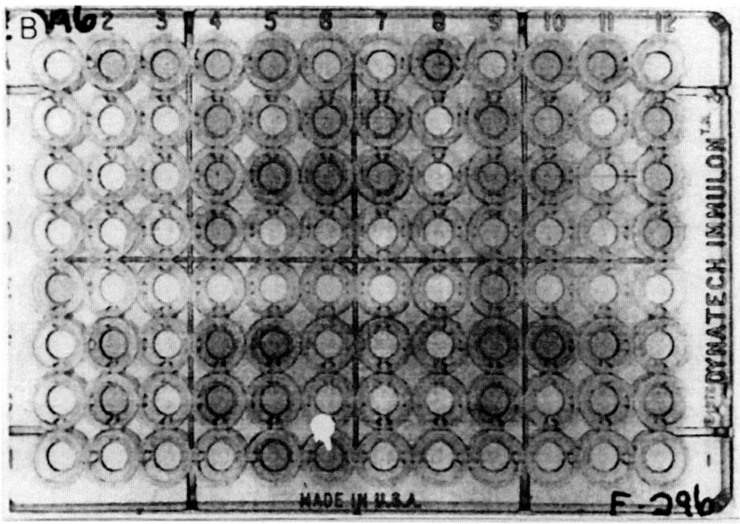

Fig. 16.3. Photographs of ELISA plate used for screening of viral diseases (From Agrios, 1997)

x. Quartz crystal microbalance (QCM) immunosensors : In this novel technique for plant virus detection, a quartz disc is coated with virus-specific antibodies. Voltage is applied across the disc, making the disc warp slightly via a piezoelectric effect. Adsorption of virus particles to the crystal surface changes its resonance oscillation frequency in a concentration dependent manner. It is, therefore, qualitative and quantitative. The developers of the technique claim that it is as sensitive but more rapid than ELISA and economical. In the first described use of QCM for plant viruses, as little as as 1 ng of particles of *Cymbidium mosaic virus* and *Odontoglossum ringspot virus* were detected in crude sap extracts.

2.3 Sedimentation analyses

The sendimentation behaviour of viruses, whether determined by conventional ultracentrifugation or by gradient centrifugation in preparative centrifuges can yield information on important taxonomic characteristics i.e. number of sedimenting components, their sedimentation coefficients and buoyant densities. Multiple components viruses such as comoviruses can be distinguished from other viruses with rather similar morphology, such as tobacco necrosis virus which sediments as a single species and this is especially useful in categorizing spherical viruses. Sedimentation coefficients and buoyant densities can be determined by rate-zonal centrifugation and equilibrium centrifugation respectively.

2.4 Analysis of coat proteins

The number of polypeptides and their molecular weight are very useful in assigning a virus to taxonomic group. Many methods are available for the isolation of viral proteins. Direct dissociation of virus followed by electrophoresis is considered to be more convenient for most types of analyses and gives comparable results. The principal technique used to determine both numbers of polypeptides and their molecular weight is SDS – polyacrylamide gel electrophoresis. Western-blot analysis, though cumbersome, is used in some cases.

The amino acid composition of coat protein of a number of viruses is known but is of little use in distinguishing between similar viruses because similar compositions may arise from quite different sequences.

2.5 Analysis of nucleic acids

i. Type of nucleic acid : Viral DNA can be distinguished from RNA by its lower buoyant density in gradients of caesium chloride salts, its differential sensitivity to Dnase and pancreatic Rnase and its base composition. A simple procedure is to treat the nucleic acid preparation with nucleases and then electrophores the products. Alternatively, gels containing native viral nucleic acid can be soaked in the enzyme solutions.

ii. Strandedness : Strandedness of the virus i.e. single stranded or double stranded(ds) can be determined by its melting behaviour (Tm) or by using nucleases and gel electrophoresis. dsRNA can be distinguished from ssRNA by its stability to Rnase. dsDNA can be distinguished from ssDNA by its sensitivity to dsDNA specific restriction endonucleases such as *Eco* RI and *Hind* III. Another method of determining strandedness depends on the effect of temperature on the hyperchromicity or a shift in the wavelength of maximum absorption of UV light following reaction of single-stranded nucleic acids with formaldehyde.

Base ratios (AT or AU/GC) and nucleotide sequences were found to be only a limited value in the virus classification.

3. MOLECULAR MARKERS

Recently, a number of nucleic acid based molecular methods have been developed for plant virus diagnosis.

3.1 PCR based methods

i. Reverse transcription-polymerase chain reaction (RT-PCR) and PCR

RT-PCR and PCR are popular techniques for detection and identification of RNA and DNA plant viruses respectively. These procedures are extremely sensitive, fairly inexpensive and require minimal skill to perform. In the case of RNA viruses, a cDNA strand complementary to the virus is made with reverse transcriptase (RT). Oligonucleotide primers, flanking part of the genome of the virus, are extended by a thermostable DNA polymerase in a series of denaturation and extension steps that exponentially increase the target DNA. For DNA viruses, the RT step is not necessary. The sensitivity and high throughput applications are major advantages of PCR methods. An RT-PCR assay of *Cucumber mosaic virus* in lupin grain was able to reliably detect one infected seed in one thousand healthy seeds.

There are a number of versions on the basic PCR technique, designed to increase sensitivity, alter specificity or allow automation of detection. The most common are listed below.

ii. Multiplex RT-PCR

Multiple species or strains are detected in a single reaction by combining primers specific for different viruses. It is important to note that the amplicons are of different lengths and that there is not cross-reactivity among them. This method was used to detect six citrus viroids and viruses.

iii. Fluorescence RT-PCR using Taqman™ technology

In this method two primers flank the sequence of interest and a third fluorescently labelled primer anneals between them. As the flanking primers extend, the labelled primer is released and fluorescence occurs. The advantages of this method are that no post reaction processing is required to detect the reaction product and that it is quantitative. A Taqman assay to detect *potato spindle tuber viroid*, a quarantine pathogen in Europe was found to be 1000 times more sensitive than a chemiluminiscent assay. The thrips, vector of tomato spotted wilt virus, were successfully screened for the virus by this method.

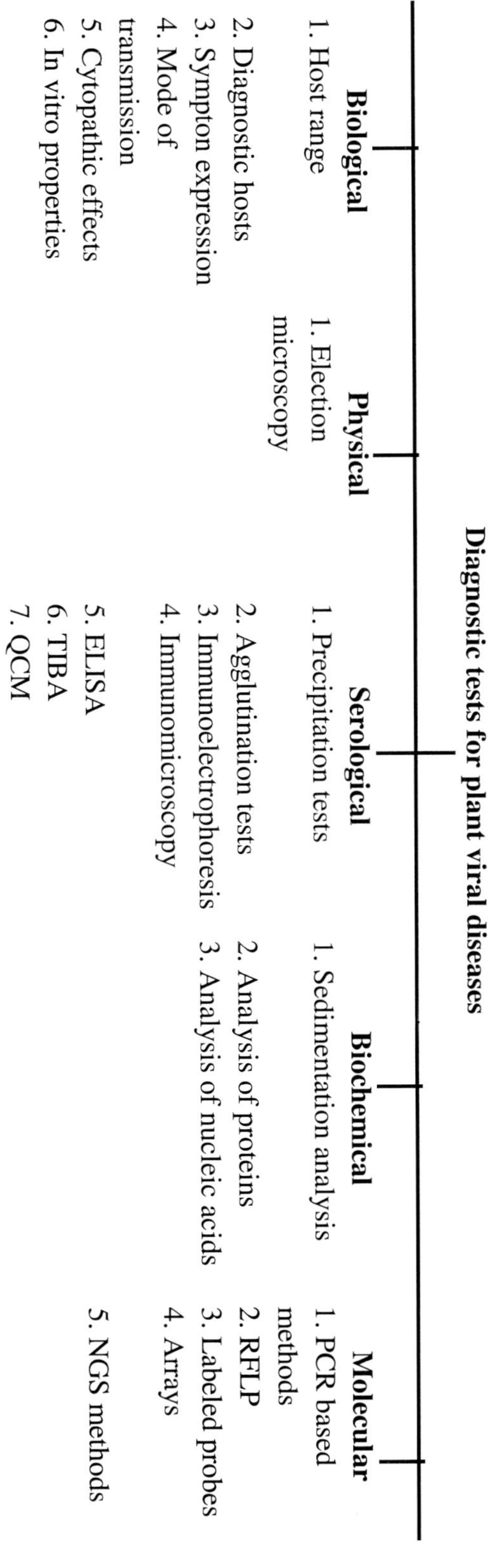

Fig. 16.1. Summary of plant viral diseases diagnostic techniques

iv. Competitive fluorescence PCR (CF-PCR)

This is an improvement over the above technique. It is used to simultaneously differentiate between virus strains and multiple virus infections. Several primer sets, each labeled with a different fluorescent marker, are added to the reaction mixture. Virus strains are differentiated with primers that differ only at the 3'end, complementary to a nucleotide position that is polymorphic between strains. Extension occurs only where 3'nucleotide is complementary. Only primers that generate amplicons fluoresce and the wavelength emitted identifies the primers that have been extended. Potatoes infected with multiple strains of *Potato virus Y* were identified using this method.

v. Immunocapture-PCR (IC-PCR)

This combines capture of virus particles by antibodies with amplification by PCR. In this method, the virus is adsorbed by the antibody bound to a surface, then removed by heating with a nonionic surfactant such as Triton X-100. The nucleic acids are then amplified using RT-PCR. This method is especially useful in concentrating virus particles from plant species, where virus titre is low, or where compounds that inhibit PCR are present For example, plum tree sap containing *Plum pox virus* and sugar cane sap containing *Sugarcane streak mosaic virus* are detected by this method.

vi. Nested PCR

In this method, two PCRs are carried out with the first reaction increasing the amount of template for the second. In the first round degenerate and low specificity primers are used. Then, an aliquot of the reaction is placed into a fresh tube for a second PCR with primers that anneal within the first amplicon. This at once increases the target molecule and dilutes inhibitors. This method has been used successfully to detect members of *Vitivirus* and *Foveavirus* species in grapevines.

3.2 Non PCR based methods

i. Restriction fragment length polymorphism (RFLP)

RFLP is used in combination with PCR to identify differences between viruses based on the presence or absence of restriction enzyme-recognition sites. After PCR amplification the amplicon is digested with a restriction enzyme(s) and the fragment sizes analyzed by gel electrophoresis. This method can be used to differentiate virus isolates.

Its effectiveness relies on polymorphisms within restriction enzyme recognition sites. RFLP was used to show that only members of subgroup 2 of cucumber mosiac virus were present in Western Australian lupin crops.

ii. Labeled probes

Nucleic acid hybridization of DNA or RNA probes has the advantage of being able to detect the nucleic acid of the virus in both forms, single stranded and double stranded. cRNA probes are preferable to cDNA probes when used to detect RNA viruses, because RNA/RNA hybrids are more stable than DNA/RNA hybrids. Probes are generally labeled with isotopes or non radioactive materials. An RNA extraction from infected tissue is blotted onto a membrane and the probe hybridized to it and detected. Orchid sap containing *Cymbidium mosaic virus* and *Odontoglossum ringspot virus* was slot blotted onto nylon membrane could be detected.

iii. Arrays

Arrays, both microarrays and macroarrays, have ben used for some years as a tool for visualizing relative change in global expression levels of mRNA, as well as single nucleotide polymorphism typing and host pathogen interaction. There are limited number of investigations on plant virus diagnosis using array technology. However, these novel techniques have recently been applied for the detection of animal and human viruses.

3.3 Diagnosis using Next Generation Sequencing (NGS) methods

The advent of next-generation sequencing (NGS) technologies has invariably led to a revolution in many fields including plant virus diagnosis. These methods do not require any previous knowledge of viral sequences and can sequence millions or billions of nucleotides in parallel, enabling the detection of all viruses, including those previously unreported. Moreover, the use of such modern techniques to understand epidemiology of viral diseases, alongside prior information about the virus and its diagnostics, is an emerging field of research that can provide holistic picture of the disease. In NGS-based methods, viral metagenomics is gaining considerable attention for diagnostics. Various sequencing platforms, like those developed for second-generation and third-generation sequencing are mostly used for diagnostics.

i. Next-Generation Sequencing (omics)-Based Methods

NGS technologies have changed the approaches to both basic and applied research in plant virology. Due to the ability of NGS platforms to generate enormous data and their prompt delivery and cost-effectiveness, followed by analysis through bioinformatics, these techniques are becoming popular in various fields of biological research. Various platforms like GS FLX by 454 Life Sciences/Roche diagnostics; HiSeq, MiSeq, and NextSeq by llumina, Inc.; SOLiD by ABI; and Ion Torrent PGM, Proton, S5, and BGISeq-500 are commercially available. These sequencing platforms were found to be more fast, sensitive and efficient in viral diagnosis compared to traditional approaches including PCR-based methods. Emerging and re-emerging viruses and diseases can also be tracked using this technology. These technologies were also found to be valuable for detection of subviral agents. Furthermore, modern NGS techniques can detect multiple viruses simultaneously, along with the possible identification of hitherto unknown viruses. Some important plant viruses and diseases diagnosed through different NGS technologies are presented in Table 16.2.

ii. Analysis Using Third-Generation Sequencing Technologies

During the last decade, third-generation sequencing technologies, also called single-molecule sequencing technologies, have gained much attention. These techniques possess many advantages over second-generation sequencing technologies. Presently, the Helicos™ Genetic Analysis System by SeqLL; LLC, USA, Single-molecule real-time (SMRT) sequencing by Pacific Biosciences (PacBio, USA), and MinION nanopore sequencing by Oxford Nanopore Technologies (ONT, UK) offer third-generation sequencing platforms. These third-generation sequencers produce sequence reads with hitherto unprecedented lengths and will help to strongly increase the quality of genome assemblies. The great advantages of these technologies are the speed of sequencing and ease of sample preparation which will facilitate sequencing and identification in the field.

In plant viral diagnosis, these techniques can provide an effective and rapid method for the detection and identification of plant viruses. In the last few years, various studies have reported the effective use of nanopore sequencing in analysing plant viral infections (Table 16.3). With the development of sophisticated bioinformatics tools, these techniques may change the current status of viral diagnosis in the field of plant virology, particularly in the areas of detection, identification, and genome sequencing.

Table 16.2: Next generation sequencing (third generation; MinJON Nanopore) based identification of causative agents associated with the viral diseases of various crops

Virus name	Host plant	Remarks
Potato virus Y (PVY)	Potato plant	Potato virus Y genetic diversity analyzed
Arabis mosaic virus (ArMV)	Potato plant	First report on ArMV in Potato plants
Sri Lankan cassava mosaic virus (SLCMV)	Cassava plant	Complete genome sequencing of SLCMV
Sowthistle yellowvein virus (SYVV)	Sowthistle plant	Historically important plant rhabdovirus
Tomato severe rugose virus (ToSRV)	Tomato weed (*Physalis angulata*)	First report on infection of *P.angulata* by ToSRV
Cowpea bright yellow mosaic virus (CABMV)	Cowpea plant	Cowpea bright yellow mosaic virus identified
Wheat streak mosaic virus (WSMV)	Wheat plant	Wheat streak mosaic virus identified
Cassava mosaic virus (CMV)	Cassava plant	Cassava mosaic virus identified
Tomato yellow leaf curl virus (TYLCV)	Tomato plant	Tomato yellow leaf curl virus identified
Watermelon chlorotic stunt virus (WmCSV)	Watermelon plant	chlorotic stunt virus identified
Tomato brown rugose fruit virus (ToBRFV)	Tomato plant	Tomato brown rugose fruit virus identified
Cucumber green mottle mosaic virus (CGMMV)	Tomato plant	Cucumber green mottle mosaic virus identified
Zucchini yellow mosaic virus (ZYMV)	Butternut squash	Zucchini yellow mosaic virus identified
Diascorea bacilliform virus(DBV)		Bacilliform virus identified
Yam mild mosaic virus(YMMV)	Water yam plant	Mosaic virus identified
Yam chlorotic necrosis virus (YCNV)		Chlorotic necrosis virus identified
Plump pox virus (PPV)	Prunus plant	Plump pox virus dentified

All tests employed in diagnoss of plant viral diseases are summarized in Fig. 16.1.

Further reading

Francki, RIB (1972). Purification of viruses. In principles and techniques in plant virology (eds. Kado. CI and Agarwal, HD), 292-235, Van Nostrand Reinhold : Princeton.

Green, SK (1991). Guidelines for diagnostic work in plant virology, Asian Vegetable Research and development center. Technical Bulletin No. 15 63p. Second Edition.

Hamilton, RI *et al.* (1981). Guidelines for the identification and characterization of plant viruses *J. Gen. Virol.* **54** : 223-241.

Hull, R (2002). Mathews Plant Virology, IV edition Academic Press, New York.

Kato-maedia, M., Gao, Q and Small, PM (2001). Microarray analysis of pathogens and their interaction with hosts. Technoreview. *Cell. Microb.* **3** : 181-187.

Koenig, R. and Paul, HL (1983). Detection and differentiation of plant viruses by various ELISA procedures. *Acta Hort.,* 127 : 147-158.

Kurstak, E (1981) (Ed.) Handbook of plant virus infections. Comparative diagnosis, Elsevier/North Holland : London.

Van Regenmortel, MHV (1982). Serology and immunochemistry of plant viruses. Academic Press, New York.

Webster, CG, Wylie SJ, and Jones MGK (2004). Diagnosis of plant viral pathogens. *Curr. Sci.* **86** (12) : 1604-1607.

Mehetre GT *et al* (2021) Current development and challenges in plant viral diagnostics. *Viruses* 13 : 421-442

17

MANAGEMENT OF PLANT VIRAL DISEASES

Viral diseases occur on all types of plants. Virtually there is no major crop plant without a viral disease. On single crop plant like peanut a large number of viral diseases are reported. Viral diseases cause serious loss of yield and quality of many important crop plants (Table 17.1). The ultimate objective of any investigation on diseased plant is to provide the affective prevention and/ or control measures. However, in designing a strategy for prevention and control of virus in a specific crop, an understanding of the epidemiology of that virus is necessary. A spectrum of control measures are conceived to mitigate the losses caused by viral diseases. Direct methods like spraying chemicals in the field are not available for viruses. In fact, the direct methods for control of plant viruses are not practically feasible because viruses are obligate parasites and depend solely on host cell's nucleic acid and protein synthesis. Hence, killing a virus *in situ* without affecting host cell is unimaginable. Another limitation is that diseases become evident only after extensive viral multiplication and cellular alterations have occured. Obviously, the only approach available to control is prevention or prophylaxis. Thus most procedures that can be used effectively involve measures designed to reduce sources of infection inside and outside the crop, to limit spread by vectors and to minimize the effect of infection on yield. Some of the methods most commonly used to control are described in this chapter.

1. EXCLUSION

Exclusion is the legal prevention of introduction of a foreign viral pathogen from other countries through quarantines and seed certification. Most countries have import and quarantine regulations aimed at preventing the entry and introduction of specific virus diseases, especially those not yet known to be widely established. It is often very difficult to detect viruses in seed and vegetative plant propagules because they rarely exhibit any significant symptoms. Therefore, at quarantine stations such seeds/planting materials are grown in virus free conditions and checked for virus infection. If no infection is observed these materials may be allowed for import. Otherwise that material is sent back or destroyed.

Table 17.1. Crop losses due to some important viral diseases (from Hull, 2002)

Crop	Virus	Country	Loss/year
Rice	Tungro	SE Asia	$ 1.5 x 10^9
	Ragged stunt	SE Asia	$ 1.4 x 10^8
	Aoja blanca	S. and C. America	$ 9.0 x 10^6
Barley	Barley yellow dwarf	UK	£ 6 x 10^6
Wheat	Barley yellow dwarf	UK	£ 5 x 10^6
Potato	Potato leaf roll	UK	£ 3-5 x 10^7
	Potato virus Y		
	Potato virus X		
Sugarbeet	Beet yellows	UK	£ 3-5 x 10^6
	Beet mild yellows		
Citrus	Citrus tristeza	Worldwide	£ 9-24 x 10^6
Cassava	African cassava mosaic	Africa	$ 2 x 10^9
Many crops	Tomato spotted wilt	Worldwide	$ 1 x 10^9
Cocao	Cocoa swollen shoot	Ghana	1.9 x 10^8 trees

Certified seed/planting materials are usually advised to the farmers for crop protection from virus diseases. The basic plant certification schemes consist of (i) known pedigree or lineage of seed, (ii) third party field inspection and (iii) quality standards relating to purity, germination, weed and seed borne diseases.

2. ELIMINATION

Many virus diseases such as sugarcane mosaic and some viruses of potato are disseminated with their vegetative parts used as seed. In such cases, care should be taken to select the plant materials from perfectly heal-thy plants. However, this method is not effective when the symptoms are not visible and when secondary infection through vector is common to occur. In such cases, attempt should be made to grow a separate crop for seed purpose in a protected area with all possible means to protect the crop from vectors.

Removal of all diseased plants as soon as they are spotted has proved advantageous in many crops like sugarcane and potato. This method has been proved to be useful in controlling sugarcane mosaic. Virus diseases sometimes can also be minimized/eliminated by treating infected plants at higher temperatures. Dormant propagative organs such as tubers or budwood which can generally withstand higher temperatures are usually dipped in hot water (35-54°C) for few minutes or hours or heat treatment for different intervals. Such treatments usually eliminate the virus from seed material and the plants arising after treatment are completely healthy. Many viruses have been eliminated by heat treatment.

Even in systemically infected plants, generally apical meristems are free from virus. Culture of meristem tips has proved an affective way of vegetatively obtaining plants free from certain viruses. Culture of single cells or small clumps of cells from virus-infected plants may sometimes give rise to virus-free plants. Virus free plants of several important tropical food crops have been obtained using meristem tip culture.

3. FIELD SANITATION AND CULTURAL METHODS

Field sanitation is a line of defense against plant diseases as it eliminates or reduces the pathogen inoculum in the field, storage area or on a plant. It also prevents the spread of inoculum from infected plant materials to healthy plants. General sanitation practices include removing diseased plants, or plant parts by pruning and roguing, destroying plants and leaf debris along with leaf litter that may contain inoculum by composting and burning; removing alternate and reservoir hosts of pathogen; keeping the fields or gardens and its surroundings free from weeds; preventing introduction of pathogens and their spread by using pathogen free tools, equipment and clothes and planting pathogen free seeds and propagating materials in a pathogen free soil.

A wide range of cultural practices are developed to minimize the viral infections in the field. The susceptible crops should be grown in isolation where vectors are less active. The crops can be covered with wiremesh or cloth barriers, but these are expensive. The susceptible seed beds can be covered with tall barrier crop to prevent the spread of vectors. Closely spaced plants tend to escape infection. Alterations of dates of sowing or changed planting dates enable a crop to escape the disease. Planting dates should be adjusted so that seedlings should emerge either after or well before the periods when the vectors are most active. Plants grown in mixed plant communities have usually a lower virus incidence. Leaving fields fallow and rotation of crop, also help in controlling the diseases.

4. VECTOR CONTROL

Morethan 70% of plant viruses are vector transmitted. Vector population in the field can be correlated with the incidence of the diseases. Thus vector control is considered as an important strategy for the vector-transmitted viruses. Chemical control of vectors is the most effective method, because it minimizes the viral transmission and also direct insect feeding damage to the crop. Insecticides provide an effective control for persistent insect-transmi-tted viruses where the vector requires several hours or days to acquire and transmit the virus. However, viruses transmitted in a non-persistent manner are not controlled because they can be transmitted by the insect before it is killed by insecticides. In case of nematode transmitted viruses, the use of nematicides and soil fumigants can be effective. Various types of oils such as vegetable, minerals, synthetic and essential oils have been tried to control virus transmission. Best efficiency has been obtained with mineral oils. Oils suppress both acquisition and inoculation mechanisms.

Non-chemical methods can be employed to control the vector population. They include cultivation of barrier crops, installing insect traps, reflective mulches, introduction of insect parasites and through avoidance. Barrier crops are useful for controlling aphid borne viruses.

Insect attracting colour traps, pheromone traps and suction traps can also be used to control the insects. Aluminium-coated or white polythene plastic reflective mulches are commonly used. These mulches are thought to act as repellent by reflecting UV-light and thus confusing the aphids in their landing attempts. Controlling virus vectors such as whiteflies and thrips through predators is widely practiced in green-houses, screen-houses and under protective cover. Adjusting the sowing time of the crop so that it does not coincide with high vector population particularly at the seedling stage is also recommended.

5. CHEMICAL CONTROL

Chemical control of viruses has not been very successful. Synthetic nucleic acid base analogues (e.g. thiouracil, 8-azaguanine) or synthetic selective agents like ribavirin, thiosemicarbozones, acyclovir etc. show inhibitory activity against viruses *in vitro*. However, they were found to seriously affect plant growth when applied to plants. Thus, they are no good for chemical control. Pretreatment of tobacco plants with virazole delayed or prevented systemic infection with tomato spotted wilt virus. A large number of other chemicals influencing virus infection have also not found any practical use. Bavistin, a well known systemic fungicide reduced virus induced symptoms but has shown no effect on virus. Chemical treatment in combination with heat treatment or meristem tip culture may have an advantage in a few instances.

Certain biologicals like plant extracts have shown to exert inhibitory effect on virus infections. Plant extracts show two types of inhibitory responses. Mostly, they reduce virus activity when co-inoculated with virus into susceptible plants. Extracts from non-host plants inhibit virus infection and development of symptoms by stimulating the production of a virus inhibitory agents in the susceptible hosts. Foliar sprays of oil or skimmed milk have also been used to prevent the spread of virus diseases by aphids.

6. CROSS PROTECTION

Infection of a susceptible plant with a mild or attenuated strain of virus sometimes helps in virus control by protecting such plants against subsequent infection by a more severe strain of the same virus. Thus, the plants may be deliberately infected with a mild strain as a protective measure against severe disease. Mild strains of tomato mosaic virus are particularly helpful in controlling infection by severe strains in tomato plants. Large scale applications of cross protection have been attempted for the control of TMV in tomato, citrus tristeza virus in citrus, papaya ring spot virus (potyvirus) in papaya, Zucchini yellow mosaic virus (potyvirus) in soybean and melon and cacao swollen shoot virus (badnavirus) in cacao. Mild strain protection may operate by competition for replication sites.

7. RESISTANCE

The viral control methods discussed above are usually only partially affective. Crop resistance to the virus or the vector is the ultimate solution for virus control. The principal aim is to produce cultivars that are able to withstand losses from serious virus diseases and at the same time have acceptable traits. Ideally, resistance should prevent entry, multiplication and movement of the virus in the host and should be effective against all strains of the virus. There are several types of resistance to infection (Table 17.2). Sources of resistance can be found in existing cultivars, primitive cultivars, or wild types, closely related species of other genera of the same botanical family. The first two sources of resistance are most desired by breeders and pathologists because they can be readily found and used immediately in a resistance breeding program. Other species as sources of resistance may be difficult to use because of genetic incompatibility and/or close linkages with other undesirable characters.

Satisfactory control has been achieved in our country through varietal resistance in sugarcane mosaic and some virus diseases of fruits and vegetables. Similary, CPRI hybrids in potato namely, VB-8, PH-12-22, 8-5 and 55-123 etc possess resistance tolerance to one or other viruses. Resistance to vector aphids has been observed in potato cultivars.

Table 17.2. Types of plant reactions to virus infection.

RESISTANCE TYPE	REACTION OF VIRUS IN THE HOST				
	Virus replication	**Virus spread in plant**	**Virus titer**	**Symptom**	**Yield reduction**
Extreme resistance (immunity)	-	-	-	-	-
Resistance to systemic spread	+ or +/-	- or +/-	low or high	+ or -	-or+
Hypersensitivitiy (quick cell death)					
local	+ or +/-	-	low	necrosis	- or +
systemic	+	+	low	necrosis	+ or ++
Latency	+	+ or -	low or high	-	- or +/-
Tolerance	+	+	high	- or +	- or +/-
Full susceptibility	+	+	high	+/++	+/++

+ = present; - = absent; +/- = limited; ++ = extensive.

8. TRANSGENIC RESISTANCE

The increased understanding of plant diseases and defense responses at the molecular level has led to an explosion of interest in controlling diseases through genetic modification of plants. Recombinant DNA technology has facilitated the introduction of foreign gene into a plant that can confer the resistance to virus infection. What could not be achieved through traditional breeding methods became possible with genetic engineering. Many approac-hes to produce virus resistant transgenic plants were made and in some cases promising results were also obtained. Some of these approaches entirely novel, for example, transferring pathogen coat protein genes in to the host.

i. Coat-protein mediated resistance

Attempts to understand the phenomenon of resistance due to cross protection led to the conclusion that expression of specific viral genes in plant might have the same effect. In 1986, this was demonstrated when the coat protein gene of TMV was expressed in transgenic tobacco plants under the control of the CaMV 35S constitutive promoter. The result was delayed symptom development in all the transgenic plants, and 10-60% of the plants failed to develop any viral symptom. The expression of this protein was believed to block the normal uncoating and replication of invading viruses, and evidence to support this was provided by use of TMV pseudoviruses (Table 17.3).

Table 17.3. Some transgenic plants that contain cloned viral coat proteins which show resistance to viruses

Plant(s)	Virus that provided the coat protein gene
Papaya	Papaya ring spot virus
Potato	Potato leaf roll virus
	Potato virus Y
	Potato virus S
Potato	Potato virus X
Rice	Rice strip virus
Tobacco	
	Tobacco etch virus
	Tobacco streak virus
	Tobacco rattle virus
Alfalfa, tomato	Alfalfa mosaic virus
Tobacco, cucumber	Cucumber mosaic virus
Tobacco, tomato	Tobacco mosaic virus
Tomato	Tomato mosaic virus

Although coat protein mediated resistance works effectively against closely related strains of viruses, in some cases requiring as little as 60% coat protein homology, it is ineffective against unrelated viruses and is therefore not the type of broad spectrum resistance sought by commercial interest.

ii. Replicase-mediated resistance

The risk of transencapsidation and lack of broad host range resistance led to search for alternative viral genes that might be effective in transgenic plants. Replicase-mediated resistance was first demonstrated against TMV, where the viral replicase was expressed in transgenic tobacco under the control of the 35S promoter. Replicase resistance has subsequently been shown to be effective against a wide range of ssRNA viruses and some geminiviruses. The resistance is generally more effective than coat-protein mediated resistance against high inoculum doses, and in one case, broad spectrum resistance against a range of viruses was demonstrated when the TMV replicase gene containing a bacterial insertion element within it was used. Replicase-mediated resistance probably works through a post-transcriptional gene-silencing (PTGS) mechanisms rather than through interference by the replicase protein. However, there have been a number of instances when resistance has been broken by a small number of point mutations within the infecting viral genome.

iii. Movement protein-mediated resistance

Because of the importance of movement proteins (MPs) for cell-to-cell movement of viruses, it has been proposed that blocking plasmodesmata and viral movement may be a more effective means of preventing spread of both related and unrelated viruses to result in a broader-spectrum resistance. Tobacco plants expressing the *Brome mosaic virus* (BMV) MP were found to have resistance against the unrelated TMV, and plants transgenic for a defective mutant of the TMV MP were resistant to a wide range of unrelated viruses. Attempts to produce transgenic plants with further MP mutations are currently underway, in an attempt to produce more effective broad spectrum resistance to viruses. Attempts have also been made to use other regions of viral genomes to make pathogen derived resistant plants such as proteases.

iv. RNA mediated resistance

The possibility that replicase mediated resistance might be conferred by post-transcriptional gene-silencing (PTGS) mechanism has led to the development of RNA mediated means of protecting plants against viruses. This mechanism is supported by experiments in which transgenic plants with enhanced resistance have been generated by expressing satellite RNAs and defective interfering RNAs associated with certain viruses. In addition, many viruses appear to possess a mechanism for suppressing PTGS. Ribozymes, which are self-cleaving RNAs based on certain satellite RNAs and viroids, have also been tested for their effectiveness in control of viral diseases. The aim has been to design ribozymes that would be expressed in transgenic plants and would bind to the viral RNA and leave it, rendering it inactive. Similarly, antisense RNA approaches have been tested against some viruses

Further reading

Dickinson, M (2003). Molecular Plant Pathology. *Bios Scientific Publishers,* London.

Duffus, JE (1971). Role of weeds in the incidence of virus diseases. *Ann. Rev. Phytopath.* **9** : 319-340.

Helgeson, JP and Deverall BJ (1983). Use of tissue cultures and protoplasts in plant pathology, Academic Press, London.

Nyland, G and Goheen, AC (1969). Heat therapy of virus diseases of perennial plants *Ann. Rev. Phytopath.* 11 : 331-354.

Pink, D and Puddephat, I (1999). Deployment of disease resistance genes by plant transformation a mix and match approach. *Trends in Plant Science* 4 : 71-75.

Tapfer, M (2002). Risk assessment of virus-resistant transgenic plants. *Ann. Rev. Phytopath.* 40 : 467-491.

Thresh, JM (1982). Cropping patterns and virus spread. *Ann. Rev. Phytopath.* 20 : 193-218.

Verma, HN, Baranwal VK, and Srivastava S (1998). Antiviral substances of plant origin. In : Plant virus disease control. Eds. A. Hadidi, R.K. Khetarpal and H. Koganezawa, APS Press. American Phytopathological Society, St. Paul. Minnesota.

Zitter, TA and Simons, JN (1980). Management of viruses by alteration of vector efficiency and by cultural practices. *Ann. Rev. Phytopath.* 18 : 289-310.

18

APPLICATIONS OF PLANT VIRUSES

Since the discovery of tobacco mosaic virus (TMV) in the 1890s, viruses have been largely viewed as pathogens. Most of the research in these 130 years has been carried out in that direction covering discovery, detection, epidemiology, disease losses, prevention, control etc.With advances made in the molecular aspects of different viruses, the misunderstanding of viruses as only the pathogens is slowly being erased out.Understanding of molecular architecture, working system coupled with biochemistry, host parasite interactions has opened a new dimensions of plant viruses hitherto unimagined. Manipulation of genomes of viruses of became possible employing the toolkits of modern molecular biology as desired. This has opened new vistas in the field plant virology. Plant viruses are proving to be ideal candidates as much as bacteria and yeasts. Biotechnological potentials of engineered plant viruses and their derived products like virus-like particles (VLPs) are rolling out in a range of fields like genetic engineering, agriculture, medicine and nanotechnology. In this chapter, several beneficial aspects of plant viruses and VLPs and possible applications are highlighted.

1. ENHANCING THE AESTHETICS OF ORNAMENTAL PLANTS

Viruses of plants have been studied primarily because they cause disease and economic losses in crops. Virus infections in host plants result in a range of symptoms that include foliar mosaic, mottle, ring spots, necrosis, malformation, curling/rolling, yellow vein, flower and/or foliage variegation, fruit malformations, and overall plant stunting. Interestingly, the same type of symptoms result in enhancing aesthetic value of infected host plant without any damaging effect. The best-known classical example is production of attractive variegated flowers produced by virus infected tulip plants. It is a well-known story to all virology students. Other well-known examples include the attractive mosaic patterns on the leaves of flowering maple (*Abutilon pictum*; Malvaceae) infected with *Abutilon mosaic virus*, the variegated flowers of camellia (*Camellia japonica*; Theaceae) infected with *Camellia yellow mottle virus*, and the yellow-veined Japanese honey suckle (*Lonicera japonica*; Caprifoliaceae) infected with *Honey suckle yellow vein mosaic virus*. In recent years, there have been other reports of plant viruses associated with unique phenotypes of ornamental plants, and some of them are being marketed as novel plant cultivars. List of selected viruses that enhance the aesthetics of some ornamental plants is presented in Table 18.1

Table 18.1. List of selected viruses that enhance the aesthetics of some ornamental plants

Virus Name	Genus	Susceptible hosts
Abutilon mosaic virus	*Begomovirus*	*Abutilon spp., Hibiscus spp., Malva spp., Gossypium spp.*
Camellia yellow mottle virus	*Varicosavirus*	*Camella japonica, C. sasanqua*
Clerodendron golden mosaic China virus	*Begomovirus*	*Salvia splendens, Clerodendrum cyrtophilum*
Cucumber mosaic virus	*Cucumovirus*	*Solenostemon spp., Viola spp., many wild and cultivated plants*
Honeysuckle yellow veinmosaic virus, Honeysuckle yellow vein virus	*Begomovirus*	*Lonicera spp., Solanum lycopersicum, Nicotiana tabacum, Capsicum annuum*
Nandina stem pitting virus	*Capillovirus*	*Nandina domestica*
Pelargonium flower break virus	*Carmovirus*	*Pelargonium peltatum, P.hortorum*
Pseuderanthemum yellow vein virus	*Begomovirus*	*Pseuderanthemum sp.*
Tulip breaking virus	*Potyvirus*	*Tulipa spp., Lilium spp.*

2. CROSS PROTECTION OF VIRAL PATHOGENS

Cross-protection is a phenomenon in which infection of a plant with a mild virus or viroid strain protects it from disease resulting from a subsequent encounter with a severe strain of the same virus or viroid. Its history started about seventy years ago, when the Dutchman Thung and the Englishman Salaman described the phenomenon independently. During the 1930s, several virologists confirmed this discovery, which was considered the first possibility to protect plants against virus infection. Plant virologists have proposed hypotheses on its mechanism, that have emerged until the 1970s, included (i) antibody formation, (ii) exhaustion of essential metabolites, (iii) limited sites for virus multiplication, and (iv) specific adsorption by new cell compounds. However, these hypotheses were re-proposed and discussed on several occasions without arriving at a consensus. The statement of molecular genetics of viruses produced new interesting "theories", fundamentally based on the interference between virus strains. A model developed by the Americans Palukaitis and Zaitlin in 1984 indicates that excess of progeny positive-sense RNA of the protecting strain would sequester the minus-strand RNA of the challenging strain. Other models involve a function of the coat protein, or gene recombination. Despite the different opinions, practical applications of this phenomenon have been well received. A number of crops (such as tomato, tobacco, citrus, cucurbits, grapevine, soybean, papaya, and so on) have been submitted to this practice. Nevertheless, during the 1980s, the idea of cross protection was dropped because of the development of new resistant or tolerant cultivars.

3. PLANT VIRUSES AS VECTORS

In molecular biology, a vector is a DNA molecule used as a vehicle to artificially carry foreign genetic material into another cell, where it can be replicated and/or expressed. The four major types of vectors are plasmids, viral vectors, cosmids, and artificial chromosomes.Viral vectors are generally genetically engineered viruses carrying modified viral DNA or RNA that has been rendered non-infectious, but still contain viral promoters and also the transgene, thus allowing for translation of the transgene through a viral promoter.

For a long time, plant viruses have been used as vectors for several purposes.The advantages of utilizing plant viral vectors include: (i) high copy number transient expression yielding rapid production of the desired product; (ii) ability to rapidly screen multiple construct variants in different host species or genotypes, thus avoiding poor constructs and delays in stable transformation/regeneration of hosts that may be difficult to transform; (iii) typically efficient expression in all susceptible hosts, avoiding position effects in different stable transgenic lines; (iv) plant post-translational modifications not possible in bacterial expression systems, enhancing functionality of expressed proteins; (v) systemic movement and expression in susceptible hosts and ability to initiate infection in established plants; and (vi) ability to express genes at particular plant growth stages by varying timing of inoculation.

Limitations of plant viral vectors include :(i) only transient, and not stable, expression usually without any transmission of desired characters to subsequent generations through seed or through breeding; (ii) crop-enhancing characters must therefore be introduced each growing season in annual crops; (iii) introduced genes may be lost over time by deletion or mutation affecting expression (more problematic with larger inserts); (iv) possibilities of adverse effects on the host or interactions with other viruses; and (v) possible transmission to other susceptible crops or wild hosts.

Autonomously replicating virus-based vectors provide alternative means to deliver genetic engineering (GE) reagents to plant cells. Among these are the RNA viruses, which for monocots include *Wheat streak mosaic virus* (WSMV) and *Barley stripe mosaic virus* (BSMV), *Tobacco rattle virus* (TRV) for dicots. Single-stranded (ss) DNA viruses, as geminiviruses, have been also widely adopted as vectors for diverse crops. These viruses can be modified to carry heterologous coding sequences, and protein expression has been achieved in important crops like wheat, barley, corn, oat, and rye. Recent development in GE technologies has urged scientists to incorporate viral vectors and utilize them for the efficient delivery of GE reagents in plant cell (Table 18.2).

Geminiviruses as vectors: Among different plant vectors, geminiviruses are much sought after vectors because of : (i) geminiviruses are able to infect a wide range of host plant species from various families, making these efficient vectors for multiple hosts at once; (ii) require only one protein, Rep (replication associated protein; RepA in case of mastreviruses), to initiate replication inside the host cell, and it can be expressed under its natural promoter within intergenic region or other user specified constitutive/inducible promoters; (iii) replicate inside host cells via homologous recombination-dependent replication, in addition to rolling circle replication, reverting host cells to S phase, suitable for homologous recombination if supplemented with SSN and complementary target sequences and (iv) replicate efficiently inside host cell and produce high amounts of replicons, in turn producing a lot of SSNs and target sequence if used as vector for GE, considerably enhancing the targeting .

Table 18.2 Important viral vectors for plant genome engineering

Virus type	Virus vector	GE platform	Plant species	Target
DNA virus	BeYDV	CRISPR and TALEN	*Solanum lycopersicum*	*ANT1*
	BeYDV	ZFN, TALEN and CRISPR	*Nicotiana tabacum*	P-*GUS:NPTII*
	BeYDV	CRISPR	*Solanum tuberosum*	*stALS1, stALS2*
	BeYDV	CRISPR and TALEN	*Solanum tuberosum*	*ALS1*
	CaLCuV	CRISPR	*Nicotiana benthamiana*	*PDS*
	WDV	CRISPR	*Triticum aestivum*	*Ubi, MLO, GFP*
	WDV	CRISPR	*Oryza sativa*	*GFP, GUS*
RNA virus	TRV	ZFN	*Nicotiana tabacum* and *Petunia hybrida*	*uidA*
	TRV	Meganuclease	*Nicotiana alata*	*DFR*
	TRV	CRISPR	*Nicotiana benthamiana*	*PDS*
	TRV	CRISPR	*Nicotiana benthamiana*	*PDS, PCNA*
	TRV	CRISPR	*Nicotiana benthamiana*	*Plant virus*
	TRV	CRISPR	*Nicotiana benthamiana*	*Plant virus*

Acetolactate synthase1 (ALS1), Solanum tuberosum acetolactate synthase1 (StALS1), green fluorescent protein (GFP), β-lucuronidase [GUS] reporter controlling gene (uidA), Promoter of GUS and neomycin phosphotransferase (P-GUS:NPTII), Bean yellow dwarf virus (BeYDV), Cabbage leaf curl virus (CaLCuV), zinc finger nucleases (ZFNs), transcription activator like effect or nucleases (TALEN), clustered regularly interspaced short palindromic repeats (CRISPR), Tobacco rattle virus (TRV), genome engineering (GE), phytoene desaturase gene (PDS), proliferating cell nuclear antigen (PCNA), ubiquitin gene (Ubi), Mildew LocusO (MLO), dihydroflavonol 4- reductase (DFR).

Geminiviruses have been engineered as vectors for the expression of heterologous proteins in plants. The cargo capacity of these viruses is quite restricted; they can be converted into non-infectious replicons by replacing genes important for infection and cell-to-cell movement with heterologous sequences. To achieve this goal, movement protein (MP) and coat protein (CP) coding sequences of geminiviruses have been removed, thereby eliminating the possibility of cell-to-cell movement as well as plant-to-plant insect-mediated transmission. Yin et al. (2015) has developed a novel use of gemninvirus vectors the "virus-based gRNA delivery system for CRISPR/Cas9 mediated plant genome editing (VIGE).” VIGE makes use of Cas9 over expression in plants (N. benthamiana so far) and transient delivery of geminivirus vectors with sgRNA targeting the gene of interest. This system can be used for generation of knock-out libraries, as an alternative to the virus induced gene silencing (VIGS).

A number of geminivirus based vectors are constructed (Table 18.3.)

Table 18.3. Geminivirus-based vectors

Types of vector	**Virus**		**Viral component**
Expression vector	Coding	Short peptide protein	*Tomato golden mosaic virus* (TGMV) *Beet curly top virus* (BCTV) *Bean yellow dwarf virus* (BeYDV) *Cotton leaf curl Multan virus* (CLCuV) *Bean yellow dwarf virus* (BeYDV) *Cabbage leaf curl virus* (CaLCuV) *Cabbage leaf curl virus* (CaLCuV) *Cotton leaf crumple virus* (CLCrV) *Cotton leaf crumple virus* (CLCrV) *Abutilon mosaic virus* (AbMV)
VIGS vector	Non-coding	Genome editing nuclease Genome editing repair templates Genome editing sg RNA mi RNA RNA sponge	*Beet curly top virus* (BCTV) *Tobacco curly shoot virus* (TCSV) *Cabbage leaf curl virus* (CbLCV) *Chilli leaf curl virus* (ChiLCV) *Cotton leaf crumple virus* (CLCrV) *Tomato yellow leaf curl China virus* (TYLCCNV) *Cotton leaf curl Multan betasatellite*(CLCuMB) *Tobacco curly shoot virus* (TbCSV)

Tobacco rattle virus (TRV) as vector:TRV has shown to be promising as a vector for GE. Tobacco rattle virus (genus *Tobravirus*, family *Virgaviridae*) is a positive single stranded RNA (CssRNA) pathogenic plant virus that infects over 400 plant species from 50 families. It is naturally transmitted by nematodes of the family Trichodoridae; and can also be mechanically and seed transmitted. TRV as a vector meets several important requirements for highly efficient and multiplexed editing: (i) can systematically infect a arge number of plant species; (ii) the virus is easily introduced into plants via Agrobacterium and delivery into growing points of the plant; (iii) the small genome size facilitates cloning, multiplexing, library constructions, and agroinfections; and (iv) the virus RNA genome does not integrate into plant genomes, (v) TRV has two genome components, TRV1 (or RNA1) and TRV2 (or RNA2). TRV1 is essential for viral movement and contain genes encoding 134- and 194-kDa replicase proteins, a 29-kDa MP and a 16-kDa cysteine-rich protein whose function is not fully known. The TRV2 genome varies among different isolates of this virus and has genes encoding the CP and non-structural proteins. These non-structural proteins are implicated in nematode transmission, but they are not essential for experimental infection. Therefore, for use as a vector, the two non-structural protein–encoding genes in TRV2 can be replaced with multiple cloning sites for inserting fragments of interest. TRV is also an efficient vector for VIGS, facilitating functional genomics in diverse plant species.

3.1 Plant viruses as expression vectors

In the last two decades, plants have become serious competitors to other production systems for pharmaceuticals, such as bacteria, yeast or mammalian cells. They are robust, inexpensive to grow and bring a low risk of contamination. The idea of edible vaccines, produced in plants generated a lot of interest but has now lost its popularity due to problems of dose and quality control which is difficult to guarantee without purification. In response, transient expression through the use of plant virus expression vectors became an attractive venue by which biopharmaceuticals could also be generated. Virus vector-based expression systems could simultaneously shorten production time and increase pharmaceutical yield, while removing concerns from the general public about the use of genetically modified organisms (GMOs). More recently, plant viruses have found a niche in the field of cancer immunotherapy, by acting as nanoparticles. Now the development of plant virus expression vectors for biopharmaceutical development has become an attractive and powerful tool.

Plant viruses have been engineered to express vaccines, monoclonal antibodies, and other therapeutic proteins (Table 18.4). Plant virus expression vectors have been designed from the genomes of both positive and negative-sense RNA viruses and single-stranded DNA viruses. Several vector design strategies, such as gene replacement, gene insertion using a duplicated subgenomic RNA (sgRNA) promoter (SGP), heterologous SGP, complementation, gene fusions, internal ribosomal entry site (IRES), and deconstructed viruses, were developed for expression of foreign proteins from virus vectors. The most common design strategies are gene replacement and the insertion of an SGP sequence. Expression vectors are based on variety of plant viruses such as tobamoviruses, potexviruses, comoviruses, geminiviruses. Some vectors have been developed merging of genetic elements from two vastly different virus expression systems. For example, TMV has been used to express the genome of the Alphavirus Flock house virus (FHV), a small insect virus that can replicate to high levels in plants without inducing apoptosis.

Table 18.4. Examples of plant viruses used as expression vectors for foreign proteins.

Recombinant protein or vaccine or VLP	Viral vector
Cholera toxin b subunit	TMV
Human anti-non-Hodgkin's lymphoma single-chain Fv (scFv) immunoglobulins	Hybrid TMV and odontoglossum ringspot virus (ORSV)
Rice a-amylase	Hybrid TMV and tomato mosaic virus (ToMV)
Assembled full-size monoclonal antibody	Combination of non-competing viral vectors TMV and PVX
Human growth hormone	Hybrid crucifer-infecting TMV (cr-TMV) and turnip vein-clearing virus (TVCV)
Plant-produced VLP developed for drug delivery	TMV
Plant-produced chimaeric virus vaccine for influenza virus	TMV
Assembled full-size monoclonal antibody	CPMV
Plant-produced chimaeric virus vaccine for human rhinovirus 14 and human immunodeficiency virus	CPMV
Plant-produced VLP developed for encapsulation of metals	CPMV
Plant-produced chimaeric virus vaccine for hepatitis C virus	PVX
Hepatitis B core Norwalk virus capsid protein (NVCP)	BeYDV

3.2 Applications of plant virus–derived vectors

Various uses of plant viral vectors can be broadly grouped under (*a*) therapeutic proteins, (*b*) epitope display and vaccines, (*c*) antimicrobial compounds, (*d*) virus-induced genome editing, (*e*) agricultural biotechnology, and (f) nanotechnology. These are discussed in the following sections.

3.2.1 Viral vectors for plant genome engineering

Genome engineering (GE) refers to the strategies and techniques developed for the targeted, specific modification of the genetic information of living organisms. The major advantage of GE is that it enables a specific sequence on a chromosome be modified, thereby increasing the precision of the gene disruption, correction or insertion, offering perfect reproducibility. GE technologies have recently evolved as promising tools for improvement of a wide range of organisms, including plants. One of the major challenges in engineering plant genome is selection of right kind of vectors that are modified in a systematic manner to deliver reagents for GE. The efficient machinery and comprehensive genome structure make viral genomes excellent choice to be used as vectors. Recent developments in GE technologies have prompted scientists to incorporate viral vectors and utilize them for the efficient delivery of GE reagents in plant cell.

i. A deconstructed version of Bean yellow dwarf virus (BeYDV) was developed and used to efficiently deliver ZFNs and a repair template to tobacco cells to achieve gene targeting (GT) at an integrated reporter gene.
ii. BeYDV based replicons were constructed for GT and insertion of a strong promoter upstream of a tomato gene that regulates anthocyanin synthesis. In this case, GT frequencies were 12-fold higher than what could be achieved using standard Agrobacterium T-DNA delivery.
iii. Gil-Humanes et al. (2017) developed replicons based on a Wheat dwarf virus (WDV) for precise genome editing of cereal crops. The WDV-derived replicons amplified and expressed heterologous proteins in wheat, corn, and rice.
iv. Recently, it has been reported that CRISPR/Cas9 mediated GT and efficient HDR (as high as 19.4%) in rice, using a WDV-based replicon system.

Another novel use of geminivirus vectors has been developed by Yin et al. (2015) the "virus-based gRNA delivery system for CRISPR/Cas9 mediated plant genome editing (VIGE)." VIGE makes use of Cas9 over expression in plants (*N. benthamiana* so far) and transient delivery of geminivirus vectors with sgRNA targeting the gene of interest. This system can be used for generation of knock-out libraries, as an alternative to the virus induced gene silencing (VIGS).

All the above mentioned studies are exciting and promising, however, certain caveats and limitations needed to be addressed. Recently developed TRV mediated CRISPR/Cas9 delivery system has the potential to bypass the laborious and time-consuming tissue culture practices to develop plants with desirable engineered traits. This might overcome the regulatory hurdles that impede the commercialization of engineered plants.

3.2.2 Viral vectors for plant genome editing

For genome editing purpose, CRISPR/Cas9 constructs can be delivered to plants in several ways, including protoplast electroporation, leaf biolistic bombardment, or leaf infiltrations with Agrobacterium strains carrying the constructs. However, there are certain limitations in these methods. As an alternative to these traditional methods, new virus-based tools have been developed to deliver CRISPR/Cas9 constructs to plants. Much like other virus-based technologies for protein expression or virus-induced gene- silencing (VIGS), viral delivery of CRISPR/Cas9 components can greatly enhance functional genomics studies geared towards the development of improved crop varieties, by circumventing the challenges of poor transformation efficiency found among crop species. A number of viruses have been adapted for delivering the Cas9/gRNA components to plants. Most of these VIGE systems were developed and tested in *N. benthamiana* due to the ease of producing viral inoculum via leaf agroinfiltrations. Most of the CRISPR/Cas9 virus delivery systems to date rely on RNA viruses; however, there are a few DNA viruses – notably geminiviruses – that also have been developed for this purpose.The CRISPR/Cas9 viral systems have great potential for developing improved crop varieties with resistance to disease, pests, and abiotic stresses. Additionally, in many studies it has been observed that the inoculated virus is not detected in progeny plants unless vegetative propagated, thus eliminates regulatory concerns.

3.2.3 Virus induced gene silencing

Virus induced gene silencing (VIGS) is an effective technology that exploits an antiviral defense mechanism in plants. It is a recently developed gene transcript suppression technique for characterizing the function of plant genes. VIGS is rapid, efficient and specific system for transient gene silencing. The major steps in VIGS includes: (i) engineering viral genomes to the appropriate viral vector to incorporate fragments of host genes that are targeted to be silenced, (ii) infecting the appropriate plant hosts and, (iii) silencing the target genes as part of the defense mechanism of the plant against virus infection. The VIGS vector is a recombinant virus engineered to be able to carry a piece an endogenous gene from the host.Over the past few years, multiple plant viruses have been modified into VIGS systems for a vast number of plant species. In addition, different techniques have been developed for VIGS experiments to deliver the virus (in any form) into the plant, such as (i) plant infection with *in vitro*-transcribed viral RNA, (ii) infection with viral particles, which are harvested from infected donor plants, or (iii) many protocols utilize modified Ti-plasmids of *Agrobacterium tumefaciens,* which include the modified viral genome instead of the original T-DNA; *A. tumefaciens* is then used to deliver the modified T-DNA into a few host cells to start the silencing reaction. Around 37 VIGS vectors have been developed for gene function studies in dicotyledonous plants, but far fewer are available for studies with monocotyledonous plants. During infection with the modified vector, the host's defense reaction will be induced against the cloned host gene; a loss of function phenotype makes it possible to identify the function of the gene. The recombinant virus is introduced into plant cells through Agrobacterium tumefaciens mediated transient expression or in vitro transcribed RNA inoculation or direct DNA inoculation. The trans gene is amplified along with the viral RNA by RNA dependent RNA polymerase generating dsRNA molecules. dsRNA is the triggering molecule of post transcriptional gene silencing.

VIGS as a reverse genetics tool for functional genomics studies presenting several advantages. Applications include in the field of functional genomics are most glaring. Silencing the gene thereby abolishing the corresponding phenotype makes it possible to identify the function of the gene. Consequently, it has fuelled research in many fields that were previously hampered by the lack of information on gene functions.

The use of VIGS has enabled the dissection of biosynthetic pathways and their regulation *in vivo. Catharanthus roseus* is an example of a species synthesizing a great diversity and number of monoterpene indole alkaloids, including quinine and strychnine. VIGS made it possible to understand the interactions between partners of symbiosis and host pathogen interactions.VIGS has been successfully used to silence genes in tomato and strawberry fruits even after the fruits were detached from the plant. This method is particularly useful if a specific target gene, to be silenced in the reproductive organs, is involved in basic metabolic processes during the early stages of plant growth to avoid deformities and also delaying early ripening.

Besides, several VIGS vectors have been used to investigate gene functions under a biotic stress in different plant species. Recently, development of a wide range of VIGS vectors with high silencing efficiency has expanded the application of VIGS to several crop species for studying abiotic stress responsive genes. VIGS is a valuable tool for functional validation of drought responsive genes identified from transcript profiling of plants exposed to drought stress. Besides, in crop plants, various investigations discovered some abiotic stress related genes using VIGS. The utility of VIGS in investigating salt stress tolerance in crop plants has also been demonstrated.

3.2.4 Virus-Like Particles (VLP) and their applications

A VLP is a version of a viral particle that mimics its morphology but is lacking infectious genomic material. However, VLPs retain the native antigenic conformation of the immunogenic proteins and the repetitive structure of the original virus particle. They thus combine both safety and the potency to elicit a strong immune response due to their size, shape and surface where they display epitopes in a dense repetitive array. Following the expression of the viral proteins many self-assemble spontaneously into VLPs, Expression and self-assembly of the viral structural proteins can take place in various living or cell-free expression systems after which the viral structures can be assembled and reconstructed. VLPs are excellent candidates for such synthetic vaccines. VLPs are highly immunogenic and are able to elicit both the antibody- and cell-mediated immune responses by pathways different from those elicited by conventional inactivated viral vaccines. VLPs have been intensively studied as immunologically active multivalent structures useful for the generation of new prophylactic and therapeutic vaccines against human or animal infectious agents, cancers and autoimmune diseases. When compared with other VLPs, plant VLP-based vaccines have additional advantages, such as flexibility in vaccine construction, the ease of VLP production and purification, stability and the low risk of pre-existing immunity. All these properties make plant viruses an attractive alternative to animal and human VLPs. In addition, certain VLPs have also been investigated as potential drug-delivery vehicles and as imaging agents for biomedical purposes.

4. VIRUS INDUCED FLOWERING

Obtaining flowers in a synchronized manner and in a particular period of time is always challenging in horticulture. The transition to flowering requires the coordination of developmental and environmental signals. Many plants experience an extended juvenile phase before being capable of transitioning to reproductive growth, Traditional breeding and agronomic practices to obtain early and synchronised flowering are costly, time consuming and in most cases difficult to achieve. The transition to reproductive growth is being accelerated through the ectopic expression of FLOWERING LOCUS T (FT), which encodes the long distance flowering signal florigen. The FT signal is well characterized in Arabidopsis (*Arabidopsis thaliana*; AtFT), and extensive research across diverse plant species shows that FT and FT-like genes coordinate the environmental and endogenous signals governing the transition to reproductive growth. Overexpressing FT generally accelerates the transition to reproductive growth, uncouples flowering from photoperiod and vernalization requirements. However, accelerated flowering through transgene overexpression can be problematic for a number of reasons. Standard transformation strategies are time consuming and require skilled labour; they are inefficient in many species and usually are applicable in a limited number of genetic backgrounds; and many important crop varieties and species remain recalcitrant.

As alternative to traditional approaches, viral vectors have been used to deliver FT orthologs to different crop plants to induce determinate growth patterns and precocious flowering. The logic in this approach is: the FT gene product, florigen, is phloem mobile and naturally finds its way into apices to influence meristem identity, and viruses use the phloem as a pathway to establish systemic infections. Therefore, in principle, coupling an FT ortholog with a virus-based vector that can amplify the inserted sequence and move it systemically will promote flowering. On per with virus-induced gene silencing (VIGS), the use of a virus to deliver sequences that promote flowering was termed VIF (McGarry and Ayre, 2012). The first demonstration of VIF used Zucchini yellow mosaic virus to deliver FT to cucurbits, stimulating flowering in short-day melon (Cucurbita moschata) under noninductive long days. Later on, it has been demonstrated in a number of ornamental and horticulture plants.

5. BIOLOGICAL CONTROL OF WEEDS USING VIRUSES

In limited cases, viruses that affect weed species have been consider ed as bioherbicide candidates. This strategy is more commonly considered for management of invasive species in broader open ecosystem rather than specifically manage areas. Examples of viruses that have been investigated for the potential to control in vasive or undesirable species include Tobacco Mild Green Mosaic Tobamo virus for control of tropical soda apple (*Solanum viarum*) in Florida, a Mosaic virus for control of mothplant (*Araujiahortorum*) in New Zealand. A patent on the former biological control agent has been filed and EPA approval for use on fenced-inpasture areas was granted in 2015 (EPA, 2015). A virus resembling Tobacco rattle virus has also been proposed as a control agent for *Impatiensglandulifera*, an in vasive weed of concern in central and Western Europe. Similarly, Óbuda pepper virus (ObPV) and pepino mosaic virus (PepMV) have been proposed as agents to reduce overall populations of the weed *Solanum nigrum*.

6. ENGINEERING METABOLIC PATHWAYS

Plant genomes can be engineered to produce compounds of nutritional, industrial or pharmaceuticalimortance. However, it is a challenging task due to extensively interconnected regulation of biosynthetic pathways. Some attempts have been made to engineer the metabolic pathways which demonstrate the potential plant viral vectors. Recently, Maler et al (2016) through a novel strategy described the use of a viral vector derived from *Tobacco etch virus* to express a whole heterologous metabolic pathway that produces the health-promoting carotenoid lycopene in tobacco tissues. The pathway consisted in three enzymes from the soil bacteria *Pantoea ananatis*. Lycopene is present at undetectable levels in chloroplasts of non-infected leaves. In tissues infected with the viral vector, however, lycopene comprised approximately 10% of the total carotenoid content. They further showed that plant viruses that express *P. ananatis* phytoene synthase (crtB), one of the three enzymes of the heterologous pathway, trigger an accumulation of endogenous carotenoids, which together with a reduction in chlorophylls eventually result in a bright yellow pigmentation of infected tissues in various host-virus combinations. Out of these studies, in addition to illustrating the potential of viral vectors for engineering complex metabolic pathways, they have also shown a yellow carotenoid-based reporter that can be used to visually track infection dynamics of plant viruses either alone or in combination with other visual markers.

7. PLANT VIRUSES AND VLPS IN NANOBIOTECHNOLOGY

Nanotechnology is a rapidly expanding field with a wide range of applications in biotechnology and medical fields.Nanobiotechnology focusses on the use of biologically derived structures with atleast one dimension smaller than 100 nm. Many features determine the suitability of nanoaricles of biological origin particularly in the case of medical applications.

Viruses, and noninfective virus-like particles (VLPs) in particular, meet the requirements to be used nanoparticles (here onwards referred to as viral nano particles (VNP) possess these desirable features. The precise size and highly symmetric nature of plant viruses make highly useful as VNPs. Using plant viruses reduces many of the risks associated with biological materials as a basis for medical nanotechnologies, and the use of noninfective VLPs results in low risk to the environment. These lower risks allow greater ease in handling, transportation and processing of VNPs, making plant virus-based particles particularly attractive platforms for a range of nanobiotechnological applications. Native viral article or VLPs can be directly used as VNPs or they can be engineered and functionalized either by genetic or chemical means, to generate synthetic nanoparticles for a wide range of applications. Deconstructed viral vectors, which can be used to produce high yields of plant virus-derived VLPs in plants rather than in a heterologous system. The first example of deconstructed viral vector use was the production of VLPs based on CPMV.

VNPs can be used as carriers of useful cargo for a plethora of nanotechnological and biomedical applications. In medical field, there are three major applications of VNP: As cargo of therapeutics, bioimaging and metallization. Cargo carrying utilizes the interior viral cavity as a vehicle for specific molecules.

There are two main strategies for loading plant viral particles with foreign cargo: the infusion technique seeks to allow diffusion of a cargo of interest into the preformed viral particle, whereas the caging strategy aims to trigger particle formation around the cargo of interest. The ability to incorporate designed cargoes into protective protein shells, and targeting of these to specific tissue types allows the use of VNPs as bioimaging agents. Using NHS ester chemistry both flexuous rod virus PVX and the icosahedral virus CPMV can be functionalized with a number of commercial fluorescent dyes for imaging in cell cultures and to mark embryo vasculature and tumor tissue in chick and mouse systems. Imaging using nonoptical methods such as MRI by loading metals such as gadolinium into VNPs is referred to as mineralization. Many VNP technological and medical applications, such as MRI, are dependent on mineralization. With a range of available VNPs with different physical characteristics (icosahedral, rigid, or flexuous rod), and by identifying the appropriate modifications required for specific metal interactions, VNPs appear to be a viable platform for the generation of a number of organic–inorganic hybrids.

Further reading

HefferonK (2017) Plant Virus Expression Vectors: A Power house for Global Health. Biomedicines 5, 44; doi:10.3390/biomedicines5030044

Venkataraman, S.,Hefferon, K. (2021) Application of Plant Viruses in Biotechnology, Medicine, and Human Health. Viruses 13, 1697. https://doi.org/10.3390/ v13091697

Steinmetz, NF (2010) Viral nanoparticles as platforms for next-generation therapeutics and imaging devices. Nanomedicine,6, 634–641.

Chung, YH, Cai, H., Steinmetz, NF (2020) Viral nanoparticles for drug delivery, imaging, immunotherapy, and theranostic applications. Adv. Drug Deliv. Rev., 156, 214–235.

Abrahamian P, Rosemarie W. Hammond, and John Hammond (2020) Plant Virus–Derived Vectors: Applications in Agricultural and Medical Biotechnology Annu. Rev. Virol. 7:513–35

Steele J FC *et al* (2017) Synthetic plant virology for nanobiotechnology and nanomedicine WIREs Nanomed. *Nanobiotechnol,* 9:e1447. doi: 10.1002/wnan.1447

Ali, Z., Abul-Faraj, A., Li, L., Ghosh, N., Piatek, M., Mahjoub, A., *et al.* (2015). Efficient virus-mediated genome editing in plants using the CRISPR/Cas9 system. *Mol. Plant* 8, 1288–1291. doi: 10.1016/j.molp.2015.02.011

Nooraei *et al* (2021) Virus□like particles: preparation, immunogenicity and their roles as nanovaccines and drug nanocarriers *J Nanobiotechnol 19:59* https://doi.org/10.1186/s12951-021-00806-7

Lozano-Duran, R. (2016). Geminiviruses for biotechnology: the art of parasite taming. *New Phytol.* 210, 58–64. doi: 10.1111/nph.13564

Zaidi SS and Mansoor S (2017) Viral vectors for plant genome engineering. *Front. Plant Sci.* 8:539. doi: 10.3389/fpls.2017.00539

PART FOUR

ANIMAL & HUMAN VIRUSES

19

CLASSIFICATION

Though human and animal diseases caused by virus are known since ancient times and even vaccination against viral diseases was developed, the confirmation of viruses as etiologic agents of these diseases was not forthcoming much later. For the first time in 1898 Loeffler and Frosch demonstrated that foot-and-mouth disease of cattle was caused by filterable agents. This was followed by the demonstration by Walter Reed and Carols Finley (1901) that yellow fever was also caused by simlar agents. The real impetus to animal viral research came from two developments : (i) successful cultivation of a virus in tissue culture and (ii) development and improvement in electron microscopy. Also, the occurence of various important human viral diseases and the global determination to eradicate the dreaded diseases like small pox provided a support to viral research. Today, the research on animal viruses covers broad areas like basic research, biological control, prevention and eradication of important diseases, understanding the cancers, exploitation of viruses as vectors in recombinant DNA technology.

1. HOST RANGE

Animal viruses are reported to infect animals belonging to all phyla, right from protozoa to mammals. Conventionally, they have been classified as invertebrate and vertebrate viruses. Among the former group, insect virus have received much attention because of their ability to cause diseases of useful insects like silk worm, and their promising role as biocontrol agents for the control of insect pests. Baculoviruses are also being used as ideal expression vectors. Vertebrate viruses are reported from the animals of all orders. Apart from human viruses,

viruses of birds, cattle and other livestock were described and some of them were thoroughly investigated. Among human viruses, vaccinia virus, retroviruses, a some oncogenic DNA viruses and recently SARS viruses have received a considerable attention.

2. MORPHOLOGY

Considerable variation is found in the morphology of viruses (Fig. 19.1 and 19.2). Majority of the animal viruses are enveloped. Nonenveloped viruses include parvoviruses, adenoviruses, papova viruses, picornaviruses and reoviruses. All these viruses are cuboidal with icosahedral geometric symmetry. In reoviruses a double icosahedral capsids are present. The overall shapes of enveloped viruses vary. Nevertheless, most of them are spherical. However poxviruses are rectangular. Similarly, bullet shaped rhabdoviruses possess helical capsid. Variations are also found in the envelope with regard to spikes and appendages. Herpes viruses consist of an amorphous layer called tegument in between the nucleocapsid and envelope. Similar additional proteinaceous layers are also present retroviruses.

3. NUCLEIC ACIDS

Animal viruses consist of all types of genomic nucleic acids. RNA containing virus families are more than DNA containing virus families (Fig. 19.1). All variations of nucleic acids i.e. double stranded, single stranded, (+) stranded and (-) stranded viruses are found. Retroviruses, with + sense RNA genome, are diploid. Hepadnaviruses possess partial dsDNA. Genome size also varies considerably in different viruses. Segmented genomes are found in some viruses (e.g. reoviruses, orthomyxoviruses). Still, in some viruses like viruses of parvoviridae exist as satellites.

4. REPLICATION

Due to wide variations in morphology, genomes, affinity for specific hosts and tissues, animal viruses adopt a variety of strategies for replication. Replications of some important animal viruses are elaborated in chapter 20.

5. CLASSIFICATION

When virologists first began to classify animal viruses, they naturally thought in terms of features such as the host preferences of each virus. Unfortunately, not all such criteria are equally useful. For example, many viruses will infect a variety of animals, and conversely a particular animal can be infected by several dissimilar viruses. Thus, virus host preferences lack the specificity to distinguish precisely different viruses. Modern classifications are primarily based upon virus morphology, the physical and chemical nature of virion constituents and genetic relatedness. Morphology is probably the most important characteristic in virus classification. Nature of the virus nucleic acid is also extremely important Nucleic acid properties such as the general type (DNA or RNA), strandedness, sense (+/-) size and segmentation are all useful.

Classification scheme proposed by David Baltimore (1971) identified 32 families of animal viruses distributed in 7 classes. The ICTV using the same criteria together with other parameters placed viruses into families and genera. The last updated ICTV classification (2005) lists 134 genera and 49 families. This classification is presented in Table 19.1 and Fig. 19.1 and 19.2.

Table 19.1. Classification of animal viruses (ICTV, 2005)

I. Invertebrate viruses

Order	Family [Subfamily]	Genus	Type species
	Ascoviridae	*Ascovirus*	*Spodoptera frugiperda ascovirus*
	Baculoviridae	*Nucleopolyhedrovirus*	*Autographa californica nucleopolyhedrovirus*
		Granulovirus	*Cydia pomonella granulovirus*
	Birnaviridae	*Entomobirnavirus*	*Drosophila X virus*
	Dicistroviridae	*Cripavirus*	*Cricket paralysis virus*
	fs infec.htm	*Iflavirus (was* ``Infectious flacherielike viruses")	*Infectious flacherie virus*
	Iridoviridae	*Iridovirus*	*Invertebrate iridescent virus 6*
		Chloriridovirus	*Invertebrate iridescent virus 3*
	Metaviridae	*Errantivirus*	*Drosophila melanogaster gypsy virus*
	Nimaviridae	*Whispovirus*	*White spot syndrome virus1*
	Nodaviridae	*Alphanodavirus*	*Nodamura virus*
	Parvoviridae		
	[Densovirinae]	*Densovirus*	*Junonia coenia densovirus*
		Iteravirus	*Bombyx mori densovirus*
		Brevidensovirus	*Aedes aegypti densovirus*
	Polydnaviridae	*Ichnovirus*	*Campoletis sonorensis ichnovirus*
		Bracovirus	*Cotesia melanoscela bracovirus*
	Poxviridae		
	[Entomopox-virinae]	*Alphaentomopoxvirus*	*Melolontha melolontha entomopoxvirus*
		Betaentomopoxvirus	*Amsacta moorei entomopoxvirus*
		Gammaentomopoxvirus	*Chironomus luridus entomopoxvirus*
	Pseudoviridae	*Hemivirus*	*Drosophila melanogaster copia virus*

Reoviridae	*Cypovirus*	Cypovirus 1
Tetraviridae	*Betatetravirus*	*Nudaurelia capensis β virus*
	Omegatetravirus	*Nudaurelia capensis ω virus*

II. Protozoa viruses

Order	Family [Subfamily]	Genus	Type species
	Totiviridae	*Giardiavirus*	*Giardia lamblia virus*
		Leishmaniavirus	*Leishmania RNA virus 1-1*

III. Vertebrate viruses

Order	Family [Subfamily]	Genus	Type Species
	Adenoviridae	*Mastadenovirus*	*Human adenovirus C*
		Aviadenovirus	*Fowl adenovirus A*
		Atadenovirus	*Ovine adenovirus D*
		Siadenovirus	*Turkey adenovirus B*
	Arenaviridae	*Arenavirus*	*Lymphocytic choriomeningitis virus*
Nidovirales	*Arteriviridae*	*Arterivirus*	*Equine arteritis virus*
	Asfarviridae	*Asfivirus*	*African swine fever virus*
	Astroviridae	*Mamastrovirus*	*Human astrovirus*
		Avastrovirus	*Turkey astrovirus*
	Birnaviridae	*Avibirnavirus*	*Infectious bursal disease virus*
Mononega virales	*Bornaviridae*	*Bornavirus*	*Borna disease virus*
	Bunyaviridae	*Orthobunyavirus*	*Bunyamwera virus*
		Hantavirus	*Hantaan virus*
		Nairovirus	*Dugbe virus*
		Phlebovirus	*Rift Valley fever virus*
	Caliciviridae	*Vesivirus*	*Swine vesicular exanthema virus*
		Lagovirus	*Rabbit hemorrhagic disease virus*
		Norovirus	*Norwalk virus*
		Sapovirus	*Sapporo virus*
	Circoviridae	*Circovirus*	Porcine circovirus
		Gyrovirus	*Chicken anaemia virus*
Nidovirales	*Coronaviridae*	*Coronavirus*	*Infectious bronchitis virus*
		Torovirus	*Equine torovirus*
		Deltavirus	*Hepatitis delta virus*
Mononega virales	*Filoviridae*	*Marburgvirus*	*Lake Victoria marburgvirus*
		Ebolavirus	*Zaire ebolavirus*
	Flaviviridae	*Flavivirus*	*Yellow fever virus*
		Pestivirus	*Bovine viral diarrhea virus*
		Hepacivirus	*Hepatitis C virus*
	Hepadnaviridae	*Orthohepadnavirus*	*Hepatitis B virus*

		Avihepadnavirus	*Duck hepatitis B virus*
	fs hepev.htm	"Hepatitis E-like viruses"	*Hepatitis E virus*
	Herpersviridae	"Ictalurid herpes-like viruses"	*Ictalurid herpesvirus 1*
	[Alphaherpesvirinae]	*Simplexvirus*	*Human herpesvirus 1*
		Varicellovirus	*Human herpesvirus 3*
		"Marek's disease-like viruses"	*Gallid herpesvirus 2*
		"Infectious laryngo-tracheitis-like viruses"	*Gallid herpesvirus 1*
	[Betaherpesvirinae]	*Cytomegalovirus*	*Human herpesvirus 5*
		Muromegalovirus	*Murid herpesvirus 1*
		Roseolovirus	*Human herpesvirus 6*
	[Gammaherpesvirinae]	*Lymphocryptovirus*	*Human herpesvirus 4*
		Rhadinovirus	*Saimiriine herpesvirus 2*
	Iridoviridae	*Ranavirus*	*Frog virus 3*
		Lymphocystivirus	*Lymphocystis disease virus 1*
Mononega virales	*Paramyxoviridae [Paramyxovirinae]*	*Respirovirus*	*Sendai virus*
		Morbillivirus	*Measles virus*
		Rubulavirus	*Mumps virus*
		Heniparvirus	*Hendra virus*
		Avulavirus	*Newcastle disease virus*
	[Pneumovirinae]	*Pneumovirus*	*Human respiratory syncytial virus*
		Metapneumovirus	*Avian pneumovirus*
	Rhabdoviridae	*Vesiculovirus*	*Vesicular stomatitis Indiana virus*
		Lyssavirus	*Rabies virus*
		Ephemerovirus	*Bovine ephemeral fever virus*
		Novirhabdovirus	*Infectious hematopoietic necrosis virus*
	Filoviridae	"Marburg-like viruses"	*Marburg virus*
		"Ebola-like viruses"	*Zaire Ebola virus*
	Bornaviridae	*Bornavirus*	*Borna disease virus*
Nidovirales	*Arteriviridae*	*Arterivirus*	*Equine arteritis virus*
	Coronaviridae	*Coronavirus*	*Infectious bronchitis virus*
		Torovirus	*Equine torovirus*
	Nodaviridae	*Alphanodavirus*	*Nodamura virus*
		Betanodavirus	*Striped jack nervous necrosis virus*
	Orthomyxoviridae	*Influenzavirus A*	*Influenza A virus*
		Influenzavirus B	*Influenza B virus*
		Influenzavirus C	*Influenza C virus*
		Thogotovirus	*Thogoto virus*
		Isavirus	*Infectious salmon anemia virus*

	Papillomaviridae	*Papillomavirus*	*Cottontail rabbit pappillomavirus*
Mononega virales	*Paramyxoviridae [Paramyxovirinae]*	*Respirovirus*	*Sendai virus*
		Morbillivirus	*Measles virus*
		Rubulavirus	*Mumps virus*
	[Pneumovirinae]	*Pneumovirus*	*Human respiratory syncytial virus*
		Metapneumovirus	*Avian pneumovirus*
	Parvoviridae [Parvovirinae]	*Parvovirus*	*Minute virus of mice*
		Erythrovirus	*B19 virus*
		Dependovirus	*Adeno-associated virus 2*
	Picornaviridae	*Enterovirus*	*Poliovirus*
		Rhinovirus	*Human rhinovirus A*
		Hepatovirus	*Hepatitis A virus*
		Cardiovirus	*Encephalomyocarditis virus*
		Aphthovirus	*Foot-and-mouth disease virus*
		Parechovirus	*Human parechovirus*
	Polyomaviridae	*Polyomavirus*	*Simian virus 40*
	Poxviridae [Chordopoxvirinae]	*Orthopoxvirus*	*Vaccinia virus*
		Parapoxvirus	*Orf virus*
		Avipoxvirus	*Fowlpox virus*
		Capripoxvirus	*Sheeppox virus*
		Leporipoxvirus	*Myxoma virus*
		Suipoxvirus	*Swinepox virus*
		Molluscipoxvirus	*Molluscum contagiosum virus*
		Yatapoxvirus	*Yaba monkey tumor virus*
Prions			*Scrapie agent*
	Reoviridae	*Orthoreovirus*	*Mammalian orthoreovirus*
		Orbivirus	*Bluetongue virus*
		Rotavirus	*Rotavirus A*
		Coltivirus	*Colorado tick fever virus*
		Aquareovirus	*Aquareovirus A*
	Retroviridae	*Alpharetrovirus*	*Avian leucosis virus*
		Betaretrovirus	*Mouse mammary tumour virus*
		Gammaretrovirus	*Murine leukeamia virus*
		Deltaretrovirus	*Bovine leukaemia virus*
		Epsilonretrovirus	*Walleye dermal sarcoma virus*
		Lentivirus	*Human immunodefi-ciency virus 1*
		Spumavirus	*Chimpanzee foamy virus*
Mononega virales	*Rhabdoviridae*	*Vesiculovirus*	*Vesicular stomatitis Indiana virus*
		Lyssavirus	*Rabies virus*
		Ephemerovirus	*Bovine ephemeral fever virus*
		Novirhabdovirus	*Infectious hematopoietic necrosis virus*
	Togaviridae	*Alphavirus*	*Sindbis virus*
		Rubivirus	*Rubella virus*

Fig. 19.1. A diagrammatic description of the families and genera of viruses that infect vertebrates.

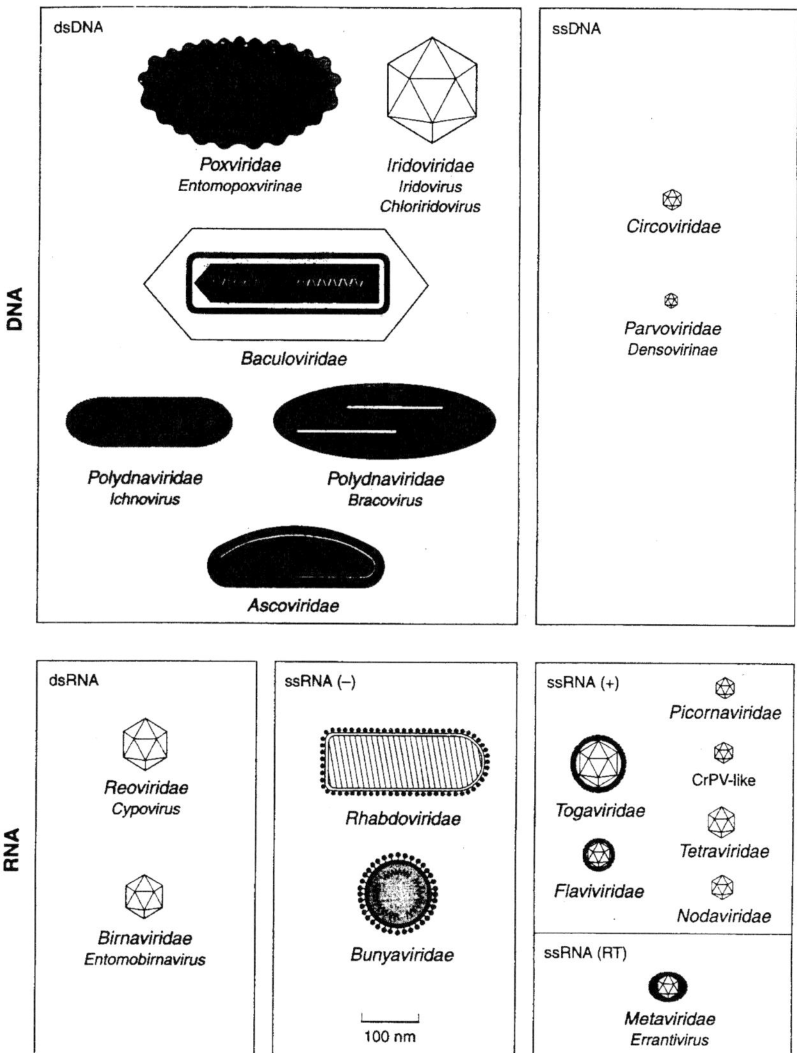

Fig. 19.2. A diagrammatic description of the families and genera of viruses that infect invertebrates.

Further reading

Cann AJ (1997). Principles of Molecular Virology, 2nd edn. San Diego, Academic Press.

Knipe DM *et al.* (2001). Fields Virology 4th edn. Lippincott Williams and Wilkins.

Granoff A and Webster RG (eds.) (1999). Encyclopedia of Virology vol. I, II and III. Academic Press, New York.

Davis BD *et al.* (1990). Microbiology 4th edn. JB Lippincott Company, EWS Square Philadelphia.

Flint SJ *et al.* (2000). Principles of Virology, ASM Press Washingotn, DC.

20

REPLICATION STRATEGIES

Animal viruses display a wide variation in morphology and biochemistry than their plant and bacterial virus counterparts. A majority of the viruses are surrounded by a complex envelope. Proteins are not only distributed in the envelope and capsid but also found associated with genome performing important functions in replication. For example, some viruses contain reverse transcriptase and other enzymes. Animal viruses possess all broad categories of nucleic acids. A wide variation can also be observed within the main categories. Some RNA genomes are segmented, some DNA genomes have unusual termini (such as inverted, repeated nucleotide sequences or covalent crosslinking of linear strands). Animal viruses uniquely interact with the surface of host cells (which do not have rigid walls). Consequently the mechanisms of entry and release of virion nucleic acids are different. As such different animal viruses have developed different strategies of replication. According to the mode of replication, animal viruses can be differentiated into virulent and moderate. Moderate viruses resemble temperate bacteriophages in their ability to establish stable relations with host cells, but the differentiation is less sharp than with bacteriophages and it is some times difficult to decide whether an animal virus is virulent or moderate.

The replication of animal viruses is very similar in many ways to that of phages. Animal virus replication may be differentiated into several stages : adsorption, penetration and uncoating, replication of virus nucleic acids, synthesis and assembly of virus capsids and release of mature virions (Fig. 20.1).

1. ADSORPTION

Replication cycle begins with the attachment of a virion, via capsid or envelope proteins, to specific cell surface macromolecules (viral receptors), frequently glycoproteins. Because the capacity of a virus to infect a cell depends greately upon its ability to bind to the cell, the distribution of these receptor proteins plays a crucial role in the tissue and host specificity of animal viruses. For example, poliovirus receptors are found only in the human nasopharynx, gut and spinal cord anterior horn cells whereas measles virus receptors are present in most tissues. This variation in the distribution of receptors for these two viruses explains the difference in the nature of polio and measles. Animal viruses utilize a variety of proteins, lipids and oligosaccharides as receptors. One class of receptors includes cellular macromolecules involved in ligand binding, endocytosis and cell recognition. For example, the receptors for poliovirus, human rhinovirus, HIV-I and EBV are members of immunoglobulin superfamily of proteins (Fig. 20.2). In contrast, the receptors for Mahoney leukemia virus is an amino acid transporter and sialic acid containing glycoproteins serve as receptors for paramyxo- and orthomyxoviruses.

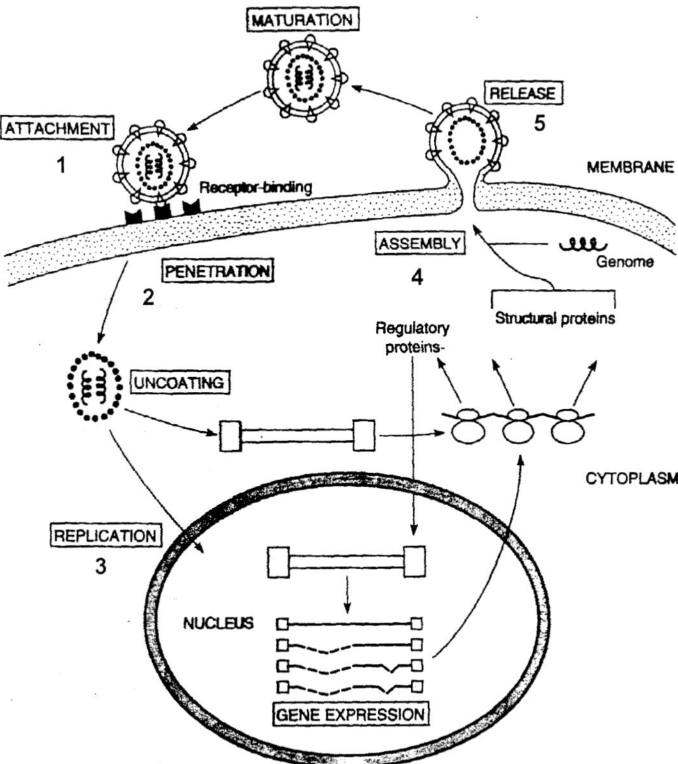

Fig. 20.1. Generalized scheme for virus replication (From A.J. Cann, 1997 Principles of Molecular Virology)

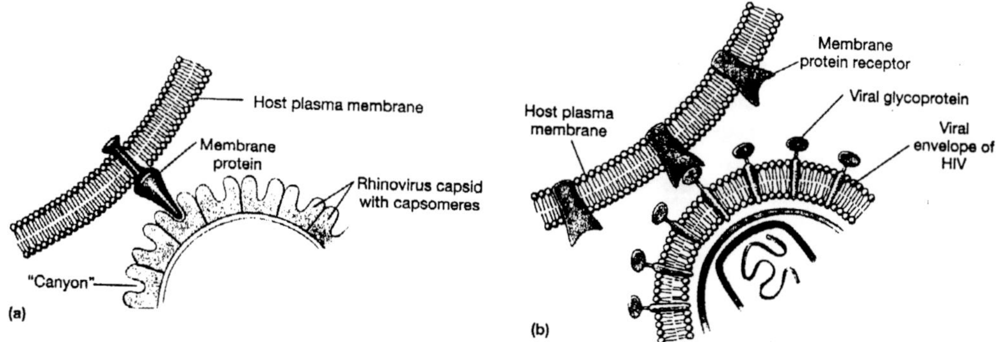

Fig. 20.2. Viral recognition of an animal host cell, (a) Rhinoviruses have ``canyons" or depressions in the capsid that attach to specific membrane proteins on the host cell membrane, (b) HIV has specific envelope spikes (viral glycoproteins) that attach to a membrane protein receptor on the surface of specific host immune defense cells.

The surface site for attachment on virus varies from virus to virus. It may simply a capsid structural protein or an array of such proteins. The atta-chment site may be found in a depression or valley or on specal projections such as the fibres or spikes. For example, in influenza virus the hemagglut- inin spikes appear to be involved in attachment to the host cell receptor site.

2. PENETRATION AND UNCOATING

Following adsorption, the virion must enter the cell (penetration) and release its genome (uncoating). The mechanism of penetration and uncoating vary with the type of virus because viruses differ greately in structure and mode of replication. Enveloped and nonenveloped viruses adopt different strategies for penetration (Fig. 20.3). Similarly, some animal viruses directly inject nucleic acid, while others ensure that a virus associated RNA or DNA polymerase also enters the host cell along with the virus genome. At present, it appears that atleast three different modes of entry are employed by animal viruses for penetration. In naked viruses such as poliovirus, upon adsorption the capsid proteins integrate with cell membrane and only the nucleic acids are released into the cytoplasm (Fig, 20.3a). In enveloped viruses, the envelope fuses directly with cell membrane and the nucleocapsid is deposited in the cytoplasmic matrix (Fig, 20.3b). The virus encoded polymerase is thus safely released into the cell along with the virus genome. It begins transcribing the virus RNA while it is still inside the capsid. Some other enveloped viruses enter the cell through engulfment by receptor mediated endocytosis to form coated vesicles. Virions, bound to their cognate receptors are transported laterally within the plasma membrane to clathrin-coated pits and enter cell as the clathrin-coated inva-ginates. Subsequently, this vesicle fuses with an endosome and, within the acidic compartment uncoating takes place (Fig, 20.3c). The acidic pH of the endosome is critical and the agents that raise the intraendosomal pH block virus uncoati-ng. Lysosomal enzymes may also aid in virus uncoating. Whatever may be the mechanism of entry, once in the cytoplasm, the viral genome may be rele-ased from the capsid or may function while still attached to capsid components.

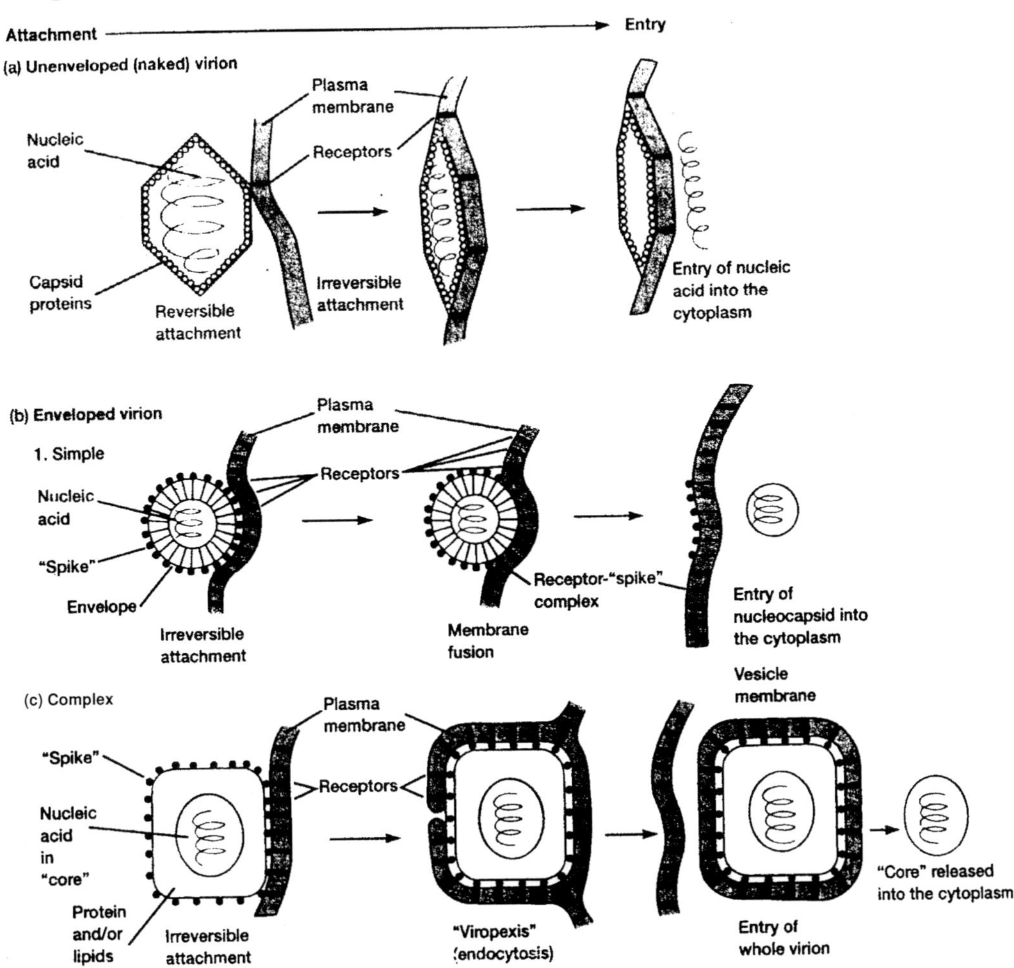

Fig. 20.3. Animal virus entry. Mechanisms of animal virus attachment and entry into host cells. See text for description of the three entry modes. (From Prescott *et al.,* 2003, Microbiology 2nd edn.)

3. REPLICATION OF VIRAL NUCLEIC ACIDS

3.1 DNA viruses

A. Replication of ssDNA viruses-Parvoviruses

Parvoviruses are cubic, naked that infect both vertebrates and invertebrates. Genomes are linear ssDNA with (-) or (+) polarity. Some viruses encapsidate (+) or (-) strands only and other encapsidate (+) and (-) strands in equivalent or different proportions. Genomic DNA is about 4800 bases and thus simplest of all DNA viruses. The small genome directs the synthesis of only three polypeptides, all capsid components. The virus resorts to the use of overlapping genes to fit three genes into such a small molecule. Since the genome does not code for any enzyme, it uses the host cell enzymes for all biosynthetic processes.

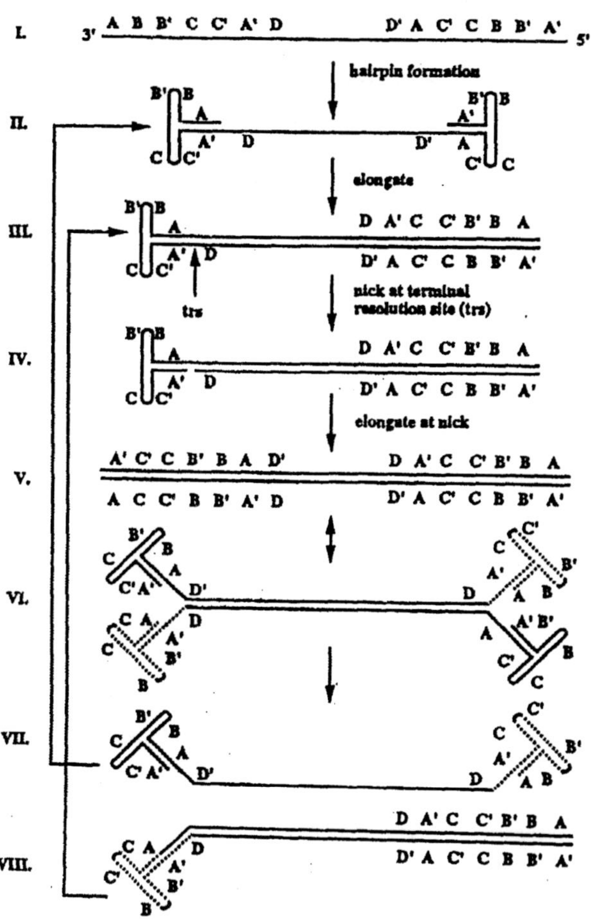

Fig. 20.4. Replication of AAV (adeno-associated virus). The terminal repeats are self-complementary and capable of forming hairpins (II). This allows for self-primed DNA synthesis from the 3'OH-group. The site-and strand-specific nick shown in IV is made by Rep proteins 68 or 78, which also possess helicase activity for conversion (isomerization) of structure V to VI. Structure VII is encapsidated during strand displacement (resulting in net virion production) or enters the template amplification pathway (VIII). (Redrawn from Berns, K.I., *et al.,* in *Virus Taxonomy, Sixth Report of the International Committee on Taxonomy of Viruses,* Murphy, F.A., *et al.,* Eds. Springer, Vienna, *Arch. Virol.* 10, 169, 1995.

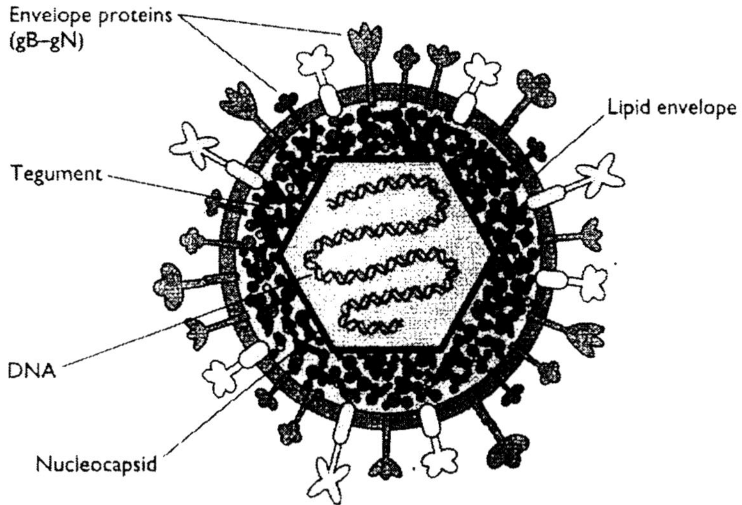

Fig. 20.5. Morphology of Herpes simplex virus

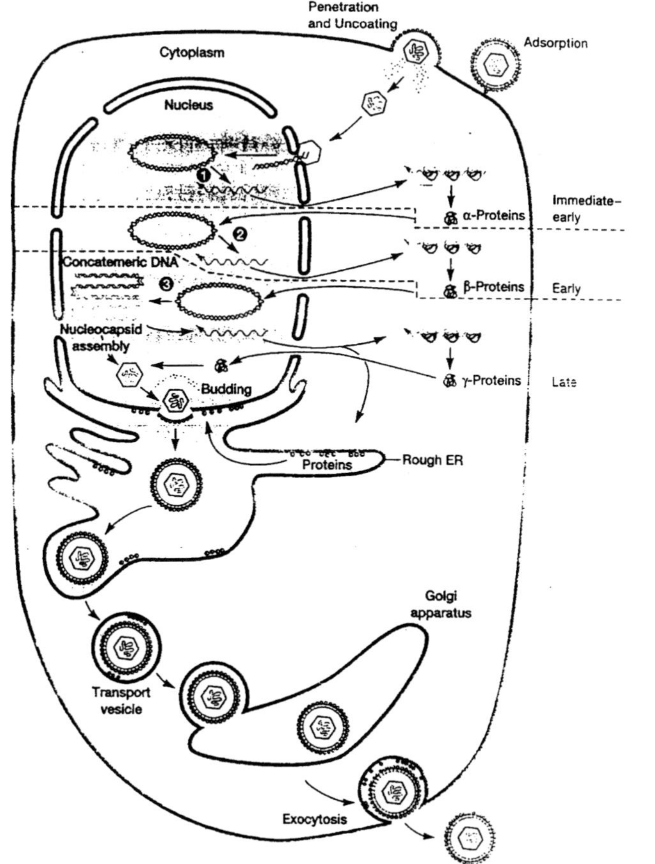

Fig. 20.6. Generalized life cycle for Herpes simplex Virus Type I.

The viral DNA replicates in the nucleus during the S period of the cell cycle. The replication proceeds via double stranded or concatemeric, interme-diates (Fig. 20.4). DNA replication is directed by unusual structure of its ends, which have terminal palindromes capable of forming hairpins. The replicative intermediates of the defective adenoassociated virus DNA includes special concatemers. They are produced in two self-priming steps. Progeny molecules are generated by strand displacement and form double stranded molecules either monomeric or concatemeric.

B. Replication of dsDNA viruses - Herpes virus, Adenovirus and Poxvirus

Herpes virus : Herpesviruses are a large group of icosahedral, enveloped double stranded DNA viruses (Fig. 20.5) that cause many important human and animal diseases. The genome is a linear DNA about 160 000 base pairs terminated by direct repeats of 300-400 base pairs often in multiple copies. It contains atleast 50-100 genes.

For adsorption, viral envelope glycoproteins attach to the cell surface receptors. Envelope and cell membrane fuse and the viral nucleocapsid is released into the cytoplasm (Fig. 20.6). Viral DNA is then uncoated and transported to the nucleus. Early transcription and mRNA are catalyzed by host enzymes. The resulting early viral enzymes are used in the DNA replication. Inside the nucleus the ends of viral DNA undergo limited exonucleolytic digestion, and then pair to form circles. In the first phase, they replicate as cyclic molecules and in the second phase they form tandem concatemers in head-to-tail connection by rolling circle mechanism. During maturation, unit-length molecules are cut from the concatemers. The late RNA transcripts are responsible for synthesis of viral capsid and envelope proteins as well as glycoproteins in the nucleic membrane. The structural proteins enter the nucleus to participate in the assembly of new viral nucleocapsids. The latter are enveloped by budding through the nuclear membrane and complete viruses are released.

Adenovirus: Adenoviruses are nonenveloped, icosahedral DNA containing viruses that multiply inside the nucleus. They induce latent persistent infections in tonsils, adenoids and other lymphoid tissues of man. Several adenoviruses are oncogenic. They serve as helpers for a group of small defective DNA containing viruses, the adeno-associated viruses which cannot replicate in their absence. Each virion contains one linear, double-stranded DNA molecule. The viral DNA has two novel features : (i) the terminal mucleotide sequences of each strand are inverted repetitions so that if the DNA is denatured, both strands form single-stranded circles through panhandles produced between the complementary ends, and (ii) a small protein of about 55,000 daltons is covalenlly linked through the terminal deoxycytosine at the 5' end of each strand. The functions of these unique terminal structures of the viral genome are important in DNA replication.

The essential features of multiplication are similar for all adenoviruses. Adsorption to susceptible cells is relatively slow and the penetration process is analogous to phagocytosis. When the pH is less than 6.0 the endocytic vacuoles membrane is ruptured, permitting the virions to find their way into cytoplasmic matrix. Uncoating of viral DNA begins immediately after virions have penetrated into the cytoplasm (Fig. 20.7A). In the cytoplasm the capsid disintegrates and the viral core migrates to the nucleus. The viral DNA is freed from the core into the nuclear pocket and free DNA enters the nucleus or the core enters the nucleus through a membrane pore.

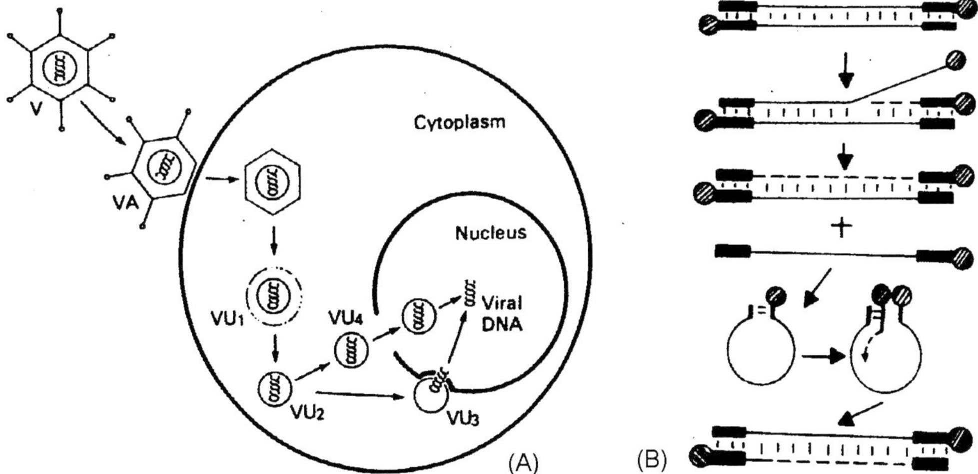

Fig. 20.7. Diagram of uncoating of the adenovirus particle. **A.** *V.* intact virion. VA, attachment of the virion to the cell, followed by penetration of the virion somewhat spherical and the viral DNA susceptible to DNase; VU2. the capsid disintegrates, leaves a viral core that migrates to the nucleus; VU3, viral DNA is freed from the core into the nuclear pocket and free DNA enters the nucleus; or VU4, the core enters the nucleus through a membrane pore, and DNA is then dissociated from proteins. **B.** DNA application. (From Davies *et al.,* Microbiology, 1990).

Replication of viral DNA is semiconservative and is brought about by proteins encoded by early mRNAs (Fig. 20.7B). Replication is initiated at either end of the molecule with the terminal deoxycytidine serving as a primer. As elongation occurs, there is an asymmetric displacement of the other strand. Viral DNA and mRNA are synthesized in the nucleus and virions are assembled in the nucleus. But viral proteins are synthesized on polyribosomes in the cytoplasm. Assembly of mature particles takes several steps that begin with procapsid followed by entry of the viral DNA and final processing. The entire replication cycle takes about 30-35 hours.

Poxvirus: Poxviruses such as smallpox (variola) virus, vaccinia virus are the largest viruses known. They can be seen with phase optics or stained preparations with the light microscope. The virus particles are somewhat rounded, brick-shaped, or ovoid and have a complex structure consisting of an internal central mass- the nucleoid-surrounded by two membrane layers. The surface is covered with ridges that may be tubules or threads. Poxviruses contain a linear double stranded DNA that carries over 200 genes. In addition to viral DNA, the viral cores contain several enzymes, primarily for transcription and modification of immediate early mRNAs : e.g. DNA dependent RNA polymerase, polyadenylate polymerase, methyl transferase and guanylyl transferase.

Unlike DNA viruses, in poxviruses the biosynthesis of the viral components and their assembly into viral particles take place entirely within the cytoplasm of the cell. The virus enters through receptor mediated endocytosis in coated vesicles : the central core escapes from the lysosome and enters the cytoplasmic matrix. The core contains both DNA and a DNA dependent RNA polymerase that synthesizes early mRNAs, one of which directs the production of an enzyme that completes virus uncoating. DNA polymerase and other enzymes needed for DNA replication are also synthesized early in the reproductive cycle. DNA replication begins about 15 hours after infection. The mode of DNA replication is imposed by the unique covalent crosslinking of the two strands. Synthesis is initiated at either end of the genome. Large circular and forked replicationg forms are found, indicating that an endonuclease cleaves the single-stranded cross links during replication. At about the same time late mRNA transcription is initiated. Many late proteins are structural proteins used in capsid contruction. The complete reproductive cycle in poxviruses takes about 24 hours.

3.2 RNA viruses

The RNA viruses are much more diverse in their replication strategies than are the DNA viruses (Fig. 20.8). Most RNA viruses replicate in the cytoplasm. However, retroviruses, orthomyxoviruses and bornaviruses are exceptions to this generalization. The circular RNA genome of hepatitis delta virus also replicates in the nucleus. Viral RNA genomes exhibit a great structural variation. RNA viruses have genomes that consist of either ds or single stranded (ss) RNA, of either positive, negative, or mixed (ambisense) polarity, of either linear or circular topology, and of either single or multiple RNA segments. Each of these variations has consequences for the pathways of genome replication, viral gene expression and virion assembly.

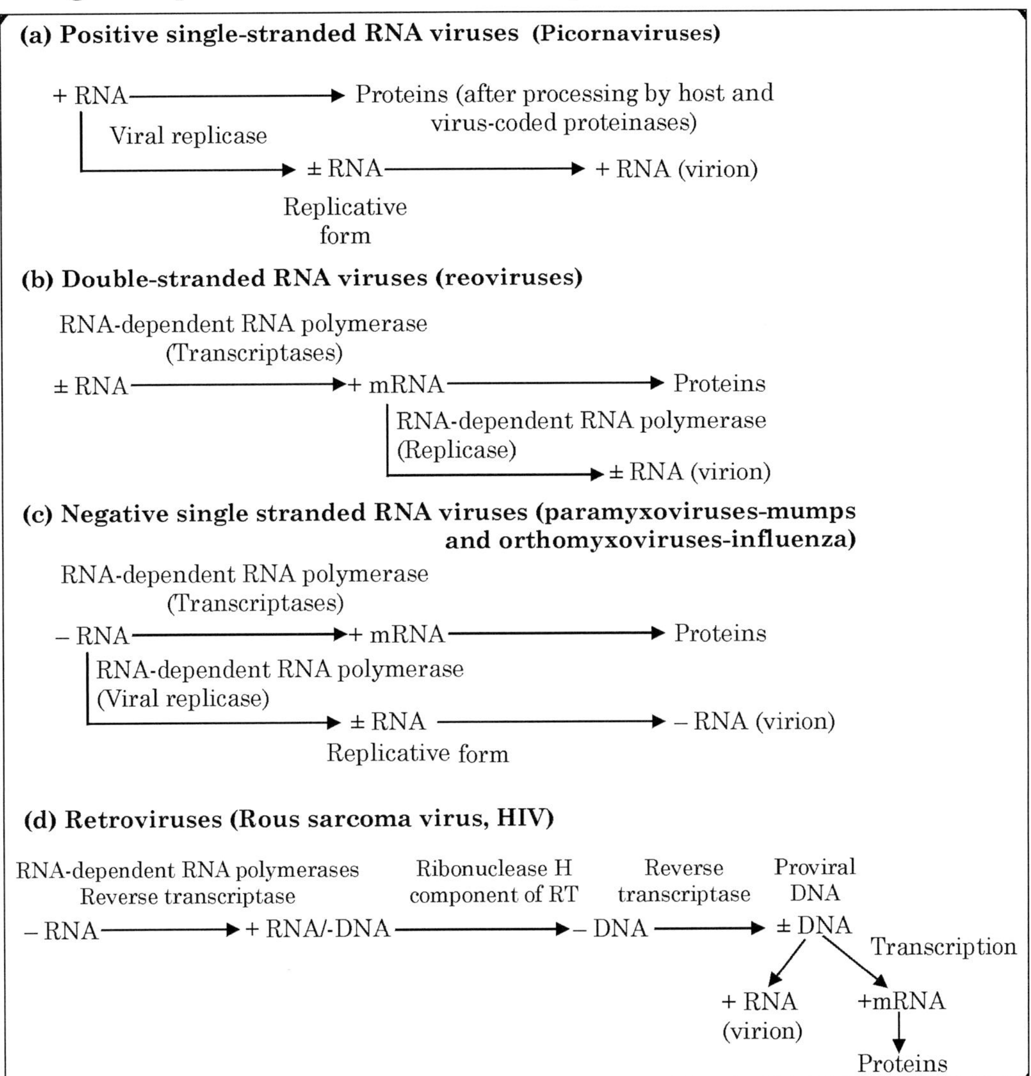

Fig. 20.8. RNA animal virus reproduction strategies.

Most RNA viruses can be placed in one of four general groups based on their modes of replication and transcription and their relationship to the host cell genome (Fig. 20.8). These modes encompass viruses with message-sense (positive strand viruses) and antisense genomes (negative-strand viruses), viruses that package their replicative form as genomes (double-stranded RNA viruses) and viruses that utilize reverse transcription.

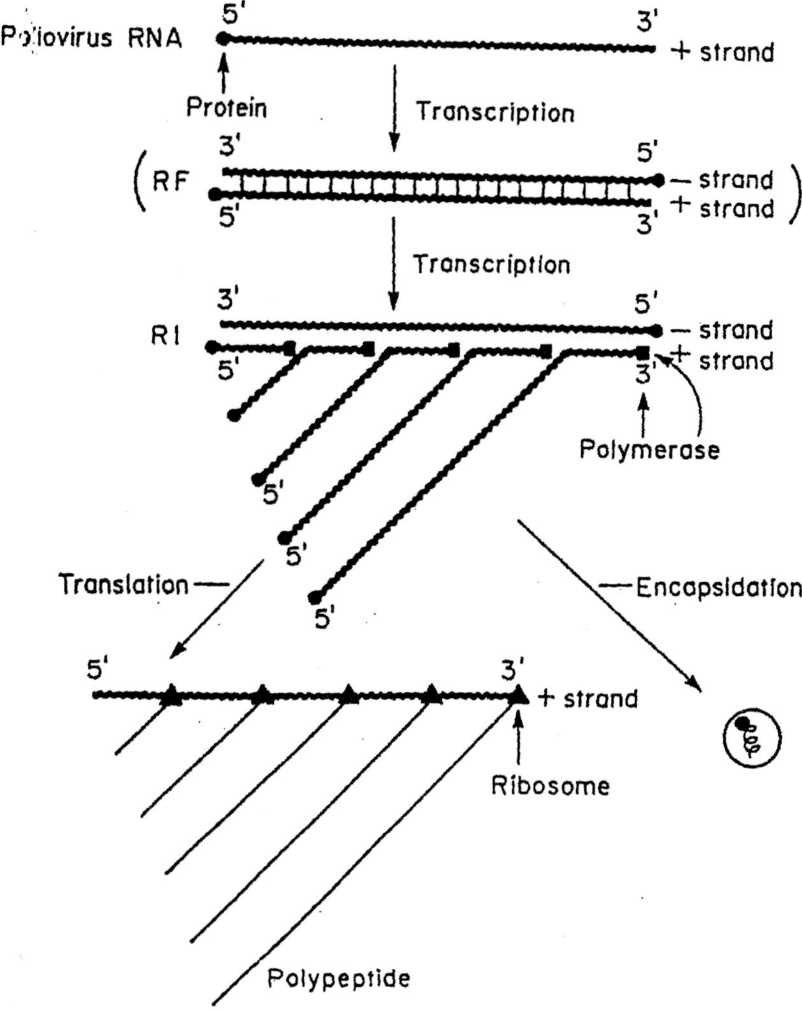

Fig. 20.9. Replication and functioning of poliovirus RNA. Parental virus RNA(+) is transcribed into a polyprotein and, by a polymerase derived from the polyprotein, into (–) strands by a peeling-off mechanism. The structure consisting of a (–) stranded template and several (+) stranded transcripts in various stages of completion is known as the replicative intermediate (RI). Progeny (+) srands are either translated or encapsidated. (From Joklik, W.K., in *Zinsser Microbiology,* 17th ed., Joklik, W.K., Willett, H.P., and Amos, D.B., Eds. Appleton - Century-Crofts, New York, 1980, 1040)

C. Positive strand RNA viruses : Poliovirus

Among the positive strand viruses, poliovirus serves as a prototype because it is most extensively studied one and provides a clear picture of strategy followed by these viruses. Following penetration and uncoating, the poliovirus genome serves directly as mRNA and is translated to yield an enormous polyprotein of 200,000 mol.wt. Initial cleavage of the polyprotein occurs cotranslationally and is mediated by the autocatalytic activity of a virus encoded protease. Subsequent cleavages yield structural (capsid) and catalytic (polymerase) proteins.

The viral RNA-dependent RNA polymerase catalyzes the synthesis of full length negative-sense copy of the genome resulting in a double stranded replicative form (I). Subsequently the negative strand serves as template and directs the synthesis of multiple plus-strands (Fig. 20.9). In the early stages of infection, when the concentration of viral proteins is low, the newly synthesized positive-strands are translated to enhance the synthesis of viral proteins. Later on, however, the newly synthesized (+) strands are utilized as genomes for progeny virions and are encapsidated to generate infectious virus particles. A small virus encoded protein (VPg) is covalently linked to the 5' ends of all picornaviruses RNAs. It is assumed that this protein plays a role in RNA synthesis.

D. Negative strand RNA viruses Para and Orthomyxoviruses

As the genome of the negative sense RNA viruses cannot be directly translated, the first event following uncoating is the synthesis of viral mRNA by a virion-associated RNA-dependent RNA polymerase using the viral genome as template. Negative-stranded viruses are divided into two classes : (i) those with unsegmented (monopartite) genomes (e.g. members of the families *Paramyxoviridae, Rhabdoviridae* and *Filoviridae*) and (ii) those with segmented (multipartite) genomes (members of the families *Orthomyxo-viridae, Bunyaviridae* and *Arenaviridae*).

Vesicular stomatis virus (VSV), a rhabdovirus, serves as a prototype for the understanding of the replication of monopartite viruses. In VSV the envelope consists of a lipid bilayer covered by external surface projections composed of a glycoprotein (G protein) and a matrix nonglycosylated M protein (M_2 protein). The nucleocapsid consists of one molecule of ssRNA many identical copies of a phosphorylated N protein and a few copies of the RNA-dependent RNA polymerase composed of a large (L) protein and a smaller phosphorylated NS (M_1) protein. The virion attaches to specific receptors of susceptible cells through the viral surface glycoprotein projections and enters the cell in endocytic vesicles. The acidic pH of the vesicle induces the viral envelope-plasma membrane fusion and the viral nucleocapsid enters the cytoplasm. The virion RNA dependent RNA polymerase transcribes the viral genome to produce five monocistronic, capped and polyadenylated mRNAs and the short leader RNA, which is identical in all the RNAs. The genome has a single promoter and the polymerase enzyme therefore enter the genome at a single site and moves sequentially down it so that production of each mRNA requires that the preceding 3` RNA be made first. Because the reinitiation does not take place every time, the downstream genes are transcribed less frequently than upstream genes. This viral gene expression is controlled by transcriptional polarity. Viral genome replication i.e the synthesis of a full length positive-sense copy of the genome and the sub-sequent generation of progeny (-) strands is brought about by the same polymerase that directs transcription. The switch between the transcriptive and replicative modes appears to be controlled by the concentration of the nucleocapsid protein.

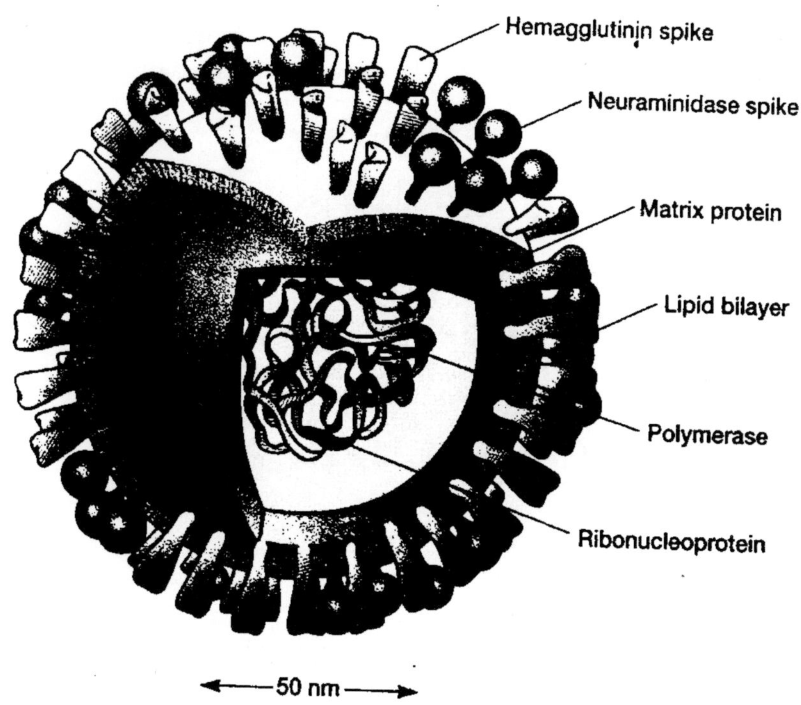

Fig. 20.10. Structure of influenza virus

Viruses with segmented genomes encode their genetic information in multiple molecules of negative-sense RNA (Fig. 20.10). For example, the genome of influenza A virus (orthomyxoviridae) is composed of eight unique segments of virion RNA, corresponding to segments of the nucleocapsid. Unlike other RNA- containing viruses influenza viruses cannot replicate in eunucleated cells. Functional but not replicative host DNA is essential for early events in mulitplication of these viruses. During replication, two classes of complementary RNAs (cRNAs) are made in infected cells : (i) polyadenylated incomplete transcripts of the virion RNAs that are associated with polysomes and serve as the mRNAs and (ii) nonpolyadenylated complete transcripts, which are the templates for virion RNAs (Fig. 20.11). Replication of the polyadenylated cRNAs requires, synthesis of viral proteins to modify the RNA polymerase, permitting production of complete transcripts. Both classes of complementary RNAs and virion RNAs are made separately for each viral RNA segment through the usual intermediary replicative forms. Both types of RNAs are synthesized in the nucleus.

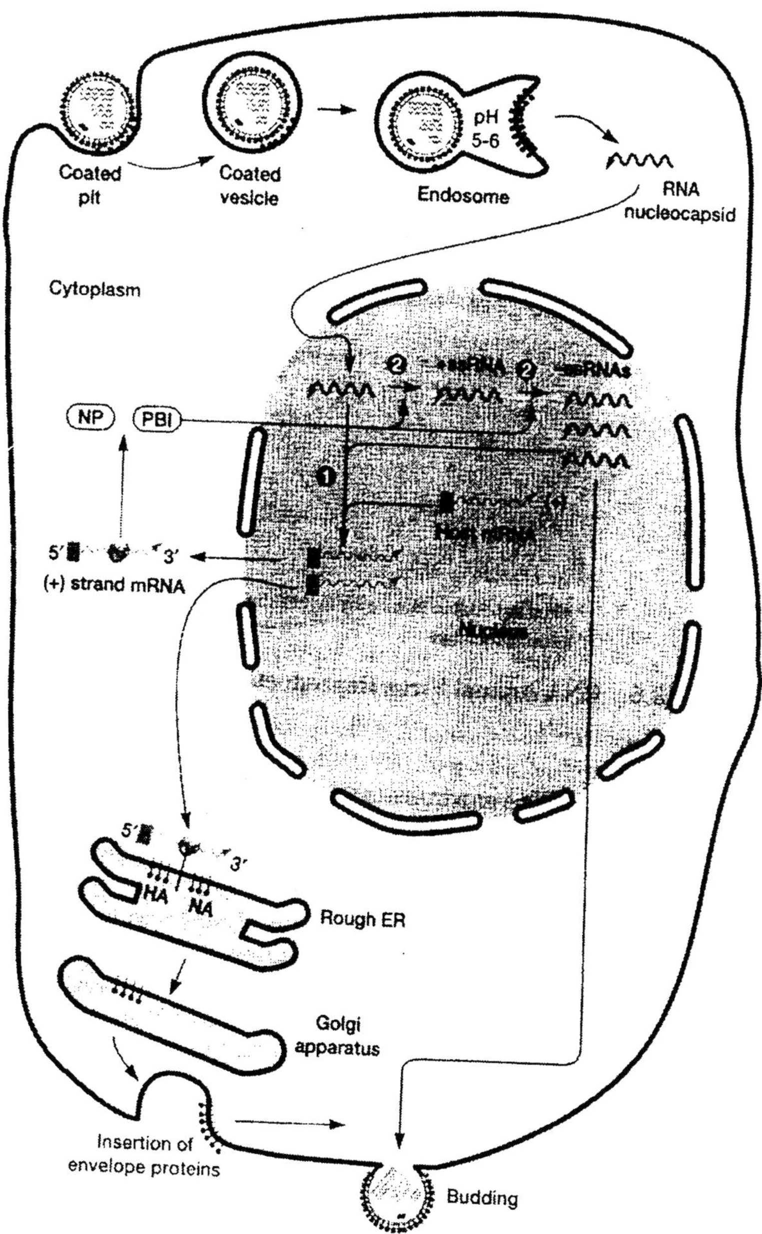

Fig. 20.11. Simplified life cycle of influenza virus. (From Prescott *et al.,* 2005 Microbiology 6th edn.)

E. Double stranded RNA viruses-Reoviruses

The virions of reoviruses have smaller sizes, icosahedral symmetry. They contain two icosahedral capsids. The genomes of reoviruses consist of 10 or 11 distinct segments, which are distributed in three size classes. These segments are distinct components of virion.

The virion RNAs donot function as messengers, but like DNA, they are first transcribed. The transcription is conservative, only the negative (-) strands serve as templates and the transcripts are processed into single stranded mRNAs. An RNA polymerase (transcriptase) contained within the core of the virion is responsible for this early transcription, which is accomplished without the synthesis of new proteins or replication of the viral genome. Unlike double stranded DNA molecules, reovirus double-stranded RNA is replicated conservatively. The negative strand of each segment is copied in great excess into plus single strands, and these then serve as templates for synthesis of the minus strand. Free minus strands are not detected because they remain associated with their templates to form the double stranded RNA segments. Replication of the viral RNA requires continuous protein synthesis to supply the viral-encoded replicase and transcriptase functions. About 75% of the newly synthesized plus strands are utilized for viral protein synthesis. As the concentration of core and capsid proteins increases, the dsRNA-protein complex exchanges nonstructural for structural proteins and forms mature virus particles.

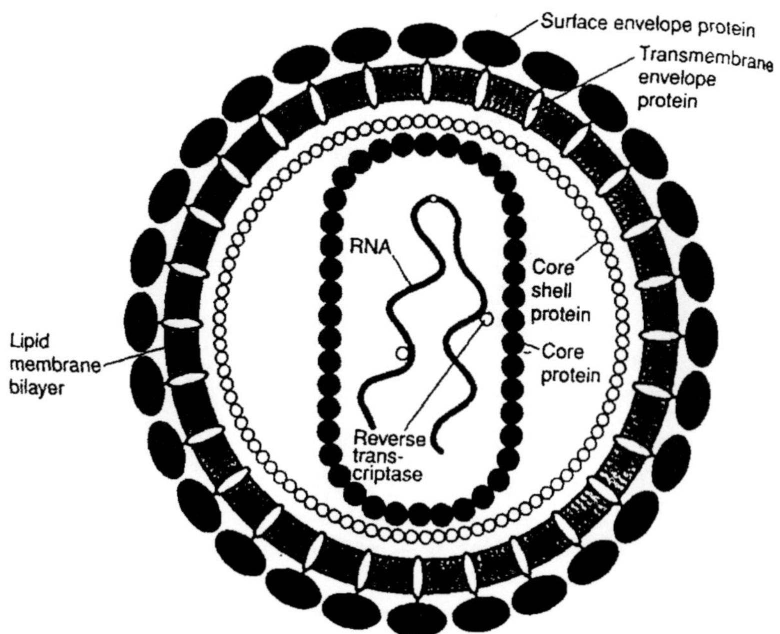

Fig. 20.12. Structure of Human immunodeficiency virus (HIV)

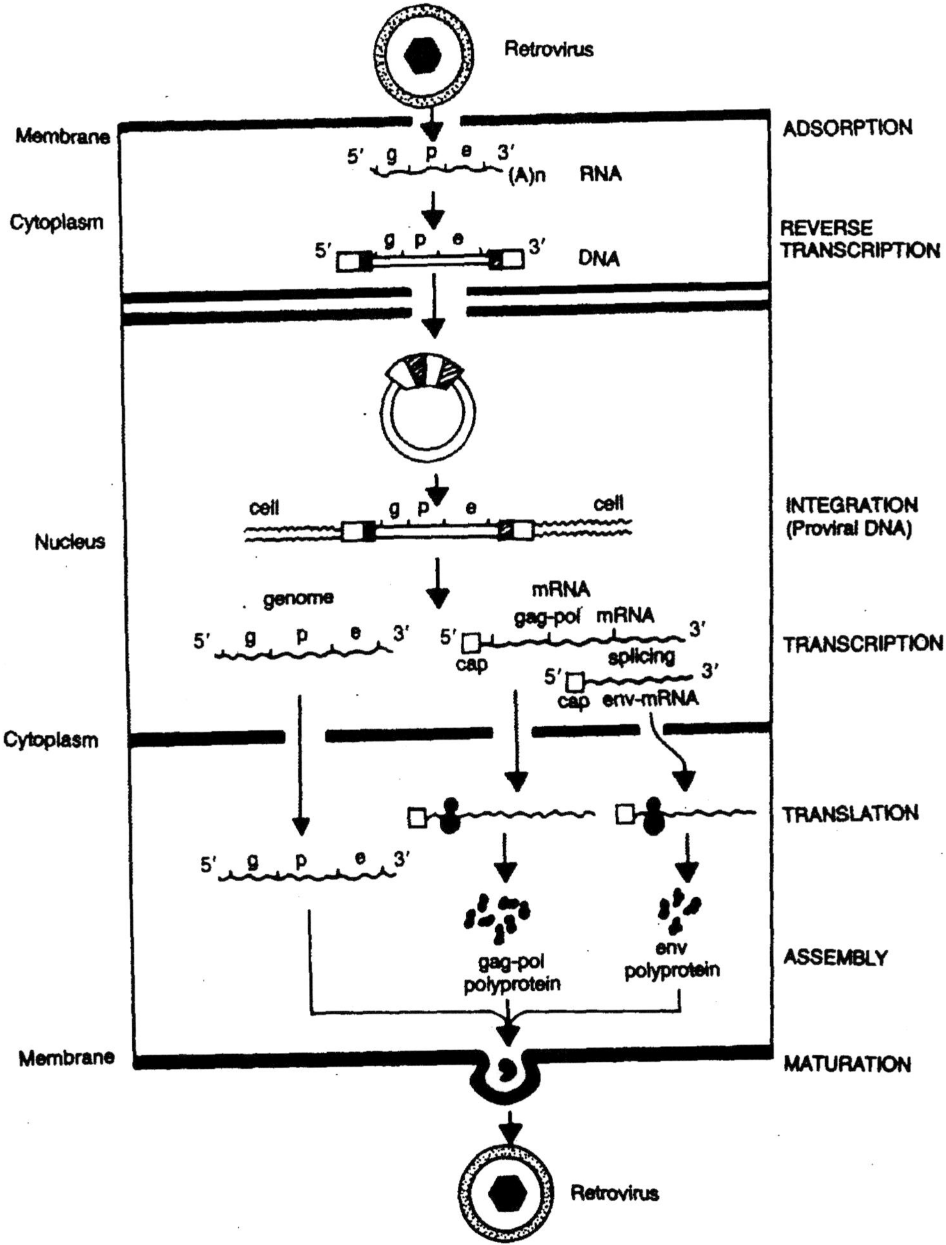

Fig. 20.13. The retroviral life cycle, from entry of the infecting virions (above) to release of progeny virions (below). (Modified from Verma I: In Becker Y (ed.): Replication of viral and cellular genomes, p. 275, Boston, Martinus Nijhoff, 1983).

5′ cap : R : U_5 : (PB) : (Leader) : gag : Prot : Pol : int : env : (PP) : U_3 : R : Poly An : 3′

F. RNA viruses that utilize a reverse transcription – Retroviruses

The retroviruses owe their uniqueness to one important ability : the conversion of genomic viral RNA into cellular DNA. The key to this ability lies in the activity of one enzyme reverse transcriptase (pol). The positive – sense strand RNA genome is between 7-10 kb long, depending on the particular retrovirus in question. This genomic RNA is capped and polyadenylated. The virion contains two copies for the RNA genome (diploid genome) (Fig. 20.12). All retrovirus genomes contain three essential genes from which structural proteins of the virus are derived. The viral RNA also contains required untranslated sequences at the 5` and 3` ends. The order of the genes and genetic elements in the virion-associated RNA genome of all retroviuses is an follows.

Retroviruses replicate their genome and transcribe mRNA using a dsDNA copy of viral genomic RNA as template. This unusual mechanism, in which single-stranded virion RNA is used as a template for dsDNA synthesis, is catalyzed by a virion associated RNA dependent DNA polymerase (reverse transcriptase). After the entry, the virion capsid is partially uncoated and a complementary DNA (cDNA) copy of the RNA genome is synthesized using reverse transcriptase (Fig. 20.13). Following cDNA synthesis retrovirus DNA is integrated into the host chromosome and in the integrated state it is termed as **provirus**. Subsequently, the provirus is transcribed by pol II to yield full-length progeny virion and one or more subgenomic mRNAs. The upstream long terminal repeats (LTR) play a crucial role in retrovirus gene expression because it contains enhancer elements which regulate pol II- catalyzed transcription. Full-length genome sized RNA can either be packaged within virions or serve as messenger for the capsid and catalytic viral proteins. A few RNAs are spliced to generate env mRNA and mRNAs encoding v-onc or regulatory proteins. The env protein is incorporated into the cells plasma membrane. Meanwhile, expression of *gag*, and *gag* : *prot* : *pol* leads to assembly of capsids in the cytoplasm. Immature virus particles bud through the plasma membrane, and final maturation of virion enzymes takes place in released virions.

G. DNA viruses that utilize reverse transcription - Pararetroviruses

There are two different genome strategies used by viruses which involve reverse transcription. One strategy as described above is used by retro-viruses that contain positive single stranded RNA/genome. Some viral groups like caulimoviruses (chapter 13) and hepadnaviruses also use reverse transcription which contain dsDNA genome. They employ a novel replication strategy via a positive-sense ssRNA intermediate and a reverse transcriptase. As this type of reverse transcription evolved parallelly and independently from retroviruses, these viruses are referred to as **pararetroviruses**.

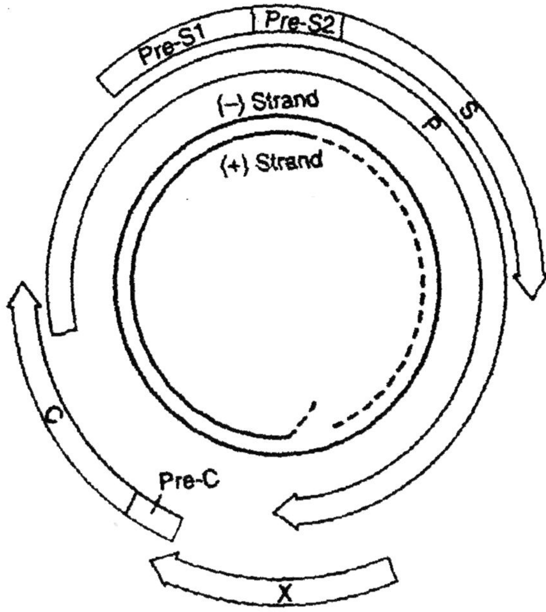

Fig. 20.14. Structure, organization and protein-coding potential of the hepatitis B virus genome.

Hepatitis B Virus (HBV) is the prototype member of the family *Hepadnaviridae*. HBV virions are spherical, enveloped, 42-47 mm diameter which contain a partially double-stranded DNA genome, plus an RNA dependant DNA polymerase (i.e. reverse transcriptase) (Fig. 20.14). Hepadnaviruses have very small genomes consisting of a (-) sense strand of 3.0 - 3.3 kb and (+) sense strand of 1.7 - 2.8 kb. On infection of cells, three major genome transcripts are produced 3.5, 2.4 and 2.1 kb mRNAs. All have the same polarity and the same 3' ends, but have different 5' ends.

Following entry and uncoating of the virion into susceptible cells, the viral DNA reaches the nucleus and utilizing the virion DNA polymerase forms complete double stranded circles, that attain a supercoiled configuration (Fig. 20.15). Host cell DNA-dependant RNA polymerase then transcribes the DNA minus strand to make multiple complete RNA copies termed the progenome RNA, as well as a smaller capped RNA. Later the RNAs are transported into the cytoplasm where the core protein is made and rapidly encapsidates the progenome RNA as well as the newly synthesized viral polymerase and the DNA terminal protein. There follows a unique reaction for DNA viruses, reverse transcription of the progenome RNA, with the DNA-linked terminal protein probably serving as a primer to initiate synthesis of the DNA minus-strand. As viral DNA is synthesized, the progenome RNA is degraded, except for a small fragment of about 20 base pairs derived from the 5' end that appears to act as a primer for the synthesis of the DNA-plus strand from the newly made minus-strand template. The progeny particles usually contain only partially double stranded viral DNA. They develop envelopes, while budding from the cell's plasma membrane.

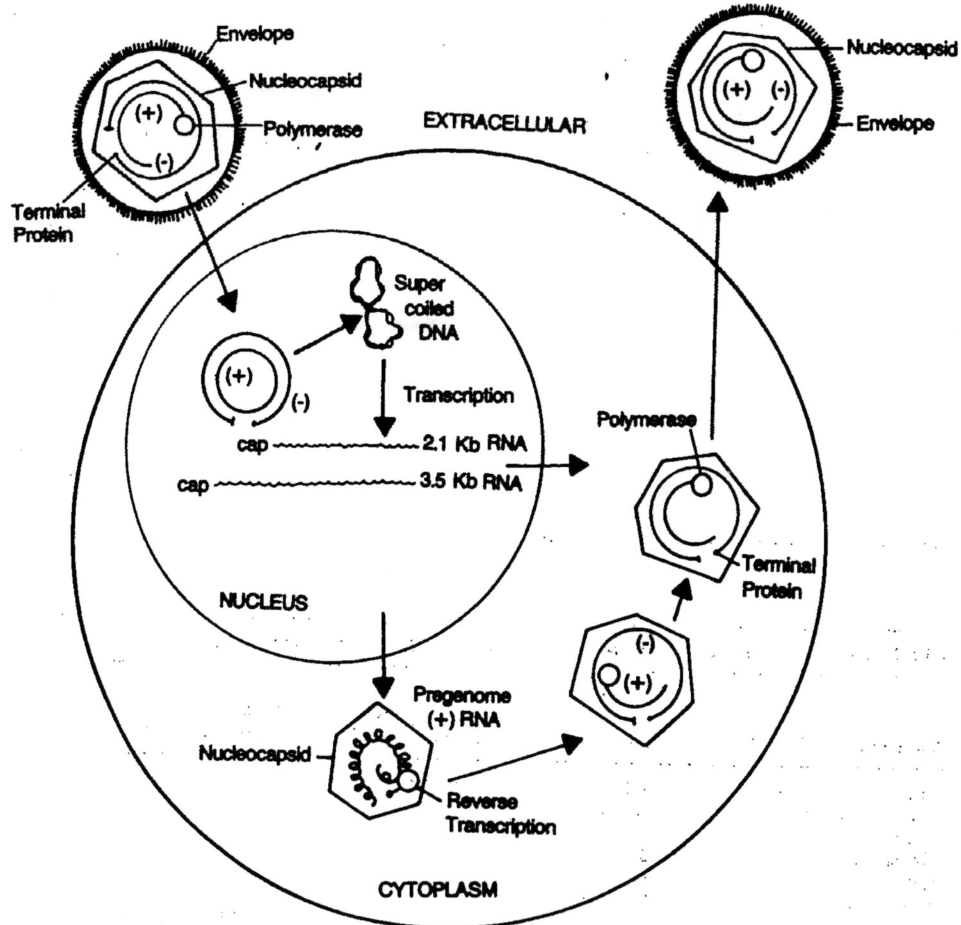

Fig. 20.15. Multiplication of hepatitis B virus in hepatocytes. (From Davies *et al.*, 1990, Microbiology)

4. SYNTHESIS AND ASSEMBLY OF VIRION CAPSIDS

Some late genes direct the synthesis of capsid proteins. Once sufficient quantities of viral proteins have accumulated in the infected cell, nucleocapsid formation and virion assembly begin and continue as long as the cells are able to support the viral components. Helical nucleocapsids form as viral proteins bind to nascent RNA transcripts and encapsidate them. During virus assembly, helical nucleocapsids migrate to cellular membranes where viral glycoproteins have concentrated. These, through concerted interaction between the nucleocapsid and the viral glycoproteins the nucleocapsid is enveloped by the cell membrane. Although virion envelopment takes place commonly at the plasma membrane (e.g. Para-, Ortho- and Rhabdoviridae), intracellular membranes (such as golgi complex, ER and in case of DNA viruses the nucleus) are used by other families. Icosahedral nucleocapsids enclose both DNA and RNA genomes. In such viruses, it is presumed that nucleocapsids form spontaneously when the concentration of capsid proteins reaches a critical level. Later the nucleic acid is inserted in some unknown way.

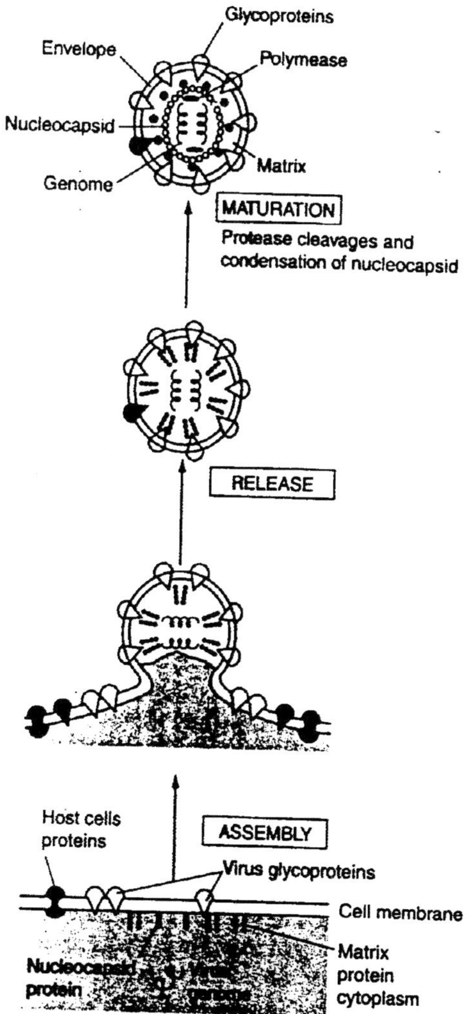

Fig. 20.16. Virus release by budding. This is the process by which enveloped virus particles acquire their membranes and associated proteins (From A.J. Cann, 1997, Principles of Molecular Virology)

5. VIRION RELEASE

The mechanism of virion release differs between naked and enveloped viruses. Naked viruses appear to be released most often by host cell lysis. In enveloped viruses the formation of envelope and the release are usually concurrent processes. First, virus encoded proteins are incorporated into the plasma membrane. Then the nucleocapsid is simultaneously released and the envelope formed by membrane budding. (Fig. 20.16) In herpesviruses, budding and envelope formation usually involves at the nuclear envelope.

Further reading

Ackermann HW, Berthiaume, L and Tremblay M (1998). Virus life in diagrams CRC Press, Boca Raton, Washington DC.

Ball, LA (2001). Replication strategies of RNA viruses P. 105-118. In Fields Virology. Knipe DM *et al.* (eds.). 4th Edition. Lippincott Williams and Wilkins.

Cann AJ (1997). Principles of Molecular Virology, 2nd edn. San Diego : Academic Press.

Chinchar, VG (1999). Replication of viruses. P. 1471-1478. In Encyclopedia of Virology. 2nd Edn. Granoff A and Webster RG (ed.) Academic Press New York.

Davis, BD *et al.* (1990). Microbiology 4th edn. JB Lippincott Company, EWS Square, Philadelphia.

Dimaio D and Coen DM (2001). Replication strategies of DNA viruses P. 119-132. In Fields Virology Knipe DM *et al.,* (Eds.). 4th Edition. Lippincott Williams and Wilkins.

Ehrenfeld E (1993). Translational regulation in virus infected cells. Semin. Virol. 4 : 199.

Levy JA, Faenkel-Conrat H and Owens RA (1994). Virology, 3rd Edn. Englewood Cliffs, NJ : Prentice Hall.

21

VIRUS-HOST INTERACTIONS

The subject of virus host cell interactions provides some of the most important insights in the field of virology. Viruses being intracellular parasites rely on their host cell for the energy, protein and nucleic acid synthesis machinery and particle assembly. Invitingly, viruses have evolved a range of mechanisms to exploit normal host cell functions. Consequently, this exploitation is associated with damage to the host cell which may be one of the major factors in the pathology and disease caused by viruses. The last few decades have witnessed a tremendous expansion of our knowledge of animal viruses. It has provided a detailed understanding of the structure and composition virus particle and genome and also the replication strategies and further the regulation of viral gene expression during infection. This information, inturn, has provided better understanding of virus host interactions during infection.

1. TYPES OF VIRAL INFECTIONS

The outcome of virus infection to a cell can be grouped into several general categories. These categories are determined by both virus and host cell involved in the infection. Infection may be of short duration and self-limited (acute) or may be long term (chronic) or even persist for the life of the host. For many viruses such as influenza, acute infection is followed by viral clearance and subsequent life long immunity against reinfection by the same viral strain. In other cases, such as HSV infections acute infection is followed by persistence of virus in a non infectious latent state from which there is periodic reactivation and shedding. A third pattern involves acute infection followed by continuous shedding of virus from infected tissues. **Productive infections** culminate in the formation of progeny and the destruction of host cell. In some cases, upon the infection the host cell is not destroyed, instead it servives and continues to produce the progeny virions. Such infections are referred to as **persistent infections**. Persistent infections but without the production of infectious virus are known as **latent infections**. In some other cases the viral genome gets integrated with the host genome or is carried as episomal DNA. Such infections are called **transforming infections.** Such infections generally cause an oncogenic alteration of the growth properties of the cell. Still in some cases, the virus is able to enter the cell but because some step essential for viral replication is absent, the replication cycle does not go to completion and no progeny virions are produced. Such infections are known as **abortive infections,** which may or may not cause cell death. Interestingly, there are some instances in which infection with a second unrelated virus can alter the type of infection, adeno-associated viruses are capable of productive, cytolytic infections only in cells co-infected with adenovirus. Similarly, Human adenoviruses can multiply in monkey cells only in the presence of SV40.

2. EFFECTS OF VIRUS INFECTION ON HOST CELL

Host cell morphology and viability

Some of the most readily recognizable affects of virus infection on host cell are those involving morphologic changes or cell death. These changes are the result of cytopathic effects (CPE). **Enders** defined viral cytopathogenicity as the capacity to induce any demonstrable departure from the normal either in morphological or functional properties of cells. Despite of a large number studies little is known of the mechanisms by which viruses induce cytopathology. Viral genes involved or implicated in cytopathogenicity have been identified. However, in most cases, the mechanisms responsible for cell destruction have not been identified.

Macromolecular synthesis

Besides gross changes in cell structure induced by many viruses, cytocidal viruses in particular usually have the ability to interfere with synthesis of host macromolecules. This interference may be caused directly by some virus function, or it may result indirectly from the effects of the virus on some other host function. For instance, poliovirus causes direct and rapid inhibition of host protein synthesis, but host DNA replication does not stop immediately after infection, suggesting some indirect inhibition mechanism. Effects on macromolecular synthesis by different viruses are summarized in Table – 21.1.

Table 21.1. Effects of animal viruses on host cell macromolecules

Function	Effect	Virus examples
Protein synthesis	Direct inhibition (viral product)	Most picorna- and togaviruses; some paramyxo- and herpes viruses
Protein synthesis	Indirect (mRNA competition)	Poxviruses, adenoviruses
RNA synthesis	mRNA inhibition-rapid	Alpha-, togaviruses
	mRNA inhibition-gradual	Picornaviruses and most cytocidal viruses
	rRNA processing- rapid	Picornaviruses
	rRNA processing – gradual	Adeno-and herpesviruses
DNA synthesis	Inhibition – gradual (indirect from lack of proteins)	Picorna-, paramyxo-, adeno-, and herpesvirueses
	Stimulation	Polyoma, SV40, retroviruses

i. Protein synthesis : The extent and timing of the inhibition of host protein synthesis vary for different viruses, most probably due to different mechanisms of inhibition. A number of viruses can rapidly and completely shut off production of host proteins. Many viruse, such as paramyxo-, papova-and retroviruses do not normally inhibit host protein synthesis during their replication. Viruses like picorna-, pox-, herpes-, adeno-, rhabdo-, reo- and orthomyxoviruses infections cause selective inhibition of the translation of host cell mRNA. Virus mediated inactivation of other initiation factors for protein synthesis has also been reported. Another viral strategy to inhibit host protein synthesis involves a direct competition of viral and host RNAs. It has also been suggested that selective translation of viral mRNA may also occur following changes in intracellular ion concentrations during infection.

ii. RNA synthesis : Viral products can directly affect the activity of cellular RNA polymerases and cause an inhibition of host RNA synthesis. Such an inhibition has been observed with VSV and polioviruses. In case of VSV, viral leader RNA and the matrix M RNA have been implicated in the inhibition of cellular polymerases at the level of RNA synthesis initiation. In polioviruses protease cleaves the host transcriptional factors required by RNA polymerase II and III. Influenza virus cleaves the cap structure and the first 10-13 nucleotides from the 5'ends of newly synthesized host mRNAs. Adenoviruses inhibit the transport of host mRNA out of the nucleus.

iii. DNA synthesis : A variety of DNA and RNA viruses are capable of affecting gene expression by directly altering the host cell genome. For example, the host cell DNA is degraded after infection with poxviruses.

Herpes-, picorna- and reoviruses cause a displacement of the cellular chromatin, while an inhibition of host DNA synthesis has been reported following infection with herpes-, pox-, adeno-, picorna-, reo-, alpha-, and rhabdoviruses. This inhibition may well be an indirect consequence of depre-ssion of host protein and/or RNA synthesis earlier during the infection, since synthesis of both these macromolecular constituents is required for normal DNA replication.

iv. Cellular functions : Not all effects of viruses on macromolecular synthesis in eukaryotes are inhibitory. Stimulation of certain macromolecular synthesis has also been observed. Polioviruses and some other cytocidal viruses stimulate membrane proliferation during infection. Certain viruses also induce specific cellular enzymes that in some cases may be needed for replications of the virus. In small DNA viruses such as SV40 specific enzyme induction has been observed.

Host cell membrane and cytoskeleton

Apart from altering membrane permeability, virus infection can cause other changes in the membranes of host cell. Insertion of viral proteins into cell membrane can induce syncytia formation by fusing infected cells with neighbouring infected cells. Syncytia formation, a characteristic feature of paramyxoviruses is also seen with herpes-, flavi-, lenti-, pox-, and corona-viruses. Flaviviruses affect internal membranes by causing the proliferation of rough endoplasmic reticulum, a site associated with the assembly of viral particles. Reo-, picorna and alphavirus infections usually produce a significant increase in vesicle formation in the cytoplasm. Cytolytic virus infections generally cause a progressive loss of integrity of the lysosomal membranes. The release of lysosomal enzymes into the cytoplasm has been described for a wide variety of viruses.

One of the common cytopathological effect, cell rounding has been correlated with alterations in the cytoskeleton. Disruption of one or more of the elements of the cytoskeletion has been reported with many viral infections. For example, microtubules are depolymerized following infection with herpes simplex virus 1 (HSV-1) and canine distemper virus. Viruses also use the structural elements of the cytoskeleton as the work benches for virion assembly and for transport of viral products within the cell.

Inclusion bodies

Some localised changes in intracellular architecture arise at sites of synthesis or assembly of viral components. These discrete foci of viral activity frequently are called viral factories, and they are often visible by light microscopy as distinct inclusion bodies, especially when the cells are stained specifically for nucleic acids or with fluorescent antibodies for viral proteins. It must be remembered here that discovery of a characteristic cytoplasmic inclusion, the Negri body, in cells infected with rabies virus provided an effective diagnostic test for this disease. Depending upon the virus, these intracellular masses may consist of either virions or unassembed viral products. Inclusion bodies may occur either in cytoplasm or nucleus. Cytoplasmic inclusion bodies are found in cells infected with pox-, paramyxo-, orthomyxo-, reo-, rubella or rabies viruses whereas nuclear inclusion bodies are found in cells infected with adeno-, and herpes viruses.

Transformation of host cells

Members of some families of animal viruses are capable of inducing transformation in infected host cells. In majority of the cases, transformation is associated with integration of the viral genome into the host cell genome or maintenance of viral DNA in an episomal state. Among the RNA viruses, only the retroviruses are capable of transforming cells. These viruses induce transformation through the action of a variety of oncogenic genes that are cellular in origin and that are not part of or necessary to the virus replicative cycle. Several DNA viruses are cause of or are associated with tumor induction in animals and cell transformation in cultured cells. These include polyoma-, adeno-, herpes-, papilloma-, hepadna- and poxviruses. The genes of DNA viruses responsible for transformation are viral in origin and required for virus replication.

3. HOST RESPONSE TO VIRAL INFECTION

Cell suicide or Apoptosis

It is evident from the above discussion that the virus by means of individual changes or by a collection of changes ultimately bring about the cytocidal effect on the host cell. An alternative to a direct cell killing is the induction by viruses of suicidal committment in infected cells. Infact, this suicide committment is advantageous to the infected multicellular organisms because it prevents the further spread of the virus. Some recent evidences reported lend support to this possibility. A cytopathic effect associated with apoptosis has been observed in human immunodeficiency virus (HIV) infections. Interestingly, lending an indirect support a noncytopathic latent infection of B cells with Epistein-Barr virus (EBV) has been associated with an inhibition of apoptosis function. These observations provide some basis for suggesting that virus-associated cell killing may involve the induction of apoptosis or some other suicide function in the infected cell.

Resistance of host cells

A major determining factor of the susceptibility of a cell to a particular virus is the ability of the viral attachment proteins to recognize and interact with specific receptors on the cell surface. Resistance to infection is generally attributed to the lack of appropriate surface receptors. Thus, infection does not take place because the viruses cannot adsorb to the cell membrane. Apart from cell surface receptors, the host range of some viruses can be determined by other factors such as host cell transcriptional regulators. It has been demonstrated that viruses from the herpes-, polyoma-, retro-, and hepadnavirus families can replicate only in cells that express the appropriate factors which permit recognition of the viral enhancers. The cells also mount active defense response to viral infections. The best characterized defense that cells have evolved for protection against viral infection is the interferon system. As discussed in chapter-24, the interferons produced in response to viral infection promote the development of an antivral state through the induction of a second group of proteins. Two of these proteins, the 2'-5' A synthetase and the protein kinase have been well characterized and evidence suggests their role in the development of the interferon-mediated antiviral state. To counter these proteins several families of viruses have evolved factors that are capable of blocking the activity of the 2'-5' synthetase and the protein kinase.

Further reading

Knipe, DM, Samuel, CE and Palese, P (2001). Virus-host cell interactions. In Fields Virology 4th edn. Knipe, DM *et al.,* eds., Lippincott Williams and Wilkins. New York.

Frankel–Conrat H and Wagner RR (eds) (1984). Comprehensive Virology, Vol. 19, Viral cytopathology. New York : Plenum Press.

Whitaker-Dowling, P and Youngner, JS (1999). Virus host cell interactions. In Encyclopedia of Virology 2nd edn. Granoff A and Webster RG (Eds.) Academic Press, New York.

Hardwick, JM (1998). Viral interference with apoptosis. Semin. Cell Dev Biol 9 : 339-349.

22

PATHOGENESIS

The term pathogenesis means the origin and progress of disease. The study of viral pathogenesis makes us understand the process by which a virus produces a disease in the host. Entire virus pathogenesis process can be identified into different steps in terms of a series of interactions between the virus and the host. Although specific steps in the pathogenesis may differ for individual virus-host combinations, the general outlines remain the same for most viruses. Different steps in virus pathogenesis include : (i) entry of the virus into a susceptible host, (ii) multiplication of the virus, (iii) spread from the site of entry to target tissues, (iv) shedding of virus into the environment and (v) transmission to a new susceptible host.

1. ENTRY

The most common natural routes of viral entry are through the skin, respiratory tract, gastrointestinal tract, urinogenital tract and conjunctiva.

Skin : Generally skin functions as an effective barrier to the entry of a number of pathogens including viruses. The outer skin layer (stratum corneum) made up of dead keratinized cells does not support viral replication. Infection is facilitated through cuts, abrasions or wounds. Viruses can also be mechanically transported across the stratum corneum by an insect or animal bite or with human-made implements like hypodermic needles. A layer of the epidermis below the stratum corneum (stratum Malphigii) contains living cells but is essentially devoid of blood vessels, lymphatics and nerves. Papilloma viruses enter this layer and cause dermal warts but rarely disseminate to produce systemic diseases. Dermis, the inner layer of the skin is luxurious in blood vessels, lymphatics and nerves. It provides the fertile ground for viral multiplication and subsequent dissemination. Infection of dermis may occur when virus is transmitted by a deep bite, as occurs with rabies virus.

Respiratory tract : In respiratory tract infections the entry of the virus is initiated by infected droplet aerosols generated by coughing or sneezing, the nose to hand to nose transmission, exchange of infected saliva through kissing or sharing of drinking glasses, tooth brushes or other utensils. Some viruses such as influenza, parainfluenza replicate within the cells of the mucosal surface, whereas other viruses such as measles and mumps viruses spread systemically after an initial phase of replication in respiratory mucosa. Host defenses against initiation of viral infection are by physical or immunological. Physical defenses include filtration and mucociliary cleara-nces. The respiratory immune system includes bronchus- associated lymphoid tissue (BALT) and the lymphoid tissue of the tonsils and adenoids. IgA and IgGs protect the upper and lower respiratory tracts.

Gastrointestinal tract : The viruses that have the propensity of infection for gastrointestinal tract are called enteric viruses. **Enteric viruses** can induce either symptomatic or asymptomatic diseases. These viruses cause different degrees of enterocyte injury. Some viruses like poliovirus donot induce gastrointestinal symptoms but produce systemic infections. The target cells for enteric viruses include the mature enterocytes of the intestinal villi and the immature dividing cells in the villus crypts. Some viruses can also infect M cells which are specialised epithelial cells overlying collections of small intestinal lymphoid tissue (Peyers patches). On the other hand, still some viruses like HIV may infect dendritic and epithelial cells of the colon and rectum. Viruses like HSV, CMV can also infect the upper gastrointestinal tract producing oesophagitis.

Enterovirus are resistant to low pH of stomach, bile salts and proteolytic enzymes of the small intestine and mucous coating of the gut. Local gastroi-ntestinal symptoms occur when viral infection involves villus epithelial or crypt cells. Destruction of the villus enterocytes results in malabsorption and diarrhea. Infection of crypt cells and spread to absorptive cells also causes diarrhea. Another, reason for diarrhea is some of the viral proteins act as secretary toxins. For example, the rotavirus NSP4 protein has toxin like actions that stimulate enterocyte hypersecretion and subsequent diarrhea. Finally, an important consequence of enterocyte infection is the shedding of large amount of virus in the stool.

Enteric viral infection typically induces production of secretary IgA. They also induce specific cytotoxic T cells and helper cells. These cells constitute part of the gut associated lymphoid tissue (GALT) system which includes regional collection of lymphoid tissue in the small intestine (Peyers patches). Cellular immunity plays a significant role in clearing enteric infection.

Genitourinal system: A number of important human viruses including HIV, HSV and papilloma viruses are venereally transmitted. Small injuries in the form of tears or abrasions in the epithetial lining of the rectum, urethra, or vagina may occur during sexual activity and permit entry of virus. Host defense factors that inhibit viral entry through these routes include cervical mucous, the pH of vaginal secretions, the chemical composition and cleaning action of urine and the presence of secretary IgA.

Conjunctiva : Though viral conjunctivitis is common, the conjunctiva rarely serves as a site of entry of viruses into the host. Local infection may occur through sharing of clothes or soaps or in the process of swimming or by direct inoculation of virus following ophthalmological procedures (tonometry, foreign body removal). Conjunctiva causing viruses include adenoviruses and enteroviruses. Enterovirus 70 (E70) commonly produces acute hemorrhagic conjunctivitis and in very rare occasions spreads from the eye to the nervous systems to produce cranial nerve palsies, myelitis or encephalitis.

2. SPREAD INSIDE THE HOST

After the successful entry, the infection may remain localised or spread from the initial entry site to other tissue and organs and thus become systemic. In many local infections, virus spread is predominantly by cell-to-cell spread in a contiguous fashion. This type of infections are typical of upper respiratory diseases caused by coronaviruses, rhinoviruses, influenza virus and the acute diarrheal disease induced by rotaviruses. More generalised symptoms like fever, chills, myalgia, malaise, fatigue, anorexia accompany these infections. Polarized release of progeny virions from the infected cells either the luminally facing apical surface or subepithelially facing basolateral surface is an important factor either limiting or favoring systemic invasion. For example, the release of virus only towards the lumen of the respiratory or GI tract would facilitate local infection of epithelial surface and conversely, release of virus from the basolateral cell surface would allow the invasion of the subepithetial mucosa and the subsequent spread of virus through lymphatics, blood vessels, or nerves.

Blood stream : Virus enters the blood stream either from the sites `primary viremia' or direct entry of virus into the blood stream like intravenous drug use, blood transfusion or by the bite of an arthropod vector. Virus in the blood stream may travel free in the plasma or in association with cellular elements. For example, enteroviruses and togaviruses are frequently found free in plasma whereas HIV, HTLV-1 and human hespes viruses infect sub-sets of T-lymphocytes. Viruses may either be simply passively transported or may actively replicate inside the carrier cells. Generally, a correlation has been observed between the magnitude of viremia generated by blood-borne viruses and their capacity to invade tissues, such as the central nervous system (CNS). The mechanism by which the blood carried viruses exit the blood stream to enter tissues is not clearly understood. In some cases, the viruses appear to directly infect endothelial cells, and then are transported across these cells into the underlying parenchyma. Also, the viruses may enter tissues inside migrating cells that are capable of emigrating across capillaries (diapedesis). Thus, endothelial cells may play an important role in determining tissue invasion by blood-borne diseases.

Nerves : Many viruses such as herpes simplex (HSV), varicella zoster (VZV), rabies and certain strains of poliovirus, reovirus and cornavirus that invade the central nevous system (CNS) spread through nerves in the infected host. Neural spread to sites other than the CNS is also common. For example, rabies virus spreads to salivary glands and VZV and HSV to the skin. Neurotropic viruses spread cell to cell through both non neural cells (Schwann cells) and through the axoplasm of neurons. Virions are transported in nucleocapid form rather than in enveloped stage. They spread through motor, sensory and autonomic nerve fibres and in both the anterograde and retrograde directions. Transneural transport also occurs specifically at synapses.

3. SHEDDING OF VIRUS

Shedding of virus from the infected host is the last step in the pathogenesis. This enables the virus transmission to other susceptible individuals. There are several routes for viral shedding. Enteric viruses are characteristically shed in high titre and for prolonged periods of time in feces. Some respiratory viruese are also shed in aerosols generated by coughing or sneezing. However, other respiratory viruses are found in saliva or nasal secretions and are transmitted by direct contact with these secretions or by spread from nose to hand to objects which are subsequently contacted by susceptible hosts. Viruses which infect genital tract are frequently released into genital secretions. A large number of viruses including HIV are transmitted through this way. Some viruses like cytomegalovirus (CMV) are shed into the milk or colostrum. Perinatal infections are caused by this manner. Although many viruses are present in urine (viruria) this method of transmission is considered rarely important. Many viruses remain in the blood and wait for the arthropod bite. In arboviruses, shedding of virus into the blood stream is an essential part of the natural life cycle of the virus.

4. TRANSMISSION

The successful propagation and survival of a virus in nature requires that it should be able to spread from the infected host to other susceptible individuals. In course of evolution, different viruses have evolved different modes of transmission. Eventhough infectious virus can be maintained for a time in a passive state, in nature, the ultimate source of a viral pathogen is an active infection in another host. The two most usual reservoirs, for human disease are other humans or other animals. Some important pathogenic viruses, their reservoir hosts, mode of transmission and diseases caused by them are presented in Table 22.1.

Table 22.1. Some pathogenic viruses and their vectors.

Virus	Mode of transmission/ Vector	Host	Disease
Poliovirus	Human – fecal contamination of water or food	Human	Enteric infection, in rare cases CNS infection (poliomyelitis)
HIV	Direct injection of virus- infected body fluids into blood	Human	AIDS
Measles	Aerosol	Human	Skin rash, neurological involvement
Yellow fever	Mosquito	Tropical monkeys	Malaise, jaundice
Dengue fever	Mosquito	Human, mosquito, primates	Mild to severe hemorrhagic disease
Ebola	Unknown	Unknown	Often fatal hemorrhagic fever
Hepatitis A	Fecal contamination of water or food	Human	Acute hepatitis
Hepatitis B	Direct injection of blood	Human	Chronic hepatitis, liver carcinoma
Hepatitis C	Direct injection of blood	Human	Acute and chronic hepatitis
Hepatitis delta	Blood, requires coinfec- tion with hepatitis B	Human	Acute hepatitis
Hepatitis E	Fecal contamination of water or food	Human	Mild acute hepatitis except often fatal to

			pregnant women
Rabies	Bite of infected animal	Vertebrates	Fatal encephalitis
Herpes simplex (HSV)	Saliva, other secretions	Human	Surface lesions followed by latency, rare encephalitis
Varicella-zoster (VZV, chicken pox)	Aerosol	Human	Rash, shingles, latency
Epstein-Barr (EBV)	Saliva	Human	Infectious mononucleosis, latency
Influenza	Aerosol	Human, many vertebrates	Flu
Smallpox	Aerosol	Human	Variola
Myxoma	Insect bite	Rabbits	Variable mortality, skin lesions
Rhinovirus	Aerosol	Human	Colds
Coronavirus	Aerosol	Human	Colds
Rubella (German measles)	Aerosol	Human	Mild rash, severe neurological involvement in first trimester fetus
Adenovirus	Aerosol, saliva	Human	Mild respiratory disease
Papillomavirus	Contact	Human	Benign warts, some venereally transmitted, some correlate with genital carcinomas
HTLV (human T-cell leukemia virus)	Injection of blood	Human	Leukemia

Like plant viruses, animal viruses are also either mechanically associated and transmitted or are involved biologically with their vectors. Of the mechanically transmitted viruses, rabbit myxomatosis virus is a good example. This virus is transmitted by a mosquito called `flying pin'. Mites are also involved in mechanical transmission of viruses. Transmission of such viruses as causing yellow fever and equine encephalitis, involves definite biological association with the respective insect vectors.

Further reading

Frankel-Conrat H and Wagner RR (eds) (1984). Comprehensive Virology, Vol. 19, Viral cytopathology. New York : Plenum Press.

Mims, C, Dimmock, N, Nash A and Stephen J (1995). Mims' pathogenesis of infectious disease, 4th edn., London : Academic Press.

Nathanson, N (ed.) (1997). Viral pathogenesis. Philadelphia : Lippincott-Raven.

Nathanson, N and Tayler, KL (1998). The pathogenesis of viral infections. In : Mathy BWJ and Collier, LH (eds) Virology, Vol I Topley and Wilson's Microbiology and microbial infections 9th Edn. London : Arnold.

Tyler, K (1999). Pathogenesis. In Encyclopedia of Virology. 2nd Edn Granoff, A and Webster, RG (Ed.) Academic Press New York.

Tyler, KL and Fields BN (20001). Pathogenesis of viral infections. In : Fields Virology Knipe DM *et al.* (Eds.) 4th edn. Lippincott Williams and Wilkins.

White, DO and Fenner, F (1994). Medical Virology, 4th Edn. San Diego. Academic Press.

23

DIAGNOSIS OF ANIMAL AND HUMAN VIRAL DISEASES

Diagnostic virology, apart from providing first hand identification of viral pathogens in specimens, is much more important in some areas of clinical medicine. For example, monitoring and management of immuno-compromised patients, HIV/AIDS, acute respiratory infections, acute or chronic viral hepatitis, serious ocular infections, gastrointestinal infections, antiretroviral therapy require the periodical diagnostic tests. Diagnostic virology has also been important in public health. For instance, surveillance of patients with acute respiratory infections is crucial for the recognition of new strains of influenza. In cases of measles or rubella regular monitoring of virus may help in the initiation of aggressive emergency immunization programs. Similarly, identification of cases of arboviral encephalitis prompt to launch the mosquito control programs. Still, detection of some viruses like caliciviruses is important in understanding and controlling food- and water borne gastroenteritis.

A spectrum of diagnostic methods are now available to detect and identify the viral pathogens, which range from simple culture based methods to molecular based methods (Fig. 23.4). Methods commonly used to detect viruses vary greatly from laboratory to laboratory. These variations depend not only on the availability of resources and equipment of each setting, but also on the availability of trained personnel.

Animal and human viral disease diagnostic methods can be broadly categorized in to five type:

1. Methods based on isolation and culture
2. Methods based on microscopy
3. Methods based on viral antigens

4. Methods based on serology
5. Methods based on molecular characteristics

1. VIRUS ISOLATION AND IDENTIFICATION

Identification of complete infectious viruses in clinical specimens through *in vitro* cultivation is considered the gold standard of diagnostic to virology. As viruses are obligate parasites, their isolation requires inoculation of either cell culture, suckling mice, embryonated eggs or other suitable host system. The host system selected largely depends on the virus to be isolated. Among different host systems, cell culture monolayer is the most convenient method for isolation of virus. This system is practiced in majority of the diagnostic laboratories. Cell cultures are categorized into : **Primary, passaged** and **continuous**. Primary cells are obtained directly from a tissue or organ and cannot be serially passaged. Passaged cells obtained from embryonic tissue can be maintained for few generations. Continuous cell lines, such as Hep-2 cells and HeLa are transformed cells and can be passaged indefinitely. Once inoculated with suitably processed specimen, the viral growth in cell cultures is most often detected based on the development of microscopically visible cytopathic effects (CPE). However, some viruses which donot produce the visible CPE are detected by other methods like hemadsorption, interference and electron microscopy.

i. Cytopathic effect : When viruses grow in cell culture, they often induce changes in the infected cells. These changes are called cytopathic effects (CPE). The CPE are often sufficiently distinctive to allow unequivocal identification of the virus. The characteristics of the cellular changes, the rapidity of both onset and progression of CPE, and the cell culture exhibiting the changes all lead to presumptive diagnosis of the causative virus.

ii. Hemadsorption : Viruses of parainfluenza group like influenza, para-influenza and mumps viruses donot induce detectable CPE. These are detec-ted by hemadsorption test. Influenza and parainfluenza viruses express a viral hemaglutinin on the surface of infected cells. To test for hemadsorption, a suspension of guinea pig erythrocytes is introduced into the cell culture previously inoculated with a clinical specimen. After an incubation of 20-30 min either at room temperature or 4°C, the monolayer is washed to remove the RBC. If a viral hemagglutinin is present in the cells, the RBC will get adsorbed in a specific pattern. Such areas can be inspected microscopically. If hemadsorption is detected, the causative virus can be identified by FA staining of cells from the culture.

iii. Interference : This assay is based on the principle that a cell culture infected by one virus may be rendered resistant to infection by other viruses to which it was originally sensitive. In this assay, a culture previously inoculated with a specimen is "challenged" with a test virus to which it is susceptible. A suitable control without the inoculation of specimen is also maintained. Interference, if present, the challenge virus grows in control but not in the culture previously inoculated with clinical specimen. Once the interference is established, the virus responsible can be identified by FA staining of cells from the culture.

iv. Plaque formation : Viruses induce plaque formation when infected cell monolayers are overlaid with a semi-solid medium. After the plaque formation, the overlay medium is removed and the monolayers are fixed and stained with crystal violet. Uninfected cell absorb the dye and stain, whereas plaques appear light which can be easily detected and enumerated. Plaque formation is also useful for the identification of virus, assay and to perform antiviral susceptibility testing.

In addition to the above methods, a variety of other procedures can be used to detect viral growth in cell culture when CPE is not helpful. Electron microscopy is very often used when there is uncertainity with regard to the identity of virus. In case a particular virus is suspected, its presence can be detected using virus-specific antibodies or nucleic acid assays.

Shell vial culture

This method was developed originally for cytomegalvirus (CMV) but has been subsequently extended to many other viruses including HSV, VZV, respiratory viruses and entero viruses. The main advantage of this method is that it enhances both the speed and sensitivity of virus detection. At first, the samples are inoculated into shell vials containing coverslips with cell monolayers. Following low speed centrifugation for 30-60 min, shell vial cultures are incubated for at least 16 hours. After the incubation period, the coverslip is removed and subjected to staining with a fluorescein or peroxidase labeled monoclonal antibody to CMV early antigens, which are expressed in the nuclei of infected cells. Thus the presence of one or more fluorescent nuclei will evidence the presence of CMV in the specimen.

2. MICROSCOPY

i. Light microscopy : Viruses cannot be directly observed under the light microscope. However, indirect evidence of viral infection may be detected. The most characteristic features of viral infection are inclusion bodies, multinucleated cells, and syncytial cells. These findings can be obtained when cultured cells are stained and mounted. The most commonly used staining is with hematoxylin and eosin (H & E), a technique routinely used in every pathology laboratory. Zenkers fixing solution is specially useful for the demonstration of virus induced intranuclear inclusions and preserves sharp details of virus infected cells. For the detection of DNA and RNA viruses, Schiff's reagent and acridine orange respectively are used for staining.

ii. Electron microscopy : Viral infections can occasionally be diagnosed using electron microscopy. Advantages of electron microscopy include speed, lack of requirement for viral viability and many different kinds viral particles can be observed. Thus, like viral culture, electron microscopy is an `open' procedure. Essentially, two basic techniques available for routine use are : (i) negative staining of virus particles and (ii) thin sectioning of virus infected cells. Negative staining of clinical materials provides the simplest and most rapid method for detection and recognition of virus particles. This sectioning, although less rapid, provides more reliable diagnosis, since the examination of virus infected cells reveals the site of viral replication.

3. VIRAL ANTIGEN BASED METHODS

The detection of viral antigens in clinical specimens by immuniological techniques has become an important method in diagnostic virology. Viral proteins or antigens are recognised by the specific antibodies used in the test. The improvements in this method of diagnosis were achieved with the availability of antibodies of any desired specificity, the monoclonal antibodies which can recognize single epitope of viral proteins. This has facilitated the increased specificity and sensitivity of this category of tests. The main advantage of antigen detection is its speed, that it can provide the diagnostic information within a few hours after one simple reaction. This has laid to widespread use of antigen detection in several areas of diagnostic virology. Common methods used for viral antigen detection include : fluorescent antibody (FA) staining, immunoperoxidase staining and enzyme immuno-assay (EIA). The enzyme immunoassay can be subdivided into solid phase and membrane immunoassays.

FA staining is widely employed for detection of cell-associated viral antigens. Two formats are available : direct and indirect (Fig. 23.1). In the direct format, a fluorescent label, fluorescin isothiocyanate (FITC) is conjugated to the antibody that is used to recognize antigen. In the indirect format, the antiviral antibody is unlabeled and is detected by a second antibody produced against the first one. The second antibody is tagged with fluorescent label. After staining, the specimen is viewed with epiillumination using UV light. The direct method is simple, nevertheless the indirect method is more sensitive and versatile. Immunoperoxidase (IP) staining is essentially similar to FA staining except that horseradish peroxidase is used in place of fluorescent label. A peroxidase specific substrate is added that changes color when acted on by the enzyme. The advantage of this method is that the staining can be viewed by the light microscopy.

EIA is a versatile and widely used method that can be applied to the detection of antigens. The principle and versions of this technique are already discussed in chapter – 16.

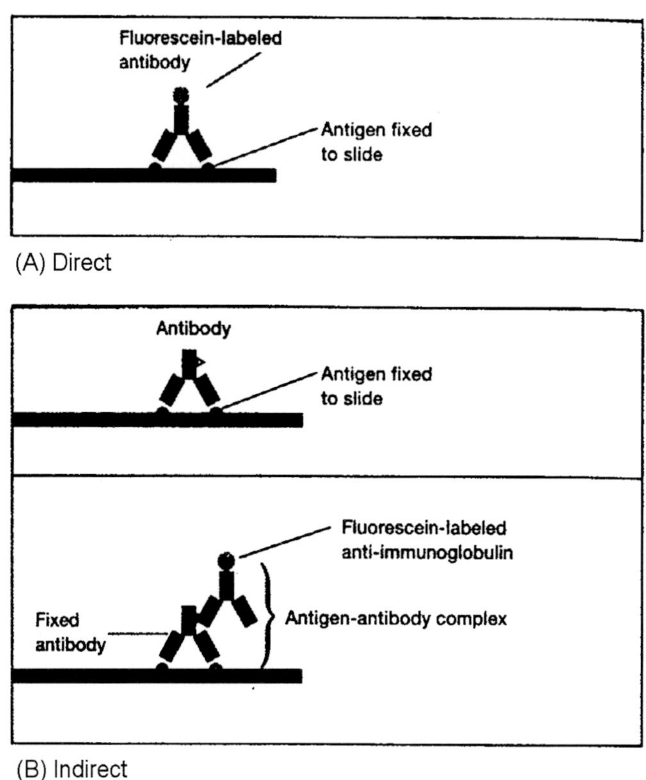

Fig. 23.1. Fluorescent antibody staining. A. Direct, B. Indirect.

4. SEROLOGY BASED METHODS

The detection of antibodies in the host to demonstrate a viral infection is one of the most successful methods of diagnosis. Viral specific immunoglo-bulins are generally produced in high concentrations that can be easily detected by several techniques. All these techniques rely heavily on the availability of specific antigens. In the past, the antigens were often obtained from viral cultures or even from infected patients. However, Now-a-days large quantities of pure antigens are produced employing recombinant DNA technology. Similarly, specific synthetic peptides containing a single epitope are used as antigens.

These developments have enhanced the sensitivity and specificity of serological methods. The principles of serological methods are almost similar to antigen detection methods.

5. MOLECULAR BASED METHODS

The last two decades have witnessed a bewildering activity in molecular biology and consequently significant achievements were made in the genotypic identification and characterization of microorganisms. These developments heralded a new era in the diagnostic microbiology. Application of genome-based molecular techniques enhance the speed, sensitivity and very often the specificity of etiologic diagnosis. The promises of these techniques are numerous : they replace the biological amplification i.e., growth in culture – by enzymatic amplification of specific nucleic acid sequences, *in situ* identification, availability of a spectrum of techniques to suit to the need. Further, the results obtained through these techniques can be delegated for a variety of purposes.

Molecular based diagnostic techniques can be broadly categorized into : i. nucleic acid hybridization methods, ii. signal amplification methods, and iii. nucleic acid amplification methods.

i. Nucleic acid hybridization

Nucleic acid hybridization chemistry has ultimately paved the way for the development of probe technology. This technology has become one of the powerful molecular techniques not only for viral identification of clinical importance but also in other disciplines of microbiology. Many nucleic acid probes have been developed and a number of commercially prepared test kits are being used in clinical laboratories to speed up the identification of important viruses directly in clinical specimens.

The principle of nucleic acid probe technology is straight forward and simple. Genomic sequences specific for a particular group of infectious agents are selected, cloned, synthesized and used as probes. Probes are single stranded oligonucleotides that typically contain less than 50 bases. Because of their unique nucleotide sequence they hybridize with DNA or RNA targets. They bind with high specificity to complementary sequences of a target nucleic acid and are capable of identifying organisms at above and below the species level. The sensitivity of the method varies directly with the size and composition of the probe as well as the nature of the original specimen. The target nucleic acid can be genomic DNA or RNA. To facilitate visual dectection or automation, probes are usually labeled with reported molecules which include enzymes, antigens or chemiluminiscent moieties or radioisotopes.

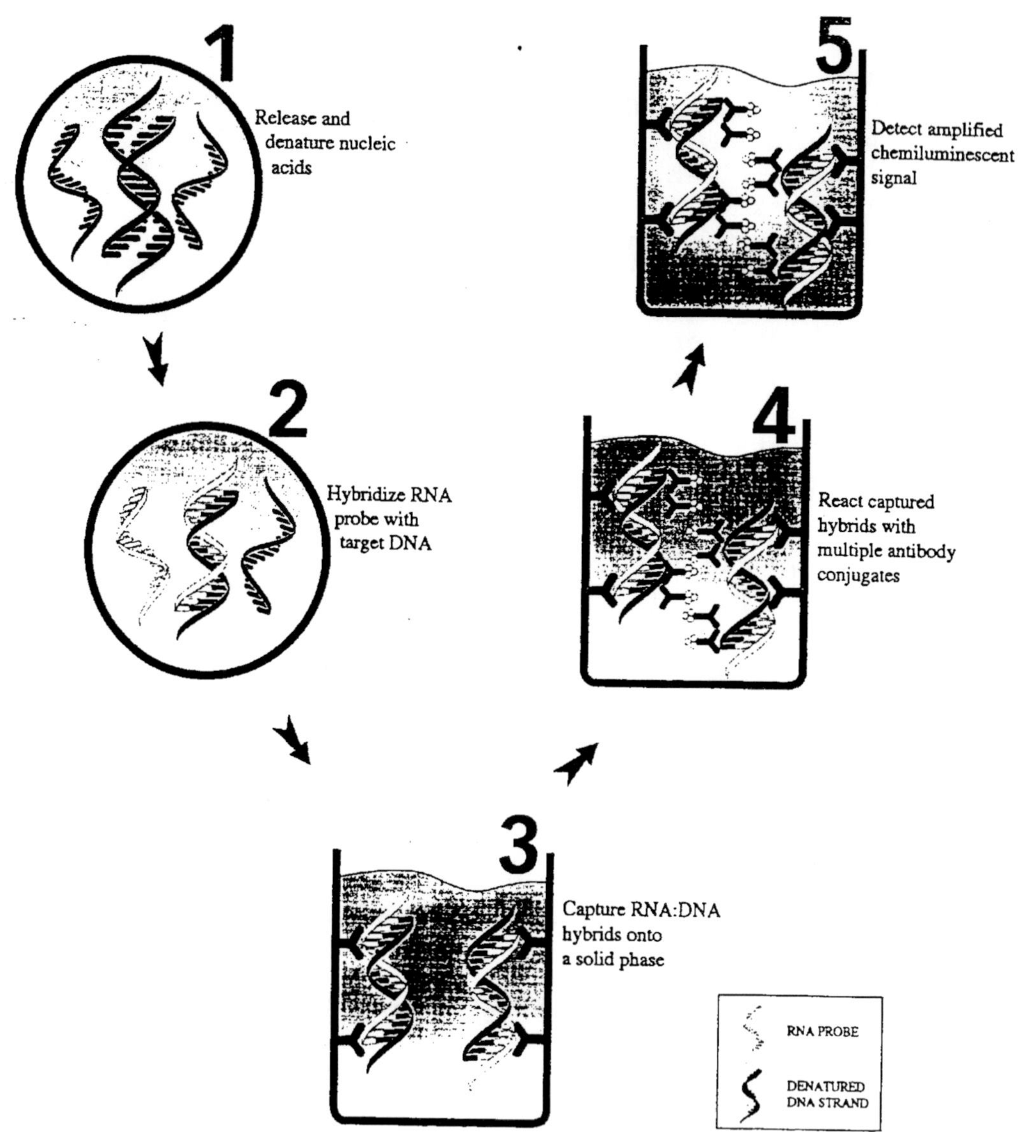

Fig. 23.2. Hybrid capture assay In step 1. target DNA is released and denatured. In step 2, RNA probe hybridizes with target DNA. In step 3, RNA: DNA hybrids are captured by an antibody to RNA: DNA hybrids that is bound to the sides of the reaction vessel. In step 4, an enzyme-conjugated antibody that recognizes RNA: DNA hybrids binds to the hybrids. In step 5, a chemiluminescent substrate is added and light is emitted if hybrids have been formed. (Courtesy of Digene Corporation, Beltsville, MD).

The probe and target nucleic acid hybridization can be performed in liquid phase, solid phase or *in situ*. In liquid phase hybridization of both probe and target are allowed to interact in liquid state allowing rapid duplex formation. The leading method in routine use for detecting hybridization in liquid phase is known as **hybridization protection assay** (HPA). The solid phase hybridization technique can also be used for the simultaneous analysis of multiple samples. Target nucleic acid bound to nylon membranes or nitrocellulose are hybridized with a probe. Hybridization and detection procedures are similar to those for the liquid format. It is a versatile technique, though somewhat less sensitive than liquid phase hybridization. Well established technique like southern and northern blot methods are good examples. *In situ* hybridization assay is another form of solid support hybridization in which whole cells or tissue sections are affixed to microscope slides and the target nucleic acids are detected directly. At present this method is routinely used for detection and typing of human papilloma virus.

ii. Signal amplification methods

The main drawback of probe hybridization is poor sensitivity which is mainly due to the limited amount of target in the clinical samples. To overcome this limitation, signal amplification methods are designed to strengthen the signal by increasing the concentration of label tagged to the target nucleic acid. Detection of target is enhanced through the use of multiple enzymes, multiple probes, two-tiered probes and multiple probes and enzymes without the amplification of neither probe nor target. A **hybrid capture system** was modulated on signal amplification principle which is a solution hybridization antibody capture assay that uses chemiluminiscent detection (Fig. 23.2). This system is widely used in the detection of HPV infection and other viral types in cervical swabs or fresh cervical biopsy specimens. Commercial test kits are also now available for the detection of blood-borne viruses like CMV, HBV and for detection and typing of HPV. The branched DNA (bDNA) probe technique, another format of signal amplification technique, developed and marketed by Chiron Corp, is a most widely used signal amplification method. This technique uses a branched multiple probe-enzyme complex to increase the signal in proportion to the amount of target in the reaction mixture. This system utilizes a series of primary probes, novel branched secondary probe, and short enzyme labeled tertiary probes. This facilitates the attachment of 60-300 enzyme molecules to each target molecule (Fig. 23.3). A chemiluminiscent substrate added reacts with the enzymes emitting light which can be quantified. bDNA method not only minimizes the contamination problems, at the same time presents high reproducible results. It is an excellent method for quantitation and therapeutic response monitoring of HCV and HIV infections.

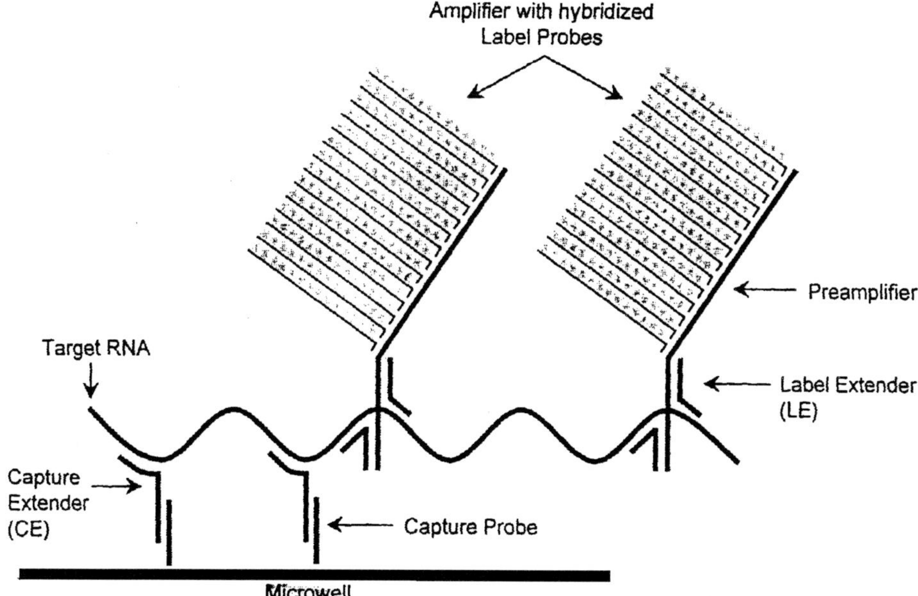

Fig. 23.3. Branched chain DNA assay as used for ultrasensitive detection of human immunodeficiency virus (HIV) (From Fields Virology)

One disadvantage of signal amplification methods is, they are less sensitive than target amplification technique. Another limitation of these methods is background noise due to non-specific binding of reporter probes. Recent advances in sample concentration and nucleotide chemistry are expected to circumvent some of these problems.

iii. Nucleic acid amplification methods (Fig. 23.4 and Table 23.1)

Polymerase chain reaction (PCR)

Nucleic acid amplification (NAA) techniques predominantly rely upon PCR and PCR derived techniques and more recently by some other alternative technique developed. PCR, first envisioned by Kary Mullis soon became a robust technology. Rapidity, reliability, reproducibility, coupled with versatility and automation has allowed this technique to make inroads into clinical laboratories with even moderate facilities. PCR based DNA amplification techniques have been developed and applied to the detection of viral pathogens, identification of clinical isolates and strain subtyping. A major problem in normal PCR is annealing of primers to the template before the denaturation of the target and polymerization resulting in non-specific amplification. In order to avoid this problem the **Hot-start PCR amplification** method is developed. This is achieved through adding taq-specific monoclonal antibodies or use of modified taq which becomes active only at 95°C.

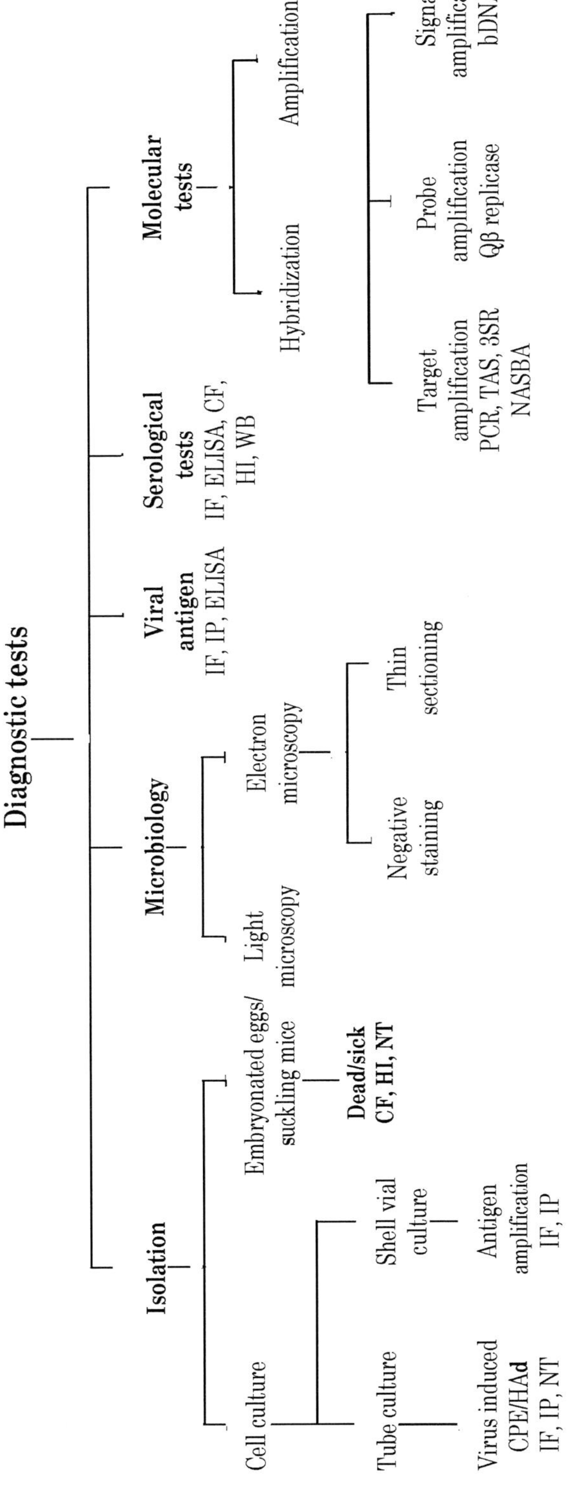

Fig. 22.4. Flow chart of diagnostic tests for viruses. CPE-cytopathic effect; CF - complement fixation test, HAd - hemadsorption test, HI - hemagglutination-inhibition, ELISA - Enzyme linked immunosorbent assay, LA - latex agglutination test, IF - immunofluorescent staining, IP - immunoperoxidase staining, NT - neutralization test, WB - western blot, PCR - polymerase chain reaction, TAS - transcription based amplification system, 3SR - self sustaining sequence replication, NASBA - nucleic acid sequence based amplification, bDNA - branched DNA

Multiplex PCR

In this PCR technique, two or more sets of primers specific for different targets are used in the same tube. Thus more than one target sequences in a specimen are amplified at the same time. Simultaneous amplification of different targets can also be achieved for a variety of purposes. For diagnostic purposes, multiplex PCR can be used for the detection of multiple pathogens from a single specimen. For example, a multiplex assay has been described to allow the simultaneous detection of enteroviruses, HSV-1 and HSV-2 and VZV in specimens of CSF from patients with encephalitis or meningitis. The agents to be detected need not to be only viruses : for example, a PCR reaction has been described for the simultaneous detection of EBV and *Toxoplasma gondii* in CSF from immunocompromised patients. Quantitative competitive PCR, a variation of multiplex PCR, is used to quantify the amount of target sequence in a specimen.

Quantitative PCR

This method is used for quantitative determination of nucleic acid target present in the sample. It is generally coupled with other versions of PCR like PCR-ELISA, cRT-PCR, real time PCR to quantify the target. Within a short time this method has become an integral part of standard medical practice to estimate the load for prognosis and treatment of many viral infections like HIV, HCV, CMV, HBV, EBV etc. Commercial assays are also now available for HIV and HCV. Because of manifold applications, quantitative methods are bound to increase dramatically in near future.

Reverse transcriptase PCR (RT-PCR)

Reverse transcriptase PCR (RT-PCR) was developed to amplify RNA targets. In this process, RNA targets are first converted to cDNA by reverse transcriptase and are then amplified by PCR. Quite naturally, RT-PCR has an important role in the detection of RNA virus infections. It has a wide application in the identification of viral pathogens and in determining the effectiveness of antimicrobial therapy. Commercial RT-PCR kits, some of which are automated, are now available for detection of HIV, HCV and enteroviruses. Now, it is widely used for the detection of SARS – CoV_2, causal agent of COVID-19.

Table 23.1. Nucleic acid amplification methods used in diagnostic microbiology

Amplification method	Enzyme(s) requirement	Temperature requirement	Nucleic acid target
Target amplification			
PCR	Taq DNA polymerase	Thermocycler	DNA, RNA
TAS, 3SR, NASBA, TMA	RT, Rnase H, RNA polymerase	Isothermal	RNA and DNA
SDA	Restriction endonucleases DNA polymerase	Isothermal	DNA
LCR	Taq DNA ligase	Thermocycler	DNA
Probe amplification			
QB replicase	QB replicase	Isothermal	DNA and RNA
CPT	Rnase H	Isothermal	DNA
RCA	DNA polymerase	Isothermal	DNA
Signal amplification			
bDNA	None	Isothermal	DNA and RNA
Hybrid capture	None	Isothermal	DNA
Invader	Cleavase	Isothermal	DNA

Transcription-based amplification (TBA) system

TBA systems, oriented primarily to amplify RNA targets were first described by Kwoh *et al.* In principle, the system involves synthesis of cDNA from RNA target and transcription of resultant cDNA to generate a vast number of RNA copies. Variations of this process are referred to as **self-sustaining sequence replication (3SR), nucleic acid sequence-based amplification (NASBA)** or **transcription mediated amplification (TMA)**. These methods make use of several enzymes (RT, Rnase H and T_7 DNA dependent RNA polymerase), primers containing T_7 polymerase binding site, a complex series of reactions, which all take place simultaneously at the same temperature (isothermal) and in the same buffer. The merits of this system over RT-PCR are that it does not require thermocycler and pretrea-tment of sample with Dnase. An automated NASBA system is commercially available from Organon-Teknika Corp for the detection of HIV-I.

Further reading

Becker, Y and Darai G (eds) (1992). Diagnosis of human viruses by polymerase chain reaction technology. Berlin : Springer-Verlag.

Evans, AS and Kaslow RA (eds) (1997). Viral infections of humans, 4th edn. New York : Plenum Medical Book Company.

Hriung, GD, Fong CKY and Landry ML (1994). Hsiung's Diagnostic Virology, 4th edn. New Haven : Yale University Press.

James, K (1990). Immunology of infectious diseases. Clin. Microbiol. Rev. 3 : 132.

Lennette EH (ed.) (1991). Laboratory diagnosis of viral infections, 2nd Edn., New York Dekker.

Murray, PR *et al.* (1995). Manual of Clinical Microbiology, 7th edn. ASM Washington, DC.

Persing DH (2004). Molecular Microbiology : Diagnostic principles and practice, ASM Press, Washington, DC.

Schmidt, NJ and Emmons RW (eds) 1989. Diagnostic procedures for viral, rickettsial and chlamydial infections, 6th edn. Washington, DC. American Public Health Association.

Storch, GA (2001). Diagnostic Virology P. 493-513. In Fields Virology. Knipe DM *et al.* (eds.), 4th edition Lippincott, Williams and Wilkins.

24

ONCOVIRUSES

Viruses cause varied effects on their infected cells. Viruses that cause lytic infection destroy the host cells. In case of enveloped viruses, release of virions by budding process may be slow and the host cell may not be lysed. The cell may remain alive and continue to produce virus over a long period of time. Such infections are referred to as **persistent infections**. Few viruses cause **latent infections**. In a latent infection, there is a delay between infection by the virus and the appearance of symptoms. A number of animal viruses have the potential to change a cell from a normal one to a cancer or tumor cell. Such viruses are referred to as cancer causing viruses or **oncoviruses** or **tumor viruses**. The study of oncoviruses has many objectives : (i) the study of how tumor viruses cause malignancy could lead to better methods of preventing or treating cancer, (ii) the possibility is that such studies could lead to better understanding of cellular regulation of growth because the tumors are formed as a result of loss of regulation.

1. DISTRIBUTION OF ONCOVIRUSES

The first tumor producing (oncogenic) virus was discovered in 1908 by Ellerman and Bang. They demonstrated that leukemias of chicken could be transmitted to other chickens by cell-free filterates. Later Rous (1911) found that a chicken sarcoma, a solid tumor can be similarly transmitted. These viruses responsible turned out to be retroviruses. The fact that DNA containing viruses can also cause cancers was demonstrated much later by Shope (1932) and Lucke (1934). Buttner (1936) demonstrated that mouse adenocarcinoma is caused by a virus. This discovery provided an ideal system for further experimental work on tumor viruses. Further impetus to investigations of tumor viruses arose from the later discovery of new group of DNA viruses that cause cancer in mice and other rodents : polyoma virus and simian virus –40. Finally, the human adenoviruses, papilloma viruses and herpes viruses were also shown to have oncogenic activity. Around this time, the study of bacterial lysogeny had clearly shown that the genetic material of viruses can become permanently integrated with that of host. These realizations have led to an explosive development of interest in viral carcinogenesis. Very soon the oncogenic effect of several viruses was also demonstrated in tissue cultures in the form of cell transformations. The discovery of many new oncogenic viruses in the 1960s revealed a perplexing distribution. The type of viruses that are capable of inducing tumors are remarkably diverse, and include viruses with RNA and DNA genomes (Tables 24.1 and 24.2). Nevertheless, certain common features have emerged from the study of neoplastic transformation by different viruses. The first is that tumor formation induced by viruses is the result of the acquisition by the host cell of the viral genome. In all cases of virally induced tumors so far described, at least the viral genome is present. Furthermore, the continued expression of atleast a part of the viral genome is required for the transformed phenotype to be present.

Francis Peyton Rous (1879-1970); Nobel Prize, 1966

2. VIRAL ONCOGENES

The study of oncogenic viruses has allowed the identification of a set of genes, which initiate the cancer process. Many of them play an important role in cancer but also in normal cell growth and differentiation. The illegitimate expression of these genes is responsible for the process of transformation and neoplasia. Martin and Vogt (1970) first identified the *src* gene in Rous sarcoma virus (RSV) responsible for transformation. Later M. Bishop and collaborators (1983) identified, a very similar gene, the cellular SRC gene (or *c-src*). The normal gene is known as *src*-proto-oncogene, and the related gene present in virus is *src* oncogene or *v-src*. *V-src* is incorporated in the RSV genome as a consequence of the recombination of nontransforming retrovirus with *c-src*. The *c-src* is highly conserved in animal species.

Table 24.1. Distribution of oncogenic viruses among animal virus families

Nucleic acid in virions	Viral group	Oncogenic viruses
RNA	Retroviruses	Oncoviruses : leukosis viruses, sarcoma viruses
DNA	Adenoviruses	Many types
	Papovaviruses	Polyoma virus, SV40, SV40-like human viruses, papilloma viruses
	Herpesviruses	Virus of neurolymphomatosis of chickens (Marek disease), Lucke's virus of frog renal adenocarcinoma, herpes simplex virus (cell transformation). Epstein-Barr virus (Burkitt lymphoma, nasopharyngeal carcinoma), cytomegalovirus, primate herpesviruses
	Hepadnaviruses	Hepatitis B virus
	Poxviruses	Fibroma virus

Table 24.2. Some viruses associated with human cancers

Virus	Genome type	Cancer
Human herpesvirus 8 (HHV8)	Double stranded (ds) DNA	Several, including Kaposi's sarcoma
Epstein –Barr virus (EBV)	dsDNA	Several, including Burldtt's lymphoma and nasopharyngeal carcinoma
Hepatitis B virus	dsDNA	Hepatocellular carcinoma
Hepatitis C virus	Single-stranded (ss) RNA	Liver cancer
Human papillomaviruses (HPV) strain 6, 11, 16 and 18	dsDNA	Cervical cancer
Human T-cell lymphotropic virus 1 (HTLV-1)	ss RNA (retrovirus)	T-cell leukemia

The possible function of *v-src* was clarified after Erikson identified its protein product. The protein expressed in normal cells by the *src* proto oncogene is similar but not identical to that expressed by *v-src* and does not cause transformation. Two reasons can be attributed to this difference : one is that the *v-src* gene differs from *c-src* gene in its 3 end and lacks a phosphorylation site which is crucial for normal control; the second is that the viral gene is more strongly transcribed. After *v-src* many other oncogenes were discovered in oncogenic viruses. More than 30 oncogenes of various origins are now known (Table 24.3).

3. MECHANISM OF VIRAL CARCINOGENESIS

Many groups of viruses induce specific alterations in the control of cell division and cell mortality, and their study has allowed the identification of many factors involved in carcinogenesis. The study of tumor viruses has provided the major impetus to our understanding of the process of tumor formation, and through this, how to control and ultimately cure certain types of cancers.

Virus-induced alterations of the genetic control of cell replication is the common feature in viral carcinogenesis. There is no common mechanism by which this occurs. It can take place through a very specific interaction between a viral gene and other genes resident in that cell. Specific viruses act on any one of the manifold genetic steps involved in the development of cancer. Table 24.3 presents the details as to how different retrovirus oncogenes affect the cellular growth. In contrast to retroviuses, in DNA viruses, transformation is the result of the activation of a gene that is not required for the completion of the viral life cycle. Transformation by DNA viruses is the result of expression of proteins intimately involved in the replication of the viral genome. Certain papovaviruses and adenoviruses share the feature of specifically blocking the function of tumor suppressor genes. Apart from the alteration of the expression of resident cellular growth control gene, the activation of a *c-onc* is also accomplished by virus induced chromosomal translocations. This type of activation is thought to be a factor in carcinogenesis by HPV. An example of further alterations involved in the development of cancer can be seen in the induction of Burkitts lymphoma by EBV. An important step involves the translocation of the cellular *myc gene*, which controls transcription, into the immunoglobulin gene region on another chromosome. This is accomplished by the integration of EBV DNA into a specific site in a cellular chromosome. But it is this translocation, along with the continued stimulation of the immune systems by the co-carcinogen and the ability of EBV to inhibit apoptosis of specific cells that contribute to the development of cancer. Continued stimulation of target cell to divide can result in response to specific tissue damage, as is the case in the formation of hepatic carcinoma as the result of chronic hepatitis B infections. This mechanism is thought to be the basis for chronic hepatitis C virus infection induced liver cancers.

Table 24.3. Oncogenes in different oncogenic viruses

Oncogene	Location	Function
Oncogenes present in DNA Viruses		
E1A	Nucleus, cytoplasm	Regulates transcription
E1B		
PV-ST	Cytoplasm	
PV-MT	Plasma membrane	Binds and stimulates $pp60^{c-STC}$ and $pp62^{c-yes}$
PV-LT	Nucleus	Initiates DNA synthesis and regulates transcription
SV40-ST	Cytoplasm	
SV40-LT	Nucleus, plasma membrane	Initiates DNA synthesis, regula-tes transcription and binds p53
Oncogenes present in retroviruses		
ab1	Plasma membrane?	Tyrosine protein kinase
erb A	Cytoplasm	Thyroid hormone receptor
erb B	Plasma and other membranes	EGF receptor Tyrosine protein kinase
ets	Nucleus	
fes	Plasma membrane	Tyrosine protein kinase
fgr	id	id
fms	id	CSF-1 receptor (tyrosine protein kinase)
fos	Nucleus	
fps (see fes)		
kit	Membranes	Tyrosine protein kinase
mill/raf	Cytoplasm	Serine/threonine protein kinase
mos	id	Serine protein kinase
myb	Nucleus	
myc	Nucleus	
ras	Plasma membrane	GTP-binding protein
raf (see mil)		
rel	Cytoplasm	
ros	Cytoplasm	Tyrosine protein kinase
sis	Cytoplasm and secreted	PDGF subunit
src	Plasma membrane	Tyrosine protein kinase
ski	Nucleus	
yes		Tyrosine protein kinase

Carcinogenesis can also occur in the complete absence of any introduction of viral genetic material into the growth-transformed cell. This is the basis for the theory that some viruses can cause cancer by a true hit and run mechanism, that is, one where the virus-mediated damage to the cell does not require the cell to have ever encountered a virus gene or genetic element.

Further reading

Hoppe-seyler F and Butz K (1995). Molecular mechanisms of virus induced carcino-genesis : the interaction of viral factors with cellular tumor suppressor proteins. *J. Mol. Med.,* 73 (11) : 529.

Laimins LA (ed.) (1996). Viruses in human cancers. *Semin. Virol.* 7 : 293.

Nevins, JR (2001). Cell transformation in viruses. In Fields, Virology 4th Edn. Knipe DM *et al.* (Eds.) Lippincott Williams-Wilkins, New York.

Reddy, EP, Shalka AM and Curran T (eds.) (1988). The Oncogene Hand Book. Amsterdam Elsevier.

Wisdom, R (1999). Transformation-Animal viruses. In Encyclopedia of Virology Vol. III Granoff A and Webster RG (Eds.) Academic Press New York.

25

PREVENTION AND CONTROL OF ANIMAL AND HUMAN VIRAL DISEASES

A. VACCINES

The development of vaccines against viral diseases is undoubtedly one of the great achievements of human endeavours. Eradication of the smallpox from globe is the best success story of the vaccines which became a torch-bearer example for launching the programmes to eradicate other viral diseases. Mortality rates due to a number of viral diseases decreased dramatically with the administration of vaccines. It is encouraging to note that many more dreaded diseases like measles, are rapidly disappearing because of effective vaccines. The world health organization (WHO) has targeted poliovirus for eradication by 2003 and measles virus shortly thereafter. Significant progress with poliovirus has already been achieved. Vaccination is not only effective but at the same time cost effective both at individual and community level. The recent pulse polio programme taken up in India under the aegis of WHO stands as a testimony for its accessibility and affordability to any category of the human community. Now serious efforts are being made to contain COVID-19 pandemic through vaccination and some success has also been achieved.

1. HISTORICAL ASPECTS

Notwithstanding the unrecorded and unscientific approaches for several centuries, scientifically speaking the science of vaccinology was borne in 1798 with the discovery of vaccination against small pox by **Edward Jenner.** **Edward Jenner** (1749-1823), a naturalist and a country doctor experimentally tested the folklore based notion that human infection with a mild cattle disease called cowpox would protect infected individuals against the human small pox. He put this idea to test on 14 May, 1796, when he injected pus from a cowpox lesion on the finger of milkmaid Sarah Nelmes under the skin of an 8-year old James Phipps, a healthy young boy. Phipps developed a fever and a lesion typical of cowpox at the site of infection. Jenner then deliberately infected Phipps with small-pox two weeks later, and the boy survived this potentially lethal challenge. The next practical vaccine for viral disease did not appear until almost 100 years later when **Louis Pasteur** (1822-1895) prepared a rabies vaccine and introduced the term **vaccination** (vacca = cow) in honour of Jenner's pioneering work. However, Pasteur prepared rabies vaccine by a different approach. The rabies pathogen was attenuated by growing it in an abnormal host, the rabbit. After infected rabbits had died, their brain and spinal cords were removed and dried. After successful testing the mixture on dogs, Pasteur tested it on a nine year old boy Joseph Meister, who had been bitten by the rabid dog. Since the boy's death was certain and absence of treatment, Pasteur agreed to try vaccination. Joseph was injected 13 times over the next 10 days with increasingly virulent preparations of the attenuated virus. He survived and the entire episode was etched out permanently in the annals of vaccinology.

Despite of early breakthroughs in the development of antiviral vaccines, the further progress in vaccine technology remained slow largely because viruses were difficult to identify, propagate and study. Indeed, the next antiviral vaccines (against yellow fever and influenza viruses) did not appear until the mid 1930s. In the early years, the production of viral vaccines was relied on infected animal organs and tissues as sources of antigens. The decade of 50's brought Enders cell culture a break through that made possible a number of human and animal virus vaccines. Intensive research during the last 70 years, brooded a variety techniques and models for the large scale production of a variety of vaccines against viral diseases. The saga of vaccine production technology pioneered by classical works of Jenner and Pasteur spans through conventional vaccines like live attenuated and killed vaccines, modern recombinant vaccines and to a new generation edible plant vaccines that is going to be a reality in 21st century.

2. WHAT IS A VACCINE?

While we want to provide a reasonably good definition, antigenic properties of the virus and immunological aspects of the host must be considered. Infact, the effective vaccination results depend on eliciting the sustainable immune response in the host without its own detriment. Consequently, virus factors, host factors, their interactions and ultimate response must be incorporated in the definition.

A vaccine is a collection of immunological determinants which are presented to the immune system as killed or live antigens which provoke protective immune response. The extension of the definition is that a vaccine renders the immunized population immune to subsequent infection to that specific or related pathogens. The smallest unit of antigenicity is known as the **antigenic determinant or epitope**. A number of properties have been identified which make a substance antigenic but the exact basis of antigenicity is still not clear. Three components of immune system participate in immune response against vaccines. These are: antigen presenting cells such as macrophages, B cells that make antibodies and cytotoxic T cells which generate cell mediated immune response. Vaccines to be most effective must induce immediate effector responses in addition to long term memory B and T cells capable of rapid anamnestic protective recall on exposure to the specific antigen.

3. IDEAL CHARACTERISTICS OF A VACCINE

A vaccine to be administered on a mass scale must be safe, efficacious and practical. From safety point of view, it must not cause disease and side effects must be minimal. The vaccinated individual must be protected from illness due to pathogen by eliciting innate, cellular and humoral responses. The protection provided by a vaccine must also be long-term, lasting many years. The other practical requirements for an effective vaccine include stability, ease of administration and low-cost.

Vaccines have been in use for the past more than half century. Conventional vaccines like live attenuated vaccines, inactivated or killed vaccines and subunit vaccines have been successfully used for many years and still used even today. However, the traditional vaccines which are currently available for veterinary and medical use are inadequate for reasons of safety, efficacy and cost effectiveness. Modern molecular methodogies have provided hope for designing better vaccines which will be devoid of the problems associated with the existing conventional vaccines. Recombinant DNA (rDNA) technology has heralded a new era in the science of vaccinology.

4. CONVENTIONAL VACCINES

i. Live attenuated virus vaccines

Virulent viruses can be made less virulent (attenuated) by growing them in cells other than those of the normal host or propagating them at non-physiological temperatures. The mutants arising from such selective condition, and which can propagate better than wild types are isolated, purified and tested and less pathogenic strains are selected. These mutants are good vaccine candidates because they no longer grow well in the natural host but replicate enough to elicite immune response. Live attenuated viral vaccines currently used include poliovirus vaccine, measles virus vaccine, measles-mumps-rubella vaccine, rotavirus vaccine, adenovirus vaccine and small poxvirus vaccine. These can be administered orally or by parenterally.

Though in use for quite some time, attenuated virus vaccines have inherent problems. For example, despite reduced capacity of attenuated viruses to spread in the vaccinated individuals, some viruses can be shed and can then infect unvaccinated individuals. Similarly, a serious problem is that live attenuated viruses may revert to virulent form as in the case of polio. The probability of reversion is more in RNA viruses. Such reversion is one of the main obstacles to develop effective vaccine which is often compared with the emergence of drug resistant mutants.

ii. Inactivated or killed virus vaccines

These vaccines also referred to as non-infectious vaccines, are made by inactivating the infectious viruses after growth in tissue culture or eggs or by using protein components of viruses only. Killed vaccines approach has the advantage of eliminating concerns about the infectious component of attenuated live virus. Generally formalin is used to inactivate the virus. A major concern for this type of vaccines is poor immunogenicity. To circumvent this problem alum is used as an adjuvant to provide the amplification of adaptive immunity. Killed vaccines are licenced for influenza, polio, hepatitis A, hepatitis B, rabies, Japanese encephalitis virus and tick-borne encephalitis (Table 25.2). Since inactivated vaccines are not effective in inducing memory, host responses as those of natural infection or live viral vaccines, the dose regimens of these vaccines incorporate 'booster' doses to assure the long-term persistence of virus specific immunity.

5. NOVEL VIRAL VACCINES

The advances achieved in molecular biology during the past four decades have generated opportunities for making novel viral vaccines. Molecular techniques are already being implemented to improve licensed vaccines and to produce new vaccines. Progress is also being made in the creation of novel adjuvants such as cytokines or immunostimulatory DNA sequences that modulate the host response to enhance antiviral immunity. Molecular approaches are being employed to create new or improved vaccines which include :

i. Synthetic vaccines

ii. Recombinant subunit vaccines

iii. Genetically altered live vaccines

iv. Vectored vaccines

v. DNA vaccines

vi. mRNA vaccines

vii. Plant and plant viruses based vaccines.

i. Synthetic vaccines

Studies carried out in the past two decades have revealed that small synthetic peptides having sequences of immunogenic epitopes on a protein will react with antibody to the intact native antigen and in some cases neutralize the biological activity. Anderer and Schlumberger (1965) demonstrated that chemically synthesized peptides corresponding to protein fragment would also produce the virus neutralizing antibody. This observ-ation led to an extensive research on synthetic antigens. Sale *et al.* (1992) demonstrated that a synthetic fragment of 20 amino acids corresponding to 89-108 of the coat protein of MS2 (a bacteriophage) elicited antibodies which reacted with the intact virus particle. These initial observations provided a ground for the subsequent exploration of peptide immunogen. The identifi-cation of genes encoding immunogenic proteins and availability of nucleic acid sequences allowed derivation of amino acid sequence of the immunogenic peptides.

There are several advantages with synthetic peptide vaccines : (i) they possess indefinite shelf-life (ii) would have precise composition, (iii) no possibility for the adventitious presence of live virus and (iv) easy to handle. A number of diseases have been targeted for synthetic vaccines which include FMDV, influenza and hepatitis B. Though significant protective immune responses have been achieved in experimental conditions, the synthetic vaccines are far from the commercialization mainly due to poor immuno-genicity, short-term immunity and high cost of vaccination. Now the focus is on improving the immunogenicity by incorporating B and T lymphocyte epitopes and better adjuvants in the synthetic peptides.

ii. Recombinant subunit vaccines

The understanding that isolated proteins of viruses can provoke protective immune response, led to application of genetic engineering in cloning and expression of critical viral genes in prokaryotic eukaryotic expression vectors. Thus, the immunogenic peptides of the viral origins could be expressed in bacteria and yeasts. Recombinant proteins can also be produced in cultured animal cells. The concept of a subunit vaccine containing only those immunogenic components necessary to elicit a protective response and excluding others not necessary for immunity, has several attractive features. Subunit vaccines lack infectivity and hence no complication of infection arising out of vaccination in immunocompromised individuals. Subunit vaccination eliminates pyrogenic, allogenic immuno suppressive and other undesirable reactogenicity. These vaccines also exhibit a great deal of stability.

The first successful recombinant vaccine was produced against hepatitis-B. This has paved the way for the development of other recombinant peptide vaccines. The other recombinant vaccine which is in advanced stage of development is against parvovirus B19. Though the recombinant subunit vaccines are quite safe, there are still considerable problems like poor immunogenicity, high cost of production, short term immunity etc. Until these problems are tackled effectively, it is unlikely that the recombinant subunit vaccines will hit the market in near future.

iii. Genetically altered live vaccines

Conventionally, attenuated or naturally found avirulent strains of viruses have been used to develop live vaccines. Now with the development of rDNA technology, it has become possible to attenuate a virus by deleting or altering the virulent genes. This kind of gene manipulation reduces its virulence or makes it unable to replicate completely and thus make a live vaccine with precisely defined modifications. A live simian immunodeficiency virus (SIV) vaccine, attenuated by deleting *nef* gene exhibited the most impr-essive protection against SIV in macaques. Investigations are in progress to generate live attenuated vaccines for the use in human including improved polio and dengue vaccines. Recently, Pirkin *et al.* (1997) demonstrated the feasibility of using genetically modified live vaccine against influenza A. Genetically altered live vaccines could be used for veterinary purpose but for human, risk-benefit factors are to be considered. The clinical trials of such products in human have not yet been carried out.

iv. Vectored vaccines

Vectored vaccine approach allows expression of heterogenous genes into an avirulent viral/bacterial organism which are harmless to vaccinated animals and induce cellular and humoral immune response against the foreign product. This approach also enables expression of more than one foreign genes encoding immunogenic proteins, thus paving the way for development of multivalent recombinant vaccines against human and animal viral infections. A wide range of viruses and bacteria have been used for expression of relevant foreign genes. The most extensively used vector has been vaccinia virus.

Vectored vaccines against viral diseases, are generally based on infectious, semi infectious viral/bacterial vectors into which the genes of interest have been cloned. Vaccinia virus has been used as a vector for development of vaccine against rabies and rinderpest. A number of relevant genes of other viruses have also been expressed in a variety of vectors and their vaccine potential has been studied extensively.

v. DNA vaccines

A relatively recent approach in vaccine development is based on the observation that when plasmid containing suitable promoters and the genes encoding immunogenic proteins are injected into tissues, the gene may be expressed and generate humoral and cellular immune responses. The same response can be generated by inoculating relevant nucleic acid into the skin or mucosa. This approach has generated a great deal of commercial interest due to several advantages.

Delivery of DNA vaccines requires expression of immunogenic proteins in tissues accessible to the immune system such as muscle or skin or mucous membranes. For effective transfection and expression of the gene within these tissues, it is imperative that the introduced DNA is supercoiled and it includes strong tissue specific promoter and the transcribed mRNA has poly A tail to ensure its stability in the host cell. Like plasmid DNA, naked mRNA encloding immunogenic protein can also be directly inoculated into host to induce immune response against the protein product. Though the nucleic acid vaccine approach appears very attractive, there are a number of concerns, including the possibility of integration of the nucleic acid, persistence and immunotolerance, which need to be addressed.

As a reference and comparison, a list of different types of vaccines developed for COVID-19 is presented in Table 25.1.

vi. Plant and plant viruses based vaccines

The recent developments in plant biotechnology have opened new vistas for the development of edible plant vaccines against human and animal diseases. The results of recent studies have suggested that genetically engin-eered plants and plant viruses could be used as vaccines. In recent years, several attempts have been made to produce various antigens and antibodies in plants. Antigens and antibodies expressed in plant can be administered orally as any edible part of the plant. The vaccine is fed raw to experimental animals or humans to prevent possible denaturation during cooking, and avoid cumbersome purification protocols. Thus, plants like tomato, banana and cucumbers are generally the plants of choice. Virus based vectors can also be used to express the gene transiently to develop the products in a short period.

Table 25.1. Different types of Covid-19 vaccines within WHO Emergency Use List (EUL)

S.no	Manufacturer	Name of the vaccine	Platform
1	Pfizer	BNT16b2/COMIRNATY Tozinameran (INN)	Nucleoside modified mRNA
2	Astra Zeneca	AZD1222	Recombinant ChAdOx1 adenoviral vector encoding the spike protein antigen of the SARA-CoV-2
3	SK BIO	AZD1222	Recombinant ChAdOx1 adenoviral vector encoding the spike protein antigen of the SARA-CoV-2
4	Serum Institute of the India	Covishield (ChAdOx1-nCoV-19)	Recombinant ChAdOx1 adenoviral vector encoding the spike protein antigen of the SARA-CoV-2
5	Sinopharm/BIBP	SARS-CoV-2 Vaccine (Vero cell), Inactivated (InCoV)	Inactivated, produced in Vero cells
6	Sinovac	SARS-CoV-2 Vaccine (Vero cell), Inactivated	Inactivated, produced in Vero cells
7	Moderna	mRNA-1273	mRNA-based vaccine encapsulated in lipid nanoparticle (LNP)
8	Janseen	Ad26 COV2.S	Recombinant, replication-incompetent adenovirus type 26 (Ad26) vectored vaccine encoding the (SARS-CoV-2)Spike(S) protein
9	The Gamaleya National Center	Sputnik V	Human Adenovirus vector-based Covid-19 vaccine

10	CanSinoBio	AdS-nCoV	Recombinationt Novel Coronavirus Vaccine (Adenovirus Type 5 vector)
11	Vecor State Research Centre of Virology and Biotechnology	EpiVax Corona	Peptide antigen
12	Zhifei Longcom, China	Recombinant Novel Coronavirus Vaccine (CHO Cell)	Recombinant protein subunit
13	IMBCAMS, China	SARS-CoV-2 Vaccine, Inactivated (Vero Cell)	Inactived
14	Sinopharm/WIBP[2]	SARS-CoV-2 Vaccine, Inactivated (Vero Cell)	No pre-submission meeting yet
15	Novavax		No pre-submission meeting yet

Plant system has the capability of producing any vaccine in large amounts and in a less expensive manner. However, the purification of the product may require the use of existing or even more cucumbersome procedures. Attention, therefore, has been paid to mainly those antigens that stimulate mucosal immune system to produce secretory IgA (s-IgA) at mucosal surfaces, such as gut and respiratory epithelia. Thus, an antigen produced in the edible part of a plant can serve as a vaccine against several infectious agents which invade epithelial membranes. The first report of the production of edible vaccine (a surface protein from *Streptococcus*) in tobacco appeared in 1990 in the form of patent application. Subsequently, a number of attempts were made to express various antigens in plants.

One of the utilities of producing antigens in plants in large amounts is in treatment of autoimmune diseases like diabetes mellitus which involve production of antibodies against glutamic acid dehydrogenase (GAD) and insulin, leading to destruction of insulin producing pancreatic cells. The antigens targeted for autoimmune response can be fed to the animals to induce immune tolerance.

Further reading

Arnon, R and R Levi (1966). In : Novel strategies in design and production of vaccines (Kohan, S and Shafferman, A eds.), Plenum Press, New York.

Sharma, AK, A Mohanty, Y Singh and AK Tyagi (1999). Transgenic plants for the production of edible vaccines and antibodies for immunotherapy. *Curr. Sci.*, 77(4) : 525.

Prasad, G and Minakshi (1997). Trends in antiviral vaccines. *Ind. J. Microbiol.,* 37: 165-176.

Plotkin, SA and WA Orenstein (1999). Vaccines, 3rd edition, WB Saunders Co., Philadelphia, PA.

Fields, BN, DM Knipe and PM Howley (2001). Fields Virology, 4th Edn., Lippincott-Raven Press, Philadelphia, PA.

Arwin, AM (2003). Viral vaccines. In: The Desk Encyclopedia of Microbiology, M Schaechter (ed.), Elsvier, Academic Press.

Pollard, AJ and EM Biker (2021) A guide to vaccinology : From basic principles to new developments. *Nature Reviews Immunology* 21:83-100

Rodriques C(2020) Impact of vaccines; health, economic and social perspectives *Front. Microbial.* 14:1-15

Plotkin SA (2003). Vaccines, vaccination and vaccinology *The Journal of Infectious Agents* 187(a) : 1349-1359.

B. INTERFERONS

Since long back virologists recognised the fact that exposure of cells to one kind of virus often rendered the cells immune to infection by another related or even unrelated virus. Similarly simultaneous occurrence of two different childhood viral diseases were never observed. For a long time the reasons for such type of phenomena remained unanswered. It was *Isaacs and Lindenmann* in 1957 unveiled the mystery surrounding these viral interference. They observed that chicken cells exposed to heat-inactivated influenza virus released a `factor' which, when added to fresh cells reduced their capacity to produce a new virus. As the factor was interfering the viral replication, they aptly termed this factor as **interferon** (IFN). Since the seminal discovery of interferons by Isaacs and Lindenmann, intensive investigations have been carried out with regard to biological effects including their action as antiviral agents. Interferons have emerged as some of the most active biological substances. Their study revealed a spectrum of effects on cells, contributing to our understanding of basic cellular immuno-and developmental biology. It has now been proved that IFN system represents the first line of defense against viral diseases and also many other infectious diseases. Thus they have been identified as potential therapeutic agents for many infectious diseases. As a part of a complex network of cytokines, IFNs display attributes beyond those of antiviral agents. With the availability of recombinant interferons, in recent years, the scope and magnitude of research on lFNs has broadened.

1. NATURE AND SOURCE OF INTERFERONS

Interferons represent the prototype of a family cytokines. Chemically, these are peptides/glycopeptides with molecular mass ranging between 15 and 25 kD. They are produced on induction by viral or non viral inducers. Most of the interferons are synthesized as inactive, larger precursor molecu-les containing N-terminal signal peptide, removal of which produces the active molecule. Interferon by itself has no direct action on viruses but it acts on other cells of the same species rendering them refractory to virus infection. Thus, interferons are species specific, in that interferon produced by one species can protect only cells of the same or related species against viral infections but not cells of unrelated species. Interestingly, the activity is not viral specific. Interferons induced by one virus (or even by non viral inducers) can confer protection against infection by the same or unrelated viruses. However, viruses vary in their capacity to induce interferon. Cytocidal and virulent viruses are poor inducers, whereas avirulent viruses are good inducers of interferons. RNA viruses are better inducers than DNA viruses.

A. Alick Isaacs (1921-1967)

B. Jean Lindenmann (1924-2015)

Interferon production is increased by increasing the temperature upto 40°C and is inhibited by steroids and increased oxygen tension. Interferon synthesis begins within about an hour of the induction and reaches peak levels in 6-12 hours. The promptness of the interferon induction, much quicker than the antibody response, suggests that interferons may play a primary role in host defense against viral infections. Interferons are inactivated by proteolytic enzymes. They resist heating at 56-60°C for 30-60 minutes and are stable over a wide range of pH (2-10). However gamma IFN is labile at pH 2.0. Interferons are poorly antigenic. Interferon assay is based on its biological activity, such as ability to inhibit plaque formation by a sensitive virus. The potency of IFN is expressed as International Units (IU) per ml. Interferons are produced by both immune cells and non-immune cells. Interferon producing immune cells include leukocytes and T lymphocytes. Non-immune cells include fibroblasts, endothelial cells, embryonic trophectoderm and pre-embryonic trophoblast.

2. CLASSIFICATION

Interferons are distinguished by several biological and molecular attributes such as the type of cells that express them, antigenicity, the 5' regulatory regions of the genes, structural coding sequences, three dimensional structure, cellular receptors and the signal transduction pathway. Broadly, interferons are categorized into three major types : Type I IFNs, type II IFNs and type IFNS III.

i. Type I interferons

These interferons exhibit certain similar biological and biochemical properties. They share a common ancestry. The hallmark of all these interferons is their stability to heat (70°C, 30 min) and acid (pH 2.0). They are produced by all cell types.

IFN-α: IFNs–α are produced primarily from leukocytes upon stimulation by dsRNA or viruses because of dsRNA formed during their replication. The primary proteins of IFN-α family contain signal and mature protein sequences of 23 and 166 amino acids respectively and produce a molecule of 19.5 kD. Disulfide bonds contribute to the heat and pH stability that characterize type I IFNs.

IFN- β : Interferons of this type are produced primarily by fibroblasts and epithelial cells following induction by dsRNA or viruses. The fully processed, glycosylated and secreted molecule of human IFN-B (HUIFN-B) contains 166 amino acids and has a molecular weight of 23 kD. The nonglycosylated molecule produced in *E. coli* has a molecular weight of 20 kD and is fully functional.

IFN – ω : It is similar to IFN- α / β in several respects. However, it is antigenically different from other type I IFNs. The mature proteins of IFN-ω contain 172 residues, six amino acids longer than IFN- α / β. IFN-ω makes upto 30% of the IFN produced by virus induced peripheral blood mononuclear cells. IFN-ω, like IFN- α / β induces an antiviral state, down regulates the cell growth and modulates the immune response. It is widely distributed among mammals (except dog).

IFN-*T* : This interferon is produced in large quantities from embryonic trophectoderm. However, its inducer is not known. Mature protein of IFN-*T* contains 172 amino acids. It is believed to be originated from IFN-ω and share upto 75% sequence homology with it. However, it is immunologically different from IFN-ω. The unique feature of this interferon is that it also functions as a reproductive hormone.

IFN- δ : It is a novel truncated type I IFN that is transiently expressed in porcine trophoblast during implantation in the uterus. The mature protein contains 149 amino acids and antigenically different from other type I IFNs. It is known to be induced by certain stress physiological conditions rather than by viruses.

ii. Type II interferon (IFN- γ)

It is also known as immune IFN. This is produced by T lymphocytes following their exposure to mitogens, specific antigens or various cytokines. The mature protein contains 143 aminoacids. The 16.8 kD mature protein is glycosylated at one or two sites. The acid and heat lability of IFN- γ distinguish it from type I IFNs.

iii. Type III interferons:

Signal through a receptor complex consisting of ILIOR2 and IFNL R1. Although discovered more recently than type I and type II IFNS, recent information demonstrates the importance of Type III IFNS in some types of virus and fungal infections.

In general type I and type II interferons are responsible for regulating and activating the immune response. Expression of type I and type II IFNs can be induced in virtually all types upon recognition of viral components, specially nucleic acids, by cytoplasmic and endoplasmic receptors, whereas type I interferon is induced by cytokines such as IL-12 and its expression is restricted to immune cells such as T cells and NK cells.

3. INDUCTION OF INTERFERONS

Many cell types are capable of producing IFN when appropriately induced by viruses, dsRNA, mitogens or other cytokines. Most RNA viruses are good inducers of IFN because they generate dsRNA during replication. Some DNA viruses also induce IFN because they generate dsRNA through convergent transcription of their genome. Even organisms like mycoplasmas, rickettsia, bacteria and protozoa can induce IFN, albeit – less efficiently.

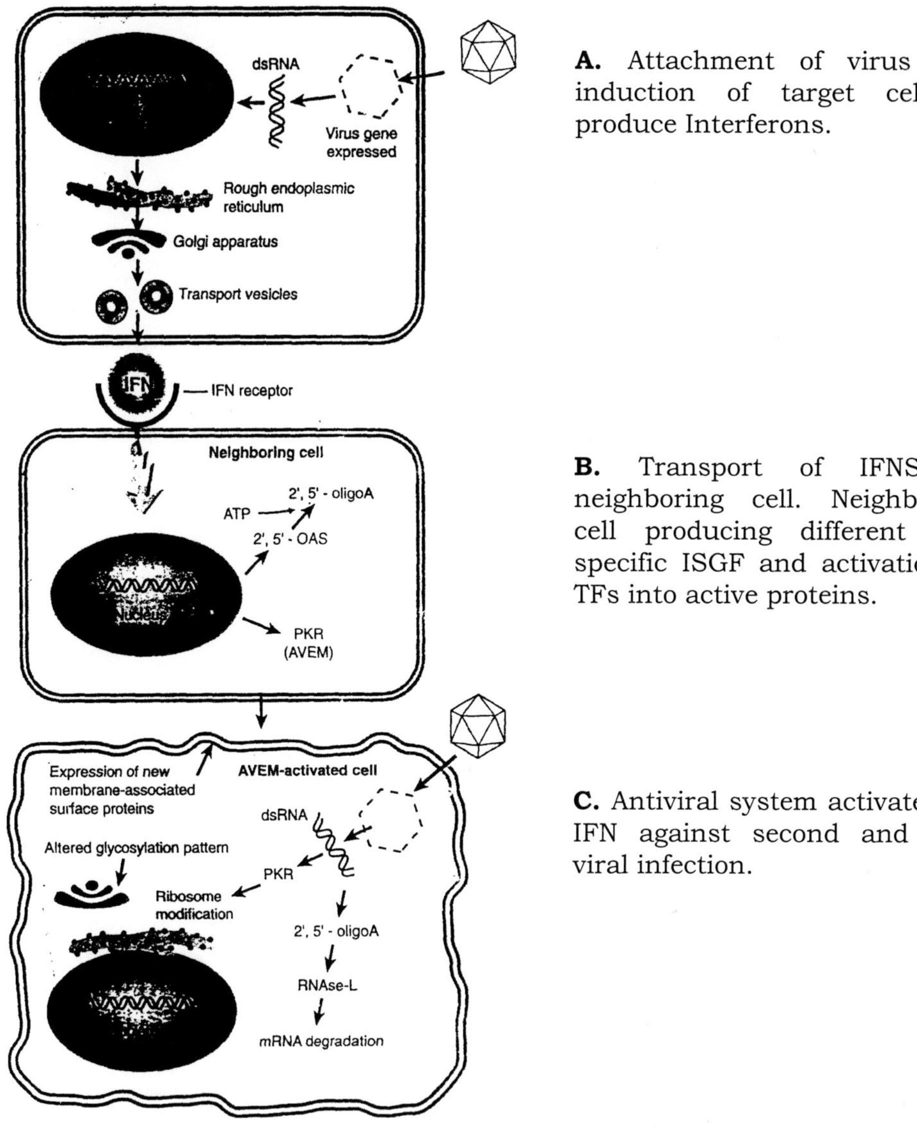

A. Attachment of virus and induction of target cell to produce Interferons.

B. Transport of IFNS to neighboring cell. Neighboring cell producing different and specific ISGF and activation of TFs into active proteins.

C. Antiviral system activated by IFN against second and new viral infection.

Fig. 25.1. The cascade of events leading to expression of interferon (IFN) and induction of the antiviral state in neighbouring cells.

In the fully competent but uninduced cell, expression of the interferon genes is tightly repressed by several protein factors bound to the 5' promoter region of the genes that act cooperatively as negative regulators. Consequent to viral infection, once dsRNA appears in the cytoplasm it is recognised by a cellular protein. Due to binding to helical

RNA, this cellular protein, protein kinase becomes activated. For HuIFN-B this activation leads to phosphor-ylation and subsequent proteolysis of inhibitors of ubiquitous transcriptional factors. Transcriptional factors salvaged from inhibitors become active and act positively allowing the expression of genes. This chain of events begins within minutes after a cell is exposed to an IFN inducer. This highly regulated process results in required yield of IFN.

4. INDUCTION OF ANTIVIRAL SYSTEM

Within hours after virus infection, newly produced and released IFN circulate in the body. The cells most proximal to the initially infected cell, and hence most vulnerable to infection, receive IFN first and in the highest concentrations. The binding of IFN to specific receptors on the cell surface starts the action phase of the IFN culminating in the development of broad spectrum of antiviral state. This entire process can be divided into three stages.

i. Receptor – IFN binding

ii. Signal transduction pathway

iii. Antiviral system activated by IFN action.

i. Receptor – IFN bindig : This first event, initiates a cascade of events that terminate in an antiviral state. The target cell consists of a number of receptors, specific sites to which IFNs bind with high affinity (Fig. 25.1). Receptors, the transmembrane proteins, distinct for type I IFN and type II IFN are responsible for species specificity. HuIFN-γ binds as a homodimer to a specific transmembrane receptor, IFNGR that contains two subunits : IFNGR-1 an affinity binding molecule and IFNGR-2 an accessory protein required for intracellular signaling.

ii. Signal transduction pathway : Within minutes after the ligand IFN binds to its receptors, a series of signal responsive events convert latent cytoplasmic transcription factors into active proteins, IFN stimulated gene factors (ISGF), including a family of IFN regulatory factors (IRF). These factors bind to specific DNA sequences namely: (i) the IFN stimulated response elements (ISRE) for type I IFNs, and (ii) the gamma activated sequences (GAS) for type II IFNs (Fig. 25.1). This is followed by transcription of the myriad IFN-stimulated genes (ISG) whose products result in the broad range of antiviral and other effects that characterize IFN action.

iii. Antiviral systems activated by IFN action : Following the successful signal transduction there is an enhanced or new expression of over 100 genes. Gene activation generally does not require *de novo* protein synthesis, enabling the cells to respond rapidly to IFN. Some genes are induced by either type I or II IFNs, other exclusively by one or other. Some of the proteins expressed during the antiviral state are: 2-5 oligoadenylate synthetase / 2-5 (A)/ Rnase L; PKR, a dsRNA-dependant protein kinase; Mx protein; ISG-15; P202; indoleamine 2, 3-dioxygenase etc.

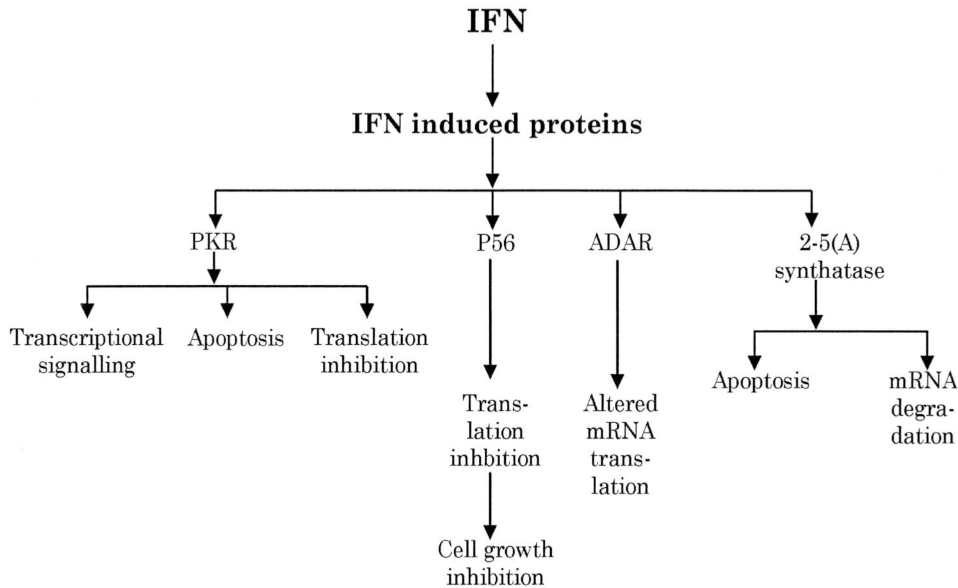

Fig. 25.2. Interferon-induced biochemical pathways affecting cell growth and metabolism.

Table 25.2 Pharmaceutical forms of interferons

Generic name	Brand name
Interferon alpha 2a	Multiferon
Interferon alpha 2b	Roferon A
Human leukocyte Interferon alpha (HuIFN- alpha- Le)	Intron A/Reliferon/Uniferon
Interferon beta 1a liquid form	Multiferon
Interferon beta 1a lyophilized	Rebif
Interferon beta 1a biogeneric (Iran)	Avonex
Interferon beta 1b	Cinnovex
Interferon gamma 1b	Betaseron/Betaferon
PEGylated interferon alpha 2a	Actimmune
PEGylated interferon alpha 2a (Egypt)	Pegasys
PEGylated interferon alpha 2b	Reiferon Retard
Ropeg interferon alpha 2b	Pegentron
PEGylated interferon alpha 2b plus ribavirin (Canada)	Besrenal, Pegetron

5. MECHANISM OF INTERFERON ACTION

IFN activated antiviral system exerts its antiviral action at different steps in the replication of different viruses and by a range of means in different cells. A variety of approaches, including gene array technology have revealed that IFNs induce synthesis of a large number of cellular proteins. The characteristics of few extensively studied IFN-induced proteins are available. Their biochemical and cellular regulatory functions are well understood and many of them play major roles in the antiviral functions of IFN. Some of these are depicted in Fig. 25.2.

6. THERAPEUTIC APPLICATIONS OF INTERFERONS

Many properties of interferons make them ideal candidates for the use in the prophylaxis and treatment of viral infections (Table 25.2). These are non-toxic, nonantigenic, diffuse freely in the body and exhibit a wide spectrum of antiviral activity. The only drawback initially was its species specificity so that interferons produced by nonhuman cells were not clinically useful. This problem was circumvented to some extent by producing interferons from buffy coat leukocytes from blood banks, with NDV or sendai virus as the inducer. Now, human interferons are available in unlimited quantities following their production by recombinant DNA technology. Even so, its initial promise as an antiviral agent has not been fulfilled. Nevertheless, some success has been achieved in the treatment of some viral infectious diseases and cancers.

i. Genital warts infections caused by human papillioma viruses have been treated successfully by intralesional, topical and parenteral therapy with IFN. It is also used as adjunctive therapy in the reoccurence of juvenile laryngeal papillomes following surgery.

ii. Recombinant human IFN- α (r huIFN- α) and IFN-β have shown significant activity against rhinovirus infections in both therapeutic and prophylactic situations.

iii. The progression of lesions from activated latent herpes viruses can be ameliorated, or their initiation prevented by IFN treatments. The added advantage is that IFNs can be used in cases of latent infections and also in immunocompromised people.

iv. Synergistic effects have been reported between IFN and viral specific antibody both in *in vitro* and *in vivo* in case of ocular infections caused by picornavirus.

v. Inspite of their discovery as antiviral agents, the major impetus to develop IFNs as products of biotechnology came from their antitumor activity. The first therapeutic use of IFN by physicians was sanctioned by the Food and Drug Administration (FDA) for the treatment of hairy cell leukemia, probably of viral origin (HTLV-II). This approval is soon followed by clinical use of IFN against hepatitis B virus (HBV), where IFN is the therapeutic agent of choice. IFN-α also emerging as the only available treatment for hepatitis C (HCV).

Further reading

Baron, S *et al.* (eds.) (1992). Interferon : Principles and Medical Applications. The University of Texas Medical Branch at Galveston, Tx

De Maeyer E and De Maeyer-Guignard J (1988). Interferons and other regulatory cytokines. New York : Wiley.

Knipe, DM, Howley PM *et al.* (eds). 2001. Fields Virology Vol I, 3rd Edn. New York Lippincott-Raven Press, Philadelphia, PA

Stark, GR *et al.* (1998). How cells respond to interferons. *Annual Rev. Biochem.* 67 : 227.

Granoff, A and Webster, RG (eds) (1999). Encyclopedia of Virology Vol II, 2nd Edn. Academic Press, New York.

Sen GC and Lengyel P (1992). The interferon system : A birds eye view of its biochemistry. *J. Biol. Chem.* 267 : 5017.

C. ANTIVIRAL AGENTS

Ever since the successful introduction of antibiotics to control bacterial infections, attempts have been made to use chemical agents, both as prophylactic and as therapeutic to control viral diseases. Until recently, it was thought that finding the antiviral agents is remote or non existent. However, only after some 50 years of work by pharmaceutical industry it has become accepted that antiviral chemotherapy is possible like antibacterial chemotherapy. The main reason for the dissapointingly slow pace for the development of antivirals is that the virus multiplication is tied so intimately to certain cellular processes that most antivirals cannot destroy the one without affecting the other. Fortunately, research on molecular biology of viruses has helped to solve the problems of differentiating the metabolism of virus and the host. Enzymes unique to viral replication are now regarded as fundamental to the logical use of antivirals. RNA and DNA polymerases, protein kinase, proteases and thymidine kinase have been identified in most groups of viruses and are considered as good primary targets. Indeed some specific inhibitors of these enzymes have been found.

1. SCREENING OF ANTIVIRAL AGENTS

Traditionally, two fundamentally different strategies have been used in discovering antiviral agents. The first method is to test a large number of compounds or specimens prepared from the culture of microorganisms against a battery of viruses. The second method of screening is to select a known or assumed metabolic pathway of the virus, particularly in relationship to its mode of replication. Obviously, the first one is high-throughput empirical screening (trial and error) and the second one is the design of tailor-made drugs. Though the second strategy appears to be attractive for many reasons, the two methods continue parellelly for the forseable future, with a recent upsurge of interest in natural plant products as potential antivirals.

2. Targets for antiviral therapy

Antivirals are targeted to interfere any stage in the replication of virus right from adsorption, penetration to release, preventing on the whole production of progeny virions. Fortunately, different sites for interference are available in the viral replication cycle. For instance, incase of HIV the following stages in replicative cycle may serve as targets for therapeutic intervention.

i. Binding to target cell

ii. Fusion to target cell

iii. Entry, uncoating and release of RNA

iv. Transcription of RNA to DNA by reverse transcriptase.

v. Accumulation of unintegrated HIV in acute infection.

vi. RNA degradation by RnaseH activity

vii. Migration of viral DNA to nucleus.

viii. Integration of HIV DNA into host DNA

ix. Transcription and translation

x. Gag-pol polyprotein cleavage

xi. Viral budding

xii. Extracellular processing by Gag-pol polyproteins.

3. CLASSIFICATION

A wide range of chemicals have been found to exhibit antiviral activity. However, their mode of actions are different. Based on mode of action, all the antivirals can broadly be divided into the following types.

1. Inhibitors of virus replication
2. Ion channel blockers
3. Inhibitors of attachment/entry of virus into target cells.
4. Inhibitors of protease activation of viral proteins.

3.1 Inhibitors of virus replication

A. Nucleoside analogues

i. Inhibitors of DNA/RNA replication

These are chemical analogs of purine and pyrimidine bases required for the synthesis of nucleic acids. Some of these analogs, or more precisely their corresponding nucleotides compete with their normal counterparts for incorporation into the newly synthesized DNA or RNA. The later thus come to have the analogs and are functionally ineffective though structurally stable. The progeny virus often has incomplete capsids and non-infectious. Some important base analogs are described below (Fig. 25.3).

Acyclovir

Acyclovir [9-(2-hydroxyethoxymethyl) guaninel], a synthetic acyclic purine nucleoside analog, is a selective inhibitor of HSV-1 and –2 and varicella zoster virus (VZV) replication. It is a prodrug - a precursor of the active antiviral compound. It has become the most widely prescribed and clinically effective antiviral drug available to date. **Valaciclovir,** is an improved version of acyclovir. Activation of the drug requires three kinase activities to be present in the cell to convert acyclovir to a triphosphate derivative, the actual antiviral drug. The first kinase activity which converts acyclovir to the monophosphate is catalyzed by viral kinase. The remaining two steps in the triphosphate formation are catalyzed by cellular kinases. The resultant acyclovir triphosphate is incorporated into DNA. As acyclovir lacks 3'-OH group of the sugar ring, the growing DNA chain terminates and further DNA replication is blocked. Further, the viral polymerase has greater affinity for acyclovir triphosphate than does cellular DNA polymerase, resulting in little incorporation of acyclovir into cellular DNA.

Ganciclovir

Ganciclovir [9- (1, 3 – dihydroxy-2-propoxymethyl) guanine] (cytovene), a guanine analog has enhanced activity against herpesviruses and CMV as compared to acyclovir. The mode of action of this drug against HSV-1 is similar to acyclovir. However, in case of CMV the action is mediated by a protein kinase, UL-97, that efficiently promotes the phosphorylation of ganciclovir to its monophosphate.

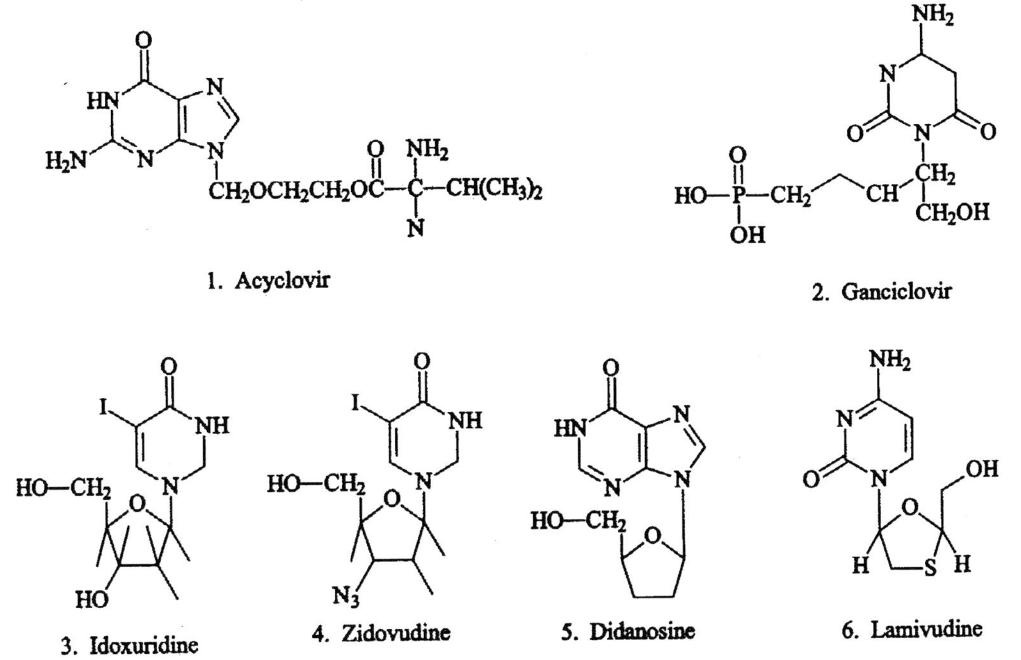

1. Acyclovir

2. Ganciclovir

3. Idoxuridine

4. Zidovudine

5. Didanosine

6. Lamivudine

Fig. 25.3. Some base analogue antiviral drugs

Cidofir

Cidofir, (S)-1-(3-hydroxy-2-phosphonomethoxypropyl) cytosine (HPMPC) is a novel acyclic phosphonate nucleoside analog and is used to treat acyclovir- and foscarnet – resistant HSV infections as well CMV retinitis. The drug has a similar mechanism of action as the other nucleoside analogs but employs cellular kinases to produce the active triphosphate metabolite.

Idoxuridine and triflurothymidine

Idoxuridine (5-ido-2'-deoxyuridine) and triflurothymidine (trifluridine) are analogs of thymidine. When adminstered systematically, these nucleo-sides are phosphorylated by both viral and cellular TK to activate triphos-phate derivatives which inhibit both viral and cellular DNA synthesis. Thus the systemic use of this drug is prevented. Both of these drugs are effective and licensed for treatment of HSV keratitis.

Penciclovir and famiciclovir

Famciclovir [9 – (4-hydroxy-3-hydroxy-methyl-but-1-yl) guanine; Famir] is the diacetyl ester of 6-deoxy penciclovir. It is the prodrug of penciclovir, and when applied orally rapidly converted to penciclovir. It's mode and spectrum of activity are similar to acyclovir. Penciclovir has a good therapeutic activity against both HSV and VZV infections.

Vidarabine

Vidarabine (Vira-A, adenine arabinoside, and 9D arabino furanosyl adenine) is active against HSV, VZV and CMV. It is a purine nucleoside analog that is phosphorylated intracellularly to its mono-, di and triphosphate derivatives. This triphosphate derivative competitively inhibits DNA polymerase of some DNA viruses. Further, vira-A is incorporated into terminal positions of both cellular and viral DNA, thus inhibiting elongation.

Ribavirin

Ribavirin (α – methyl-1-adamantane methylamine hydrochloride) was reported to have broad-spectrum antiviral activity against many DNA and RNA viruses. Despite its long history, its mechanisms of antiviral action is not clear. Ribavirin monophosphate is a competitive inhibitor dehydrogenase and such inhibition leads to reduced GTP pools in the cell. It also inhibits initiation and elongation by virus specific RNA polymerase and interferes with 5' capping of RNA. Now a days, it is widely used for the treatment of COVID.

ii. Inhibitors of reverse transcriptase

Zidovudine

Zidovudine (3' – azido – 2', 3' – dideoxythymidine; Azidothymidine and Retrovir) is a pyrimidine analog with an azido group substituting for the 3' hydroxyl group on the ribose ring. The triphosphate derivative completely inhibits HIV reverse transcriptase and also functions as chain terminator. It also inhibits a variety of other retroviruses. It was the first drug to be licensed for the treatment of HIV and still is used in combination with other drugs.

Didanosine

Didanosine (2', 3' – dideoxyinosine; ddl and videx) is a purine nucleoside with inhibitory activity against both HIV-1 and HIV-2. Didanosine is activated by intracellular phosphorylation. Its conversion to triphosphate derivative is more complicated than that with other nucleoside analogs, because it requires additional enzymes. The triphosphate metabolite is a competitive inhibitors of HIV reverse transcriptase and is also a chain terminator. The spectrum of activity of didanosine is enhanced by synergism with zidovudine and stavudine as well as protease inhibitors.

Zalcitabine

Zalcitabine (2', 3'-dideoxycytidine, ddc and Hivid) is a pyramidine analog which is activated by cellular enzymes to its triphosphate derivative. The enzymes required for activation of zalcitabine are cell cycle independant. Synergy has been reported between zidovudine and zalcitabine as well as with saquinavir.

Stavudine

Stavudine (2', 3' – didehydro, 3'-deoxythymidine, d4T and Zerit) is a thymidine analog with significant activity against HIV-1. Its mechanism of action is similar to that of zidovudine. It is either additive or synergestic with other nucleoside and non nucleoside reverse transcriptase inhibitors.

Lamivudine

Lamivudine is the (-) enantiomer of a cytidine analog with sulfur substituted for the 3' carbon atom in the furanose ring [(-) 2', 3' dideoxy, 3' – thiacytidine]. It shows significant activity against HIV-1, HIV-2 and HBV. The triphosphate derivative is a competitive inhibitor of the viral reverse transcriptase. It is used in the treatment of HIV in combination with other reverse transcriptase and protease inhibitors. It is also licensed for the treatment of chronic hepatitis B.

Abacavir

Abacavir is a carbocyclic synthetic, nucleoside analog. It is phosphory-lated by cellular enzymes to its active metabolite carbovir triphosphate which is an analog of deoxyguanosine–5' –TP. Carbovir TP then inhibits the activity of HIV reverse transcriptase. It is licensed for the treatment of HIV infections in humans.

B. Non-nucleoside inhibitors

Forcarnet

Forcarnet, a pyrophosphate analog of phosphonoacetic acid has potency against all herpes viruses and inhibits the DNA polymerase by blocking the pyrophosphate binding site, inhibiting the formation of the 3', 5' phospho-diester bond between primer and substrate and preventing chain elongation. It also inhibits HBV polymerase and the reverse transcriptase of HIV.

Fomivirsen

Fomivirsen (5' – GCG TTT GCT CTT CTT-30') is the first antisense oligonucleotide licensed for the treatment of CMV retinitis. The drug binds to the mRNA or the immediate early 2 gene of CMV. The drug is approved for patients intolerant to the medications.

Nevirapine

Nevirapine (11-cyclopropyl – 5, 11 – dihydro-4- methyl – 6H – dipyrido [3, 2 – b; 2', 3' – el] is a reverse transcriptase inhibitor of HIV-1. It is metabolized by liver microsomes to hydroxymethyl – nevirapine.

Delavirdine

Delavirdine (1-[5-methanesulfonamido-1H-indol – 2- yl – carbonyl]-4- [3-(1-methylethyl amino) pyridinyl] piperazine) is a second generation bis (heteroaryl) piperazine licensed for the treatment of HIV infection. It inhibits the DNA polymerase, however, the exact mode of action is not clearly known.

Efavirenz

Efavirenz Sustiva [(S) – 6 – chloro – 4 – (cyclopropyl ethynyl) – 1-4 – dihydro – 4 – (trifluromethyl) – 2 H – 3, 1 benzoxazin – 2- one] is a non nucleoside reverse transcriptase inhibitors which can be administered once daily. Activity is mediated predominately by non-competitive inhibition of HIV reverse transcriptase. However, HIV-2, reverse transcriptase is not inhibited by efavirenz.

3.2 Ion channel blockers

Amantadine (1-adamantane amine hydrochloride, symmetrol is a tricyclic amine which is effective against all influenza A variants. Amantadine shows a narrow spectrum of activity. Rimantidine is the α – methyl derivative of amantadine (α-methyl-1-adamantane methyl amine hydrochloride). It is 5-to 10 fold more active than amantadine and has the same spectrum of activity and mechanism of activity. The mode of action of these drugs relates to the influenza A virus M2 protein, a membrane protein which is ion channel for this virus. By interfering with the function of M2 protein, amantadine and rimantadine inhibit the acid-mediated association of the matrix protein from the ribonuclear protein complex with endosomes. This event occurs early in the virus life cycle.

3.3 Inhibitors of attachment/entry of virus into the target cell

Oseltamivir

Oseltamivir [ethyl (3R, a 4R, 5S) – 4 – acetomido – 5 – amino – 3 – (1- ethyl propoxy) – 1 cyclohexene – 1 – carboxylate] is a selective neuramindase inhibitor. It inhibits both influenza A and B virus. The drug inhibits viral replication by targeting the neuraminidase protein via binding in a competitive fashion to the enzyme, rendering the virus incapable of entry and subsequent replication.

Zanamivir

Zanamivir (5-acetylamino – 4 – [amino-iminomethyl (amino] – 2, 6 anhydro – 3, 4, 5- trideoxy – D- glycero – D-galactonon – 2 – enoic acid) is also a neuraminidase inhibitor. It binds competitively to influenza neuramidiase of both influenza A and B. Zanamivir is licensed for the treatment and prevention of influenza A and B.

Pleconaril

Pleconaril is active against many rhinoviruses and enteroviruses. This compound causes a conformational change in VPI proteins so that the virus cannot disassemble. Pleconaril is under investigation for chronic enterovirus infections of the CNS in the immune deficient proteins.

3.4 Inhibitors of protease activation of viral proteins

Saquinavir

Saquinavir is a hydroxyethylamine derived peptidomimetic HIV protease inhibitor. It inhibits both HIV-1 and HIV-2 and synergestic with other nucleoside analogs and other selected protease inhibitors.

Indinavir

Indinavir is a peptidomimetic HIV-1 and HIV-2 protease inhibitor. It inhibits 90% HIV isolates at concentrations of 100 nM. It is recommended in combination with nucleoside analogs.

Other peptidomimetic protease inhibitors include Ritonavir and Nelfinavir

Amprenavir

Amprenavir is a hydroxyethylamine sulfonamide peptidomimetic drug. It acts by binding to the active site of HIV-I protease, preventing the processing of viral gag and gag-pol polyprotein precursors and resulting in the formation of immature non-infectious viral particles. Amprenavir is licensed for the treatment of HIV infections.

Fuseon, Enfuvirtide

T-20 is an inhibitor of fusion of HIV-1 with CD_4 cells that consists of a 36-amino acid synthetic peptide with N-terminus acetylated and the C-terminus is a carboxamide.

4. LIMITATIONS OF ANTIVIRAL DRUGS

Clinical use of the currently available antiviral drugs is limited by toxic side effects. There are also some general limitations inherent in antiviral chemotherapy. First, the more selective the antiviral drug the narrower its antiviral activity spectrum. Second, since antiviral drugs target steps in virus replication, the latent phases characteristic of some viral (i.e. herpesviral) infections are not amenable to chemotherapy. This is particularly relevant for herpesvirus and retrovirus infections. Thus, eradication of latent virus infections is not feasible currently. Third, antiviral drug treatment should be started early, before irreversible tissue damage occurs. Such timely treatment is not possible without early and accurate diagnosis, which is difficult for many viral infections (such as infections of the respiratory tract). Fourth, perhaps inevitably for a specific antimicrobial agent, there is the risk of emergence of drug-resistant virus strains. This has been most dramatically demonstrated in AIDS patients from whom drug-resistant human immunodeficiency virus strains have been isolated against virtually all the drugs that have been tried so far. In the same patients treated with acyclovir for HSV or VZV infections, acyclovir-resistant HSV or VZV strains have been detected. The fact that these HSV and VZV mutants, as well as the resistant HIV strains, retained full susceptibility to other antiviral agents is reassuring. These findings indicate that on the one hand, combination therapy should be studied, particularly for the prevention of HIV drug resistance, and that, on the other hand, attempts to develop new antiviral agents should continue unabated.

Further reading

Abu-ata, O, Slim, J, Perez, G and Smith, SM (2000). HIV therapeutics : Past, present and future. *Adv. Pharmacol.* 49 : 1-40.

Balfour, HH, Jr (1999). Antivirals (non-AIDS) *N. Engl. J. Med. 340* : 1255-1268.

Declereq, E (2002). New anti-HIV agents and targets. *Med. Res. Rev.,* 22 : 531-565.

Emery, S and Cooper, DA (2003). Antiviral agents. In ``Antibiotic and chemotherapy". R.G. Finch *et al.,* (eds.) pp. 473-493. Churchill Livingstone, Philadelphia.

Galasso, G, Whiteley, RJ and Merigan, TC (Eds.) (1997). ``Antiviral agents and viral diseases of man" Lippincott – Raven, New York.

Whitley, RJ (2003). Antiviral agents. In ``The Desk Encyclopedia of Microbiology" Schaechter, M. (ed.) Elsevier Academic Press.

Inouye, RT, Panther, LA, Hay, MH and Hammer, SM (2002). Antiviral agents. In Clinical Virology. Richman DD *et al.* (eds.) 2nd Edn., pp 171-242, ASM Press, Washington, DC.

PART FIVE

MISCELLANEOUS VIRUSES

&

SUBVIRAL AGENTS

26

PHYCOVIRUSES

In comparison to other viruses, investigations on algal viruses were much delayed. Reports on algal viruses started appearing from early 1970's. Now, viruses or virus-like particles (VLPs) have been reported in about 44 taxa of eukaryotic algae. However, most of these reports described isolated accounts of microscopic observations. The virus particles were not characterized because they were difficult to obtain in reasonable quantities. Chlorella viruses are the first viruses to be characterized. Later certain marine algae are also described (Table 26.1). The algal viruses are similar in some respects to other large DNA viruses, in particular African swine fever virus, iridoviruses, poxviruses and herpes viruses. However, they also have some unique and interesting properties.

Taxonomy

The algal viruses are polyhedral (100-190 nm) with large DNA genomes. The chorella virus genomes are linear nonpermuted molecules with crosslinked hairpin ends. In contrast, brown algal viruses have circular ds DNA genomes. Both MpV- and CbV - virus genomes have linear dsDNA molecules. Despite the differences in genome structure, phylogenetically all algal viruses are more closely related to each other. They form a distinct phylogenetic group suggesting a common ancestor.

All the algal viruses are included in the family *Phycodnaviridae* (ICTV, 2005) with four genera (Table 26.2).

Table 26.1 Representative algal viruses

Virus	Host	Genome size (kb)	DNA structure	%G+C	Level of methylation	
					5mC[a]	6mA[b]
CA-4B	Chlorella NC64A	330	Linear	40.0	0.12	ND
PBCV-1	Chlorella NC64A	330	Linear	40.0	1.9	1.5
IL-3A	Chlorella NC64A	330	Linear	40.3	9.7	ND
NC-1A	Chlorella NC64A	340	Linear	40.5	7.1	7.3
SH-6A	Chlorella NC64A	330	Linear	41.1	12.6	10.3
XZ-6E	Chlorella NC64A	330	Linear	41.7	21.2	15.2
NYs-1	Chlorella NC64A	350	Linear	42.1	47.5	11.3
CA-4A	Chlorella NC64A	330	Linear	40.3	39.8	19.6
NY-2A	Chlorella NC64A	380	Linear	41.5	44.9	37.0
CVA-1	Chlorella Pbi	>300	Linear	46.1	43.1	ND
CVB-1	Chlorella Pbi	>300	Linear	45.4	42.7	17.7
CVM-1	Chlorella Pbi	>300	Linear	46.0	41.9	10.1
EsV	*Ectocarpus siliculosus*	320	Circular	50	1.0	3.0
EfasV	*Ectocarpus fasciculatus*	320	Circular	-	-	-
FsV-178	*Feldmania* sp.	178	Circular	-	-	-
FsV-158	*Feldmania* sp.	158	Circular	-	-	-
MpV-SP1	*Micromonas pusilla*	-	-	-	-	-
CpV-Pw1	*Chrysochromulina brevifilum*	-	-	-	-	-

[a]Percentage of 5mC per C. [b]Percentage of 6mA per A. ND, not detected.

Morphology

Chlorella viruses are generally considered as representative of algal viruses. The chlorella virus PBCV-I is a polyhedron (175-190 nm) with a multilaminate capsid. Electron micrographs indicate that the virions contain flexible hair-like appendages that extend from atleast some of the vertices. Virus contains about 64% protein, 21-25% ds DNA and 5-10% lipid. The virus contains at least 50 structural proteins, ranging in size from 10-200 kD. Major capsid protein (Vp54) comprises about 40% the total PBCV-1 protein and is one of the viral glycoproteins.

Genome structure

PBCV-1 genome is 330740 bp long linear and non permuted with covalently closed hairpin ends. The termini consist of 35 nucleotide long, incompletely base-paired, covalently closed hair pin loops that exist in one of two forms (flip and flop). Each hair pin loop is followed by an identical 2221 bp inverted repeat sequence, after which the genome diverges. The genome contains 702 ofen reading frames (ORFs) and also cluster of ten tRNAs located in the middle of the genome. Certain genes are interrupted by one or two introns. An unusual feature of the chlorella virus genomes is that they contain methylated bases.

Table 26.2 Classification of algal viruses.

Family (subfamily)	Genus	Type species
Phycodnaviridae	*Chlorovirus*	*Paramecium bursaria Chlorella virus I*
	Prasinovirus	*Micromonas pusilla virus SP'1*
	Prymnesiovirus	*Chrysochromulina brevifilum virus PW1*
	Phaeovirus	*Ectocarpus siliculosis virus I*

The EsV and FsV viruses that infect filamentous brown algae have circular dsDNA genomes. FsV contains two genomes packed in separate particles. The EsV dsDNA contains several single stranded gaps. Unlike the chlorella viruses the FsV and EsV genomes are integrated into the host chromosome(s) in certain stages of the algal life cycle.

Replication

PBCV-1 attaches rapidly, specifically and irreversibly to cell walls. Following attachment, the host wall dissolves at the point of attachment and viral DNA enters the cells, the viral DNA is rapidly transported to the host cell nucleus. Early transcription is detected within 5-10 min. post infection. DNA synthesis and late virus transcription begin after 1 hour of infection. The virus DNA synthesis requires both energy and deoxyribonucleotide triphosphates from the host.

At approximately 2h post infection host cells change cytologically. The nucleolus disintegrates and chromatin heterochromacity increases. Membranous material, which may function as templates for capsid assembly, appear in the virus assembly centres. This is followed by the accumulation of more dense material on exterior membrane surface which gives rise to virus capsids at the periphery of the virus assembly centres. Complete capsids assemble prior to DNA packaging, which probably enters through an opening in the capsid. By 4-5h postinfection, cell contains many filled virions which are distributed throughout the cytoplasm. PBCV-1 release, by cell lysis, begins at 4h post infection and is completed by 8-10 h post infection. Thus the virus is completely assembled inside the host and does not aquire a membrane by budding through the host membrane.

Virus replication has not been detailed for any of the other algal viruses.

Further reading

Fuhrman, JA and Suttle EA (1993). Viruses in marine planktonic systems. Oceanography 6 : 5.

Muller DG, Kapp M and Knippers R (1998). Viruses in marine brown algae. *Adv. Virus Res.* 50 : 49.

Van Etten, JL (1999). Algal viruses. In: Encyclopedia of Virology Vol. I Granoff, A and Webster, RG (Eds.) Second Edn. San Diego Academic Press.

Van Etten JL, Lane LC and Meints RH (1991). Viruses and virus-like particles of eukaryotic algae. *Microbiol. Rev.* 55 : 586.

27

MYCOVIRUSES

Mycoviruses are widely distributed in fungi but their effect on fungal properties is still not well understood excepting few viruses like yeast L-A virus, smut virus and chest nut blight fungus virus. They are typically latent and are associated with symptomless infections of their host. The fungal viruses were first detected in 1948 as `La France' disease of cultivated mushrooms in Pennsylvania, and its study led to the first description of fungal viruses in 1962. Now more than 100 species of fungi belonging to all major taxa are reported to contain mycoviruses (Table 27.1). But biological significance of mycoviruses could not be established as these viruses are not amenable to conventional infectivity assay. Thus, conclusive correlations between the occurence of viruses and specific symptoms in fungi could not be established. Mycoviruses share some of the characteristics of plasmids but their morphology, nucleoprotein composition and possession of virion associated RNA polymerase activity is quite different from plasmid. They are more like a virus. As they have not proven to be infectious, some workers prefer to call them virus like particles (VLPs).

1. PROPERTIES

Most of the mycoviruses are isometric particles of 25-48 nm diameter. However, other forms like bacillform particles, rod shaped etc. are also known. The genome is either monopartite or multipartite. In later case, the segments are separately encapsidated into capsids. The genomic nucleic acid is linear molecule of uncapped dsRNA of 4.7 to 6.7 kbp in size. No dsDNA mycoviruses have been isolated from fungi. Some viruses contain additional satellite dsRNAs which encode killer proteins. These satellites are encapsidated separately in capsids encoded by the helper virus. Antisera against these viruses have been developed but titres are low and hence these can not be used for ELISA test.

Table 27.1. Some important mycoviruses and their hosts.

Virus	Host species	Genome size (kb)
dsRNA viruses		
Totiviridae		
L-A	*Saccharomyces cerevisiae*	4.6
M_1, M_2, M_{28}	*S. cerevisiae*	1.6-1.8
L-BC	*S. cerevisiae*	4.6
Hv190S	*Helminthosporium victoriae*	4.5
P1-H, P4-H, P6-H	*Ustilago maydis*	2.6-6.1
AfV-S, AfV-F	*Aspergillus foetidus*	
AnV-S, AnV-S	*A. niger*	
YIV	*Yarrowia lipolytica*	
GgV-87-1-H	*Gaeumannomyces graminis*	4.0, 2.9, 1.9
	Zygosaccharomyces bailii	5.0
	Blastomyces dermatitidis	
Partitiviridae		
	Atkinsonella hypoxylon	2.2, 2.1, 1.8
	Penicillium chrysogenum	
	P. stoloniferum	
	Agaricus bisporus	
	Rhizoctonia solani	2.2, 2.0
Hypoviridae		
CHV1-EP713	*Cryphonectria parasitica*	
Reoviridae		
	C. parasitica	1 to 3 (11 segs)
ssRNA replicons		
Narnaviridae		
20S RNA (=T dsRNA)	*Saccharomyces cerevisiae*	
23S RNA (=W dsRNA)	*S. cerevisiae*	
NB631 RNA	*Cryphonectria parasitica*	

2. CLASSIFICATION

Mycoviruses are reported from all major taxa of fungi including yeasts. The last updated ICTV classification (2005) lists seven families and 10 genera. This classification is presented in table 27.2.

Table 27.2. Classification of mycoviruses (ICTV, 2005).

Order	Family	Genus	Type Species
	Barnaviridae	*Barnavirus*	*Mushroom bacilliform virus*
	Hypoviridae	*Hypovirus*	*Cryphonectria hypovirus 1-EP713*
	Metaviridae	*Metavirus*	*Saccharomyces cerevisiae Ty3 virus*
	Narnaviridae	*Narnavirus*	*Saccharomyces cerevisiae 20sRNA narnavirus*
		Mitovirus	*Cryphonectria parasitica mitovirus-1 NB631*
	Chrysoviridae	*Chrysovirus*	*Penicillium chrysogenum virus*
	Partitiviridae	*Partitivirus*	*Atkinsonella hypoxylon virus*
	Pseudoviridae	*Pseudovirus*	*Saccharomyces cerevisiae Tyl virus*
		Hemivirus	*Drosophila melanogaster copia virus*
		Rhizidiovirus	*Rhizidiomyces virus*

3. TRANSMISSION

There are no natural vectors for direct and efficient transmission of mycoviruses. Transmission of these viruses in nature appears to occur during cell division and production of spores by their fungal hosts (serial transmission) or following cell fusion (lateral transmission). Lateral movement might occur following hyphal anastomosis and heterokaryon formation. Host range of these viruses would largely be decided by the vegetative compatibility of the fungal hosts.

Attempts to infect mycelial cultures using purified virus have so far failed. The rigid cell wall in fungi acts as the barrier. Injecting purified preparations in the cell has met with some success in the case of mushroom viruses. Electroporation has been found successful in transmitting some of the viruses.

4. HOST-VIRUS INTERACTION-PATHOGENESIS

Killer system

The killer systems in the yeast, *Saccharomyces cerevisiae* and the smut *Ustilago maydis,* represent the only known examples in which specific fungal phenotypes have been correlated with the presence of specific dsRNA molecules of mycoviral origin. The killer systems of these strains produce an extracellular proteinaceous toxin capable of killing sensitive strains. Three distinct killer specifications are recognized in *U. maydis.* The toxins produced by these strains are distinct from each other. All yeast killer strains contain two dsRNA segments, designated as L and M. These two segments are separately encapsidated in identical capsids.

Antibiotic and mycotoxin production

Present evidence strongly points out that mycoviruses are not directly involved in antibiotic production. It has been noticed that in *Penicillium chrysogenum,* virus titres and penicillin production are not related. Role of mycoviruses in toxin production is variable. In some cases, e.g. *Helminthosporium maydis* the production of toxin (pathogenicity) is associated with presence of virus particles. However, no such correlation was observed in case of *Periconia circinata* causing root and crown rot of sorghum.

Hypovirulence

The possibility that viruses may impair the pathogenicity of pathogenic fungi was first suggested in France in take-all fungus of cereals, *Gaumannomyces graminis.* Weakly pathogenic isolates of this fungus were reported to contain virus particles, whereas strongly pathgenic isolates were virus free. Hypovirulence has been reported in several fungal pathogens. This phenomenon has been studied in detail in *Endothia parasitica,* the chestnut blight fungus. This fungus contains two types of strains : (i) virulent strains; and (ii) less virulent strains, also known as hypovirulent strains or TH (transmissible hypovirulence) strains. These TH stains are not only less virulent themselves but they also prevent more aggressive strains from attacking the host when both forms are present together. It has now been established that dsRNA is the cause of TH.

5. VIRAL DISEASES OF MUSHROOMS

The symptoms of mushroom viral diseases are highly variable and are influenced by environmental and cultural conditions. It leads to malforma-tions of sporophores, degeneration of mycelium in compost and overall yield reduction. Generally six viruses are associated with this disease out of which three are most common. However, the role of individual viruses in the disease development has not yet been established. Some viruses remain symptomless reasons for which are not known.

Further reading

Ghabrial, SA (1994). New developments in fungal virology. *Advances in virus Research* **43** : 303-338.

Saksena, KN and Lemke PA (1978). Viruses in fungi. *Comp. Virol.* 12 : 103-143.

Buck, KW (1979). Replication of double stranded RNA mycoviruses. In : Lemke PA ed. Viruses and Plasmids in Fungi, New York Marcel Dekker 93-160.

Murphy, FA, Faquet, CM, Bishop DHL *et al.* (eds.) (1995). Virus Taxonomy : New York : Springer-Verlag 515-520.

Kottin, Y, Leibowitz, MJ (eds.) Viruses of fungi and simple eukaryotes. New York : Marcel Dekker, 209-242.

Kotta–Loizou I and Coutts RHA (2017) Mycoviruses in aspergilli : A comprehensive review. *Front. Microbiol.* 8 : 1699

28

GIANT VIRUSES

1. INTRODUCTION

The serendipitous discovery of the giant virus of amoeba — Acanthamoeba polyphaga mimivirus (APMV; more commonly known as mimivirus) — in 2003 has changed the field of virology. Andrew Lwoff (1950) for the first time introduced the technical definition of viruses as simple entities that depend on a cellular host to replicate, and consisted of a nucleic acid enclosed in symmetrical protein shell known as capsid. Until recently, no virologist thought of questioning or modifying that definition. However, the discovery of Acanthamoeba polyphaga mimivirus (APMV), in 2003, a giantvirus demolished this paradigm and initiated debates with regard to the definition and classification of viruses. APMV size was unprecedented, being on the scale of small bacteria or archaea cells. Initially it was mistaken for a bacterium and recognized as a virus only ten years after its isolation. Unlike any previously identified virus, APMV could be seen with a light microscope with very large genome (>1 Mbp) encoding almost a thousand proteins. Giant viruses have several unique and novel biological features, which include: they can be infected by parasitic viruses (termed virophages), contain mobile DNA elements, (termed transpovirons), and they have a defence mechanism against virophages termed the mimivirus virophage resistance element (MIMIVIRE). Many more intriguing features of this virus, prompted virologists to investigate many such viruses. In the following years, many additional giant viral species have been identified and their genomes have been fully sequenced.

Major differences between classical viruses and giant viruses are shown in Table 28.1

Table 28.1: Differences between classical viruses and giant viruses

Classical viruses	Mimiviruses and other giant viruses
Virus size <0. 2 mm	Bigger; visible by light microscopy
Viruses contain only one type of nucleic acid: DNA or RNA	Giant viruses contain both (genomicDNA+ messenger RNAs)
No binary multiplication	True
Obligate parasitic life style	True
Energy metabolism enzymes absent	Found in an alga-infecting mimivirus
Components of the translation apparatus absent	Translation apparatus present, upto 70, and 20 tRNAs
No ribosomal proteins or genes	True
Capsid present	Presence of a capsid; Absent in pandoraviruses

All giant viruses belong to the dsDNA group, as do all NCLDV families. All known giant viruses belong to the phylum Nucleocytoviricota.This phylum includes families, Poxviridae, which infects vertebrates (for example, smallpox virus) and invertebrates, the aquatic viruses Iridoviridae and Phycodnaviridae, and the vertebrate virus Asfarviridae. Metagenomic data collected from oceanic data suggests that giant viruses are common in nature and many more are awaiting discovery. The presence of giant viruses in almost any environment, including extreme niches and man-made sites (e.g., sewage and wastewater plants), suggests that the ecological role of these fascinating entities.

Given their distinctiveness in morphology, ecology, genome size and gene uniqueness, a new name has been proposed for the giant viruses—*giruses.*

2. UNIQUE FEATURES OF GIANT VIRUSES

Although giant viruses exhibit several phenotypic and genotypic characteristics of viruses, they also share many features with microorganisms.

i. The distinctive features of giant viruses are:

ii. Giant virus particles are visible using a light microscope,

iii. They have larger genomes than small bacteria and contain more diverse genes than other viruses.

iv. Giant viruses are more complex than other viruses, as their capsid contains two types of nucleic acid — a DNA genome and some mRNAs— as well as 114 proteins.

v. Giant viruses contain translation factors and aminoacyl-tRNA synthetases (which is unprecedented for a virus), and six tRNAs (tRNA is rarely detected in other double-stranded DNA (dsDNA) viruses).

vi. The mobilome of giant viruses is more diverse than that of other viruses and includes several self-excising introns and inteins, as well as pro-virophages and transpovirons.

vii. Unlike many viruses, giant viruses can enter host cells through phagocytosis that does not require specific viral–host interactions, owing to its large size.

3. HISTORY

Chlorella viruses are the pioneer viruses with genomes larger than 300 kb that provided us with the new concept of giant viruses.The first giant viruses to be described were chloroviruses of the family Phycodnaviridae. These were discovered in 1981 by Meints and his colleagues. The first chlorovirus was initially called HVCV (Hydra viridis Chlorella virus), since it was first found to infect Chlorella-like algae. Since the discovery of chlorella viruses, the microbiology community became interested in other large viruses predominantly residing in aqueous environments.

Subsequently, other giant viruses that infected marine flagellates were described. The first of these, then named BV-PW1, was isolated and described in 1995. However, it was not recognized as such until its sequenced genome was released as Cafeteria roenbergensis virus (CroV) in 2010. Subsequently, the giant virus Acanthamoeba polyphaga mimivirus was characterized (which had been mistaken as a bacterium in 1993), and then sequenced. It was named as Mimivirus for microbe-mimicking virus. Mimivirus is the largest virus ever discovered.Megavirus was isolated from a water sample collected in April 2010 off the coast of Chile, by Jean-Michel Claverie and Chantal Abergel. Pandoraviruses, other giant viruses of amoebae, which were considered to be amoebal eukaryotic parasites before genome sequencing were discovered in 2013 by a team of French scientists led by husband and wife Jean-Michel Claverie and Chantal Abergel.It is the second largest in physical size of any known viral genus.Most giant viruses have only been discovered and characterized in the past few years. Since 2013, the number and diversity of giant viruses of amoebae have expanded considerably, and, as of 2016, two virus families, the Mimiviridae and Marseilleviridae, have been described. Some important discoveries of giant viruses are presented in Table 28.2

Table 28.2: Some important discoveries of giant viruses in chronological order

Year	Discovery
2003	Acanthamoeba polyphagamimivirus (APMV), the first giant virus of amoeba; founder of family Mimiviridae
2008	Sputnik, the first virophage
2009	Marseillevirus, founder of Marseilleviridae
2011	Mega virus chil iensis, prototype virus of lineage C
	Lasennivirus prototype virus of lineage B
2013	Pandoravirus salinus, founder of a new putative family of giant viruses of amoeba
	Largest viral genome and gene content identified to date
	First giant virus without capsid protein identified
2014	Pithovirus sibericum, founder of a new putative family of giant viruses of amoeba
	Largest virus isolated to date, but genome smaller than Mimivirus
	Faustovirus, founder of a new putative family of giant viruses of amoeba
2015	Mollivirus sibericum, founder of a new putative family of giant viruses of amoeba, Replication in the nucleus
2016	The MIMIVIRE system, mimiviruses defend themselves against virophages
2018	Tupunavirus, discovered to contain largest set of translation components in the virosphere
	Tetraselmis virus I, Genes encoding energy production enzymes in a mimivirus

4. DISTRIBUTION AND HOST RANGE

Giant viruses have been isolated from numerous environmental niches, ecosystems and distant geographic locations, revealing their global distribution and diversity. These are isolated from different hosts, including amoebae, algae, invertebrates and mammals. Most giant viral genomes have been obtained from large-scale metagenomic sequencing projects covering aquatic ecosystems (e.g., oceans, pools, lakes and cooling wastewater units); others have been sequenced from samples extracted from underexplored geographical and ecological niches (e.g., the Amazon River, deep seas and forest soils). Currently available evidence suggests that the representation of giant viruses is underexplored, especially in soil ecosystems and unique ecological niches. Giant DNA viruses have now been detected from a range of bacteria, protists and lower animals, and are no longer thought to be restricted to a specific host or phylogenetic clade. Many more virus–host systems, most of them reported in the last five years, still await isolation, characterization and classification.

5. CLASSIFICATION

All giant viruses with i. genome dsDNA ii. genome encoding roll major capsid proteins (MCP) and containing the vertical jelly fold forming pseudo hexameric capsids are classified in Realm VI Varidnaviria, Kingdom Bamfordvirae and Phylum Nucleocytoviricota

Realm 6: Varidnaviria

Kingdom: Bamfordvirae	
Phylum: Nucleocytoviricota	2 classes
Class 1: Megaviricetes	3 orders
Order 1: Algavirales	1 family
Family: Phycodnaviridae	6 genera
Order 2: Imitervirales	1 family
Family: Mimiviridae	2 genera
Order 3: Pimascovirales	3 families
Family: Ascoviridae	2 genera
Family: Iridoviridae	2 subfamilies
Family: Marseilleviridae	1 genus, 2 species
Class 2: Pokkesviricetes	2 orders
Order 1:Asfavirales	1 family
Family: Asfarviridae	1 genus
Order 2: Chitovirales	1 family
Family: Poxviridae, 2 subfamilies	

6. ORIGIN AND EVOLUTION

The origin of the NCLDV group that includes all reported giant viruses is controversial. Upon discovery of mimivirus, some researchers, noticing its huge number of genes with no detectable resemblance to any cellular genes—86 percent of the total coding sequences in mimivirus—concluded that NCLDVs should be considered a fourth domain of life along side the Archaea, Bacteria and Eukarya. It has also been suggested that some NCLDV genes may have arisen from the same ancient gene pool from which sprang the prokaryotes and eukaryotes.Two interesting yet contrasting hypotheses suggest that given the size and complexity of NCLDV genomes, either the ancestor of modern NCLDVs may have given rise to the eukaryotic genome itself, or decay of a eukaryotic genome may have produced the genome of the NCLDV family. Horizontal gene transfer between virus and host has also played an important role in the evolution of NCLDVs (and their host cells), beginning far back in biological history.

A driving point to the placement of giant viruses in the tree of life is the presence of numerous previously unknown virus-encoded gene families. A recent reconstruction of NCLDV evolution suggests a common ancestor that likely contained a minimum set of 47 genes that have left foot prints in modern viral genomes. It is suggested that the present NCLDVs then evolved by losing some genes, duplicating others and acquiring genes from their hosts and other organisms. The giant viruses, like other viruses, have grabbed some genes from their hosts in the course biological evolution. Interpretation of viral phylogenetic reconstructions is therefore a difficult subject. Yet, few evidences are available to make certain vague interpretations. Analysis of genomes of 45 giant viruses identified five genes common to all of the NCLDV viruses and 177 additional genes that are shared by at least two of the virus families.

7. STRUCTURE AND GENOMES OF GIANT VIRUSES - APMV AS A MODEL

APMV and other giant viruses of amoebae have several structural and genomic features that had not been described in viruses before their discovery.

Structure : The APMV capsid is ~500 nm in size and is covered by fibrils that are ~120–140 nm long and 1.4 nm thick. These fibrils are morphologically unique among viruses).They form a dense layer, are extensively glycosylated, and enable the attachment of APMV to their host cells through glycans. The capsid comprises proteins that have a double jelly-roll fold and is icosahedral except at one vertex, which is covered by a unique five branch starfish-shaped structure (termed 'stargate') that is without fibres. Beneath the capsid, and surrounded by an inner lipid membrane and fibres, is a spherical lipid bilayer compartment that is ~340 nm in size. and contains the genome and proteins. This nucleocapsid has a large depression that faces the 'stargate', which creates a cavity and indicates that the nucleocapsid has a fixed position relative to the external capsid.

Genome: APMV has a double-stranded DNA (dsDNA) genome that is 1.2 Mb in length and contains 979 genes that putatively encode proteins with a coding density of 89%. The genome is also AT rich, comprising 72% AT nucleobases. Several APMV genes are not found in viruses other than giant viruses of amoebae, including those that encode translation factors and aminoacyl tRNA synthetases; some of the genes encoding aminoacyl tRNA synthetases are expressed and encode functional proteins. In addition, the APMV genome encodes four different tRNAs. Other genes that are unique to APMV and other giant viruses of amoebae encode proteins that are involved in nucleotide synthesis, amino acid metabolism, protein modification, lipid or polysaccharide metabolism, DNA repair or protein folding. In addition, the APMV genome contains four major groups of genes, including core genes that are shared with poxviruses, ascoviruses, iridoviruses, asfarviruses, phycodnaviruses and other giant viruses of amoebae. Moreover, mimivirus genomes harbour significant proportions of duplicated genes, of so-called ORFans (genes with no known homologue in sequence libraries) accounting for about 48% of all genes, and of genes that are predicted to have been involved in lateral exchanges of sequences with other organisms, including their amoebal host. Promoters, introns and inteins have also been predicted. The APMV genome also contains early and late gene promoters, and mRNAs are expressed as polyadeny lated transcripts that most often end with short palindromic sequences thatform hairpin-like structures. Recoding events, including a frameshift and a read through, were described for a gene that encodes a translation termination factor in the APMV genome. In addition, introns and inteins were detected in a few conserved genes, including those that encode the major capsid protein and the DNA polymerase and a 'mobilome' that is unique to APMV was identified. Finally, a few mRNAs were detected in mimivirus capsids.

Morphological and genomic features of some giant viruses are presented in Table 28.3

8. INFECTION CYCLE OF GIANT VIRUSES

The replicative cycles of giant viruses in amoebal cells show several similarities, including phagocytic entry, DNA release and replication in 'viral factories'. However, differences exist between giant viruses in the duration of their replication cycles, the involvement of the amoebal nucleus in virus replication, and the assembly and release of virus particles.

Table 28.3: A list of representative giant viruses and their morphological and genomic features

Family	Shape	Size (nm)	Genome size (kb)	DNA	tRNAs	RNA polymerase subunits	Replication cycle
Pandoraviridae (Phycodnaviridae)	Ovoid with a pore	1200 × 500	2770	Linear	3	4	Nuclear and cytoplasmic
Phycodnaviridae	Icosahedral with a spike	170	331	Linear	11	0	Nuclear and cytoplasmic
Mimiviridae	Icosahedral (with fibres)	390 (630)	1181	Linear	6	9	cytoplasmic
Pithoviridae	Ovoid with a capped pore	1500 × 800	610	Circular	0	4	cytoplasmic
Marseilleviridae	Icosahedral	250	368	Circular	0	3	Cytoplasmic, involving the nucleus
Ascoviridae (Iridoviridae)	Bacilliform or allantoid	400 × 130	157	Circular	0	4	Nuclear and cytoplasmic
Iridoviridae	Icosahedral	175	106	Linear	0	2	Nuclear and cytoplasmic
Asfarviridae	Icosahedral	200	170	Linear	0	7	Nuclear and cytoplasmic
Poxviridae	Brick- shaped	310 × 240	195	Linear	0	9	cytoplasmic

The replicative cycle of APMV, which lasts about 12 h and occurs in the amoebal cytoplasm It has several features that have not been found before in viruses. APMV is internalized through phagocytosis before its genome-containing internal core is released through the 'stargate' into the cytoplasm, through the fusion of viral and phagosome membranes;transcription may be initiated in these cores. Subsequently, the genome is released from the core and is replicated at high levels, which generates an early cytoplasmic replicative centre that is thought to be engulfed by a membrane layer of the endoplasmic reticulum. Replicative centres then merge into a single virus factory that contains, from the inside to the outside, zones that are involved in replication, membrane biogenesis, capsid assembly and DNA packaging, and fibre acquisition. The virus internal layer seems to originate from multivesicular membrane structures that bud from the endoplasmic reticulum and become open membrane sheets.The major capsid protein is assembled around these sheets while acting as a scaffold. Genome packaging occurs through a transient aperture that is distinct from the 'stargate'. Finally, the layer of fibres is assembled. The involvement of the amoebal nucleus in the life cycle of APMV has been debated. Indeed, the size of the amoebal nucleus decreases approximately twofold during APMV replication, even though this process occurs in a viral factory in the cytoplasm; amoebal nuclear factors may be involved.

9. The presence and significance of giant viruses in humans

Giant viruses have been detected in human samples in several studies especially through metagenomic analyses. Various other techniques have contributed to identify the presence of giant viruses in humans, including serologies, immunohistochemistry and immunofluorescence techniques, PCR and fluorescence in situ hybridization. These involved sequences related not only to mimiviruses and virophages but also to marseilleviruses and more recently discovered giant viruses. Although the presence and clinical significance of giant viruses in humans are still controversial issues, they represent an emerging field that deserves broader analyses and increased consideration in future studies. Further studies are necessary to gain a better knowledge on the presence and significance of giant viruses in humans.

Further reading

Aherfi S, Colson P, LaScola B and Raoult D (2016) Giant viruses of amoebas: An Update. *Front. Microbiol.*7:349. doi: 10.3389/fmicb. 2016.00349

Yamada T. (2022) Giant viruses in the environment: Their origins and evolution, *Current Opinion in Virology*, doi:10.1016/j.coviro.2011.05.008Available

Colson *et al* (2017) Mimivirus: leading the way in the discovery of giant viruses of amoebae. *Nature Reviews* 15: 243-253

Colson *et al* (2019) Giant mimiviruses escape many canonical criteria of the virus definition.Clinical Microbiology and Infection 25: 147e154

Brandes N and Linial M (2019) Giant Viruses—Big Surprises. *Viruses* 11 : 404-415; doi:10.3390/v11050404

Sun T *et al* (2020) Host range and coding potential of eukaryotic giant viruses. *Viruses* 12,1337-1356; doi:10.3390/v12111337

29

VIROPHAGES

1. INTRODUCTION

Two fascinating discoveries were made in the first decade of the present 21st century: i discovery of the first giant virus, Acanthamoeba polyphaga mimivirus (APMV), belonging to the family *Mimiviridae* in 2003, and ii. its associated virophage, Sputnik in 2008. Until then virologists know many satellite viruses and their corresponding co-infecting helper viruses (Chapter...). In physical association, though the virophages appear to be close relatives of satellite viruses, the research during the last decade conclusively proved that in many aspects they are entirely different from satellite viruses and proved to be entirely different entities.Virophages have been defined as the first viral parasites to infect another viral entity.The interaction between virophage, coinfecting giant virus and host cell are interesting in that it exhibits inhibitory effect on co-infecting virus as it replicates using the machinery of giant virus. Thus they have a parasitic relationship with the co-infecting virus. At the same time, virophage prevent the lysis and improves the survival of the host. In this respect they are analogous to temperate phages. To date, about 18 virophages have been isolated from different environmental sources. All of them have been assigned under the family *Lavidaviridae*.Virophage genomes display highly variable gene content and are most closely related to members of polintons (a widespread group of eukaryotic large DNA transposons). Virophages are geographically widespread, genetically diverse and appear to be commonly associated with giant virus-infected protist populations. Their discovery has significantly contributed to expand our knowledge regarding the diversity, evolution, and complexity of viruses. Research in the virophage field is in its initial phase and further discoveries that could challenge the current concept of a virus which is likely to change in coming years.

2. HISTORY

Virophages have been defined as obligatory parasites of giant viruses. Therefore, their description was subsequent to the isolation of their viral hosts. Giant viruses were discovered in 2003, and since then, they have sparked sustained interest. They appear to be the most complex of the known viruses based on genomic and structural analyses.

APMV was the first giant virus discovered. It has several structural and genomic features that had not been described in viruses before its isolation. In 2008, giant viruses were shown to be themselves the prey of another group of viruses that were named virophages. The discovery of virophages dates back to the isolation of the second giant virus of amoebae, mimivirus strain and close APMV relative, *Acanthamoeba castellaniimamavirus* in 2008. Following the discovery of the Sputnik virophage, other virophages, very similar to Sputnik, were isolated. Sputnik 2 was isolated in association with Lentillevirus, a mimivirus from the lineage A that has been cultivated from a contact lens rinsing liquid of a patient with keratitis (Table 29.1). Unexpectedly, a new class of mobile genetic elements, that have been named transpovirons, was also discovered in the Lentillevirus genome. In contrast, the third Sputnik strain was found without its natural host mimivirus in a soil sample collected in Marseille, France. Rio Negro virophage (RNV), the fourth Sputnik strain,was isolated from the Negro River in Amazonia, Brazil. A giant virus from the Mimivirus genus was isolated from the same sample.The Guarani is a Sputnik-like virophage isolated in 2019 from a water sample collected in Pampulha lagoon, Belo Horizonte, Brazil

Although the first virophage was reported in 2008, 11 years ago, co-culture strategies contributed to isolate only ten strains of virophage. In contrast, 57 novel complete or partial virophage genomes were reported from different metagenomic datasets. Metagenome assemblies not only revealed the diversity of the virophage group, but also suggested a relevant ecological impact of these small viruses and their potential role in the stability of the microbial network. The first virophage detected by metagenomics was the Organic Lake virophage (OLV). OLV is also the second virophage discovered after Sputnik.

Table 29.1 Chronological order of major discoveries in the virophage field

Year	Discovery
2003	Discovery of APMV, the first giant virus
2008	Discovery of Sputnik, the first virophage
2011	The first provirophage Sputnik2 and transpovirions
	Discovery of Mavirus, the first marine virophage and organic Lake Virus (OLV), the first metagenomic virophage
2013	Discovery of Sputnik3
2014	Rio Negro Virophage (RNV)
	Discovery of Zamilon and its host specificity
2016	First description of the MIMIVIRE system and the host genome integration- reactivation of Mavirus
2018	Platanovirus saccamoeba virophage
2019	Discovery of Guarani

3. VIROPHAGES AND SATELLITE VIRUSES

Virophages as well as satellite viruses are associated with another virus, giant viruses, or helper viruses, respectively. They can exert a negative effect on their associated virus which varies significantly depending on the virus and multiple infection parameters. Satellite viruses also share some other aspects with virophages, such as subcellular localization or overlap with their helper viruses regarding polyadenylation sites. On the other hand, virophages are genetically distinct from satellite viruses. While the latter have short ssRNA genomes (satellite viruses infecting plants and arthropods) or short ssDNA genomes (adeno-associated viruses), virophages possess considerably longer dsDNA genomes encoding genes that are conserved in the PRD1–adenovirus lineage. Unlike satellite viruses, virophages encode multiple morphogenesis proteins (two capsid proteins, maturation protease, DNA-packaging ATPase) and DNA replication proteins. Virophages probably depend on the giant virus-provided transcription apparatus in a way that is similar to how other small dsDNA viruses depend on the cellular transcription system. Thus, while several phenomena are shared between virophages and satellite viruses (dependency on another virus, variable hypovirulence), their larger particle size and coding potential, genetic and infection properties, and relatedness to viruses of the PRD1–adenovirus lineage clearly set virophages apart from those much simpler and genetically distinct groups of satellite viruses. Virophages can be viewed as eukaryotic dsDNA viruses that became evolutionarily adapted to giant dsDNA viruses, and that use the viral transcription enzymes in the cytoplasm instead of accessing the host nucleus for cellular transcription.

4. CLASSIFICATION

Classification of virophages has been the subject of intense debate within the scientific community. Krupovic et al. (2011) argued that virophages should be classified within the group of classical satellite viruses. They suggested that virophages are no more than a subgroup of satellite viruses with a double-stranded DNA genome. On the basis of available data, the ninth report of the International Committee on Taxonomy of Viruses for Virus Taxonomy (ICTV, 8 November 2018), classified virophages in the group of satellite viruses. However, this classification proposal has met with criticism. In 2015, Fischer and Krupovic proposed the creation of the family Lavidaviridae (Large virus-dependent or -associated virus), which includes two genera (Sputnikvirus and Mavirus). This taxonomic proposal has been approved by the 10th report of the ICTV, which now considers virophages as a new divergent family of double-stranded DNA satellite viruses infecting protists that are not subviral agents.

The family ***Lavidaviridae*** (Large virus-dependent or -associated virus) with the two genera, *Sputnikvirus* and *Mavirus*, has been established by the International Committee on Taxonomy of Viruses for classification of virophages. It is the sole family under order ***Priklausovirales*** (from Lithuanian *priklausomas*, "dependent"), which in turn is the sole order under class ***Maveriviricetes*** (from Maverick transposons).

Realm :*Varidnaviria*

Kingdom:*Bamfordvirae*

Class:*Maveriviricetes*

Order:*Priklausovirales*

Family : *Lavidaviridae*

Genus *Sputnikvirus*

Species - *Mimivirus-dependent virus Sputnik*

Species - *Mimivirus-dependent virus Zamilon*

Genus *Mavirus*

Species - *Cafeteriavirus-dependent mavirus*

Unassigned genus

Organic Lake virophage (OLV)

5. HOST RANGE

Virophages require a co-infecting virus in order for them to replicate. The virophages do not have the necessary enzymes to replicate on their own. Virophages use the giant viral replication machinery to replicate their own genomes and continue their existence. The host range for virophages include giant viruses with double stranded DNA genomes. They are assigned to mimiviruses composed of a wide variety of viruses isolated from various ecosystems. Based on the conserved genes, such as family B DNA polymerase encoding gene sequences as well as phylogenomic analyses, mimiviruses of amoebae were found primarily to belong to three lineages, designated A, B, and C. APMV is the pioneer member of lineage A, while Moumouvirus and Megavirus chiliensis are the prototype members of lineages B and C, respectively.

The information on virophage host cell range is extremely low. Currently, the only known cellular hosts for virophages are phagotrophic protists.Only two such virophage–virus–host systems are known in culture (*Acanthamoeba* sp. and *Cafeteria* sp.). Most of the environmental virophage genomes that have been assembled from metagenomic sequences, or host-integrated virophage genomes are assumed to be associated with giant viruses of the extended *Mimiviridae*family, which mostly infect phototrophic protists.

6. MORPHOLOGY

The virophage particles that have been examined so far are 50–75 nm in diameter and possess icosahedral symmetry (Fig.29.1). The Sputnik capsid structure is composed of the major capsid protein (MCP) encoded by the V20 gene, and the minor capsid protein (mCP) encoded by the V18/V19 gene. The MCP contains a double jelly-roll fold and forms trimeric capsomers (hexons) with pseudohexagonal symmetry that build the 20 faces of the icosahedral particle whereas the mCP is a single jelly-roll protein that forms pentameric capsomers (pentons) occupying the 12 vertices.The mature capsid consists of 260 hexons and 12 pentons that are arranged in a lattice with a triangulation(T) number of 27 ($h = 3$; $k = 3$).

The capsid proteins of different virophages show low levels of conservation on the amino acid level (e.g. ~40% for the MCPs of Sputnik and mavirus). The mCP in particular is highly diverse, which hinders sequence similarity-based identification of novel virophages. Despite the high sequence divergence, all MCPs and mCPs are assumed to adopt the double and single jelly roll folds, respectively, and preliminary data shows that the three-dimensional structure of the mavirus MCP is very similar to that of Sputnik. The virophage genomes assembled from a sheep rumen metagenome apparently lack a penton gene. In addition to the main capsid components (MCP and mCP), other virophage-encoded proteins may be present in the mature virion, such as the mavirus MV13 protein, which could have lipase activity.

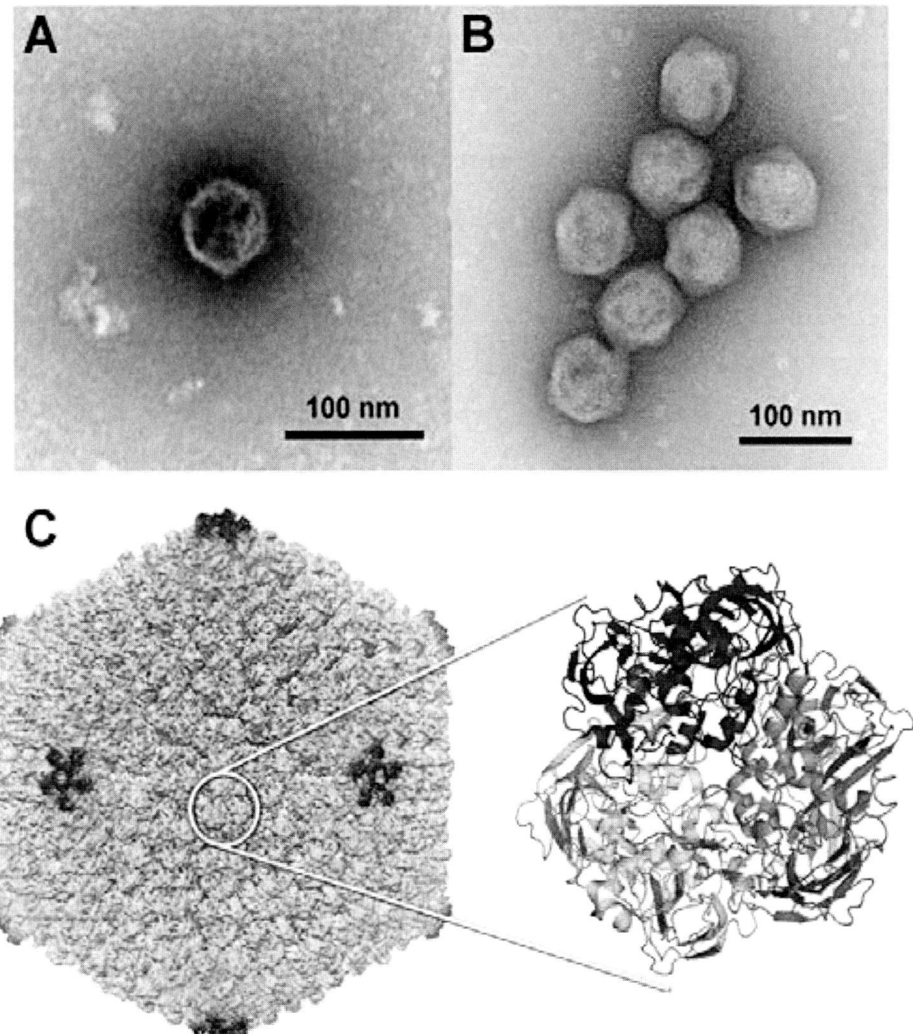

Fig. 29.1 The structure of lavidavirus capsids. A) Negative stain electron micrograph of a Sputnik particle(courtesy of M. Gaia and B. La Scola, Univ. Aix-Marseille, France). B) Negative stain EM image of CsCl gradientpurifiedmavirus particles (U. Mersdorf and M. Fischer, Max Planck Institute for Medical Research, Germany).

7. GENOME

All virophages described so far known to possess dsDNA genomes ranging in size from 17 to more than 30kbp. Genome organization of virophages and their properties are precised in Table 29.2. Cultured virophages (Sputnik, Mavirus, Zamilon) have circular genomes while those that were assembled from metagenomic sequences are either circular or linear. The G+C content of genomes of these virophages ranges between 27% and 37%, with the exception of YSLV 5 (51.1% G+C). Similarly, low G+C contents are found in their host giant viruses, e.g.mimivirus (28% G+C), CroV (23%), PgV-16T (32% G+C), and OLPV-1 (30% G+C). In contrast, the G+C contents of the nuclear genomes in the eukaryotic hosts are typically higher, i.e 59 to 70 percent. Virophage genomes contain from 16 to 34 protein-coding sequences, which are for the most part spaced in a non-overlapping manner, similar to other dsDNA viruses and bacterial genomes.

Nearly, all virophage genomes contain four conserved genes that code for a major capsid protein (MCP), a minor capsid or penton protein (mCP), a cysteine protease (PRO), and an FtsK-HerA-likeATPase.These proteins are involved in virion formation. MCP–mCP gene pair is highly conserved and found in all virophages. In most virophages, these two genes are syntenic which indicates these virophage species are of relatively recent divergence.The MCP codes for a double jelly-roll fold, whereas the mCP gene encodes the singlejelly-roll penton protein of virophage capsids. PRO product is responsible for C-terminal processing of the MCP in at least Sputnik and mavirus. FtsK-HerA-like ATPase belongs to the FtsK-HerA superfamily. These NTPases usually form hexamers at a unique vertex of the viral capsid and pump the genome through the central pore of this hexamer. Many of the non-conserved genes found in virophage genomes exhibit diverse evolutionary affiliations, including eukaryotes, bacteria, different viral families, and diverse mobile genetic elements.

8. VIROPHAGE, GIANT VIRUS AND HOST CELL INTERACTIONS

The relationship between virophages, giant viruses, and their host cells is complex tripartite relationship.This relationship becomes evident with the discovery of several virophage sequences integrated into giant viruses and host cells genomes. Endogenized virophage sequences appear to play an instrumental role during the tripartite host cell–giant virus–virophage interactions, mainly as being involved in different host defence strategies.

Table: 29.2 Properties of partial and complete virophage genomes

Virophage	Length (bp)	% GC	Topology	ORFs	Cultured or Sequenced	Associated giant Virus(es)	Host organism
Sputnik	18.342	27.04	Circular	21	In culture	Mimi-,Moumou-, Megaviruses	*Acanthamoeba* sp.
Sputnik 2	18,338	27.04	Circular	21	In culture		*Acanthamoeba* sp.
Sputnik 3	18,338	27.04	Circular	21	In culture	Mimi-,Moumou-, Megaviruses	*Acanthamoeba* sp.
Sputnik strain Rlo	1788 (partial)	33.45	Unknown	Unknown	In culture	Mimi-,Moumou-, Megaviruses	*Acanthamoeba* sp.
Zamilon		29.67	Circular	20	In culture		*Acanthamoeba* sp.
	17,276					Samba virus (SMBV)	
Zamilon 2		32	Unknown	15	Sequence only		Unkonwn
Mavirus (exogenous)	6717 (partial)	30.26	Circular	20	In culture	Unknown	*Cafeteria Roenbergensis*
		30.67	Integrated	20	In culture		
Mavirus (endogenous)	19,063	26.73	Unknown	22	Sequence only	Cafeteria roenbergensis virus (CroV)	*Cafeteria Roenbergensis*
Ace Lake Mavirus (ALM)	20,190	36.51	Unknown	26	Sequence only	Cafeteria roenbergensis virus (CroV)	Unknown
	17,767	33.37	Unknown	28	Sequence only		Unknown
Organic Lake virophage (OLV)	26,421					Unknown	Unknown
						Unknown	
Yellowstone Lake virophage (YSLV)1	27,851					Unknown	

i. Virophage–giant virus interactions

Virophages target the virus factory and the replication machinery of their host and also integrate their own genome in that of their viral hosts. Genomes of most virophages possess integrase genes. These genes encode integrases that are able to physically link their genomes to those of other organisms.The tyrosine recombinase found in Sputnik virophages may catalyse the occasional integration into mimivirus genomes. Sputnik 2 may also exist as a provirophage in the genome of its host virus.Similarly, the PgVV genome also integrates at multiple sites within its associated Phaeocystis globosa virus 16T. Integration as a provirophage in a giant virus genome has clear advantages for the smaller virus genome, as it will automatically co-infect and co-localize with its supporting giant virus.

Inhibitory or other-wise effect of virophages on the production of giant virus particles during a coinfection, appear to vary considerably. For instance, mavirus is a potent inhibitor of CroV and coinfected cells often, produce no CroV virions at all. On the other hand, negative impact of Sputnik on mimivirus is less pronounced. The Zamilon virophage on the other hand has no significant negative impact on the replication of a coinfecting mimivirus. The reasons for these variations are not known.

ii. Virophage–host cell interactions

Ever since the virophages were discovered and their genomes were understood, it was believed that virophages integrate into host genomes. This was based on the presence of integrases in Sputnik and mavirus, and also the genetic relatedness between mavirus and the endogenous eukaryotic maverick/polinton elements. Mavirus/ALM and Sputnik/Zamilon encode two different types of integrases and members of the genus *Sputnik virus* have a bacteriophage lambda-like tyrosine recombinase. The mavirus-like virophages encode a retrovirus RVE-type integrase. Integration process physically ties the virophage to one of its two host components and increases its chances to encounter conditions favourable for replication. At the same time, integrated viral DNA is maintained and replicated by the host cell, whereas free virions are exposed to various environmental parameters that may destroy infectivity.

9. REPLICATION CYCLE

Replication process of virophages can be differentiated into three phases viz. i. Entry, ii. Genome release and DNA replication and, iii. Assembly and release of virions (Fig. 29.2)

i. Entry

Virions enter the host cell either by endocytosis, or phagocytosis.Two different modes of entry have been proposed. Sputnik strains are supposed to use the paired-entry mode where they infect the host cell simultaneously with their giant virus host, attaching to its capsid fibrils. The second mode of entry, observed with mavirus, virophages infect its host cell independently of the giant virus (CroV) through clathrin-mediated endocytosis.

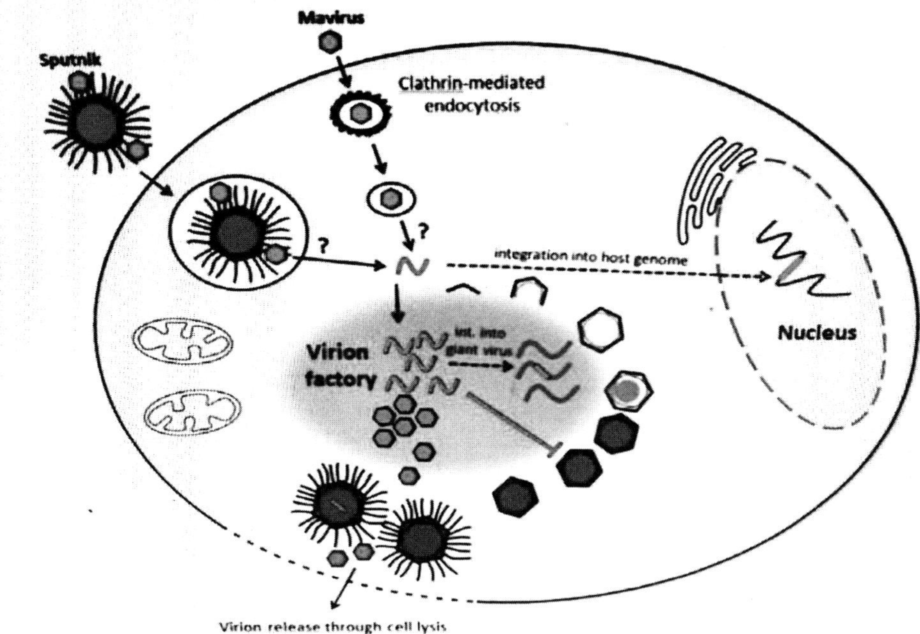

Fig. 29.2 A simplified diagram showing the replication cycle of virophage (source Fischer M G 2021)

ii. Genome release and DNA replication

After entry, the virophage genome is targeted to the developing virion factory of the coinfecting giant virus.The next step is release of the capsid contents, notably the genome. The Sputnik particle capsid is composed of 260 trimeric capsomers and 12 pentameric capsomers assembled into a T = 27 lattice. Sputnik genome delivery occurs after a loss of one or more pentameric capsomers. The virion contents that are released comprise transcripts which are used at the early stage of infection, immediately after the virus uncoating, to initiate the parasitic pathway of Sputnik with its giant virus.

Virophages are defined as bonafide parasites of giant viruses. Sputnik, Zamilon, mavirus, and presumably all other virophages cannot replicate in their host cells without co-infection of their respective host viruses. Virophage replication depends on the availability of cytoplasmic transcription. They depend on the transcription and DNA replication machinery of their virus hosts, rather than on that of the cell host. Mimivirus virophage encodes its own DNA polymerase and it has been thought that genome replication is governed by the virophage-encoded replication machinery. However, no RNA or DNA polymerase genes were identified among the 21 Sputnik genes, thus it suggests that Sputnik virophages depend on the transcription–replication complex of the giant virus. On the other hand, late replication of genetic material and virions has been experimentally demonstrated for Guarani. In any case, replication steps probably occur inside the giant virus factory.

Recently, Fischer and his colleagues demonstrated that the Mavirus virophage can integrate at several loci in the nuclear genome of its cellular host *C. roenbergensis*. This process is allowed by the presence of a retroviral integrase packaged in the virophage particle. Provirophage expression then leads to a massive production of Mavirus particles from the giant virus factory. Mavirus-produced particles do not seem to have any protective role for their host cell at this stage. However, this integration mechanism does not seem very widespread in amoebae.

iii. Assembly and release of virion

At the end of the replication cycle, four virophage-encoded proteins, including MCP, mCP, the ATPase, and the cysteine protease, are involved in the virion assembly and maturation. MCPs and mCPs assemble into virion-like icosahedral particles. This assembly steps do not require the presence of a specific giant virus or cellular initiation factor. After the assembly, the packaging ATPase packages the virophage genome inside the virion, and then the cysteine protease processes the MCPs and prepares virions for their maturation. After an eclipse phase, which lasts 2–4 h for Sputnik, newly synthesized virophage particles are visible within the virion factory.

Finally, both virophage and giant virus particles are released upon host cell lysis.

10. ECOLOGICAL IMPORTANCE OF VIROPHAGES

Ecologically, knowledge about naturally occurring virophage population dynamics and interactions with their giant virus and eukaryotic hosts is limited. Co-infection of virophages with their respective giant viruses leads to decreased giant viruses production and better survival of host-cell populations. Although the molecular mechanisms underlying virophage inhibition of giant viruses in natural ecosystems are unknown, modelling approaches indicate that lavidaviruses influence the population dynamics of giant viruses and protists alike, with consequences for nutrient cycling and primary production. Further, virophage infections could lead to increased frequencies of algal blooms and higher secondary production by reducing host algal cell mortality. Virophage metagenomes have been assembled from different habitats, including oceans, lakes, soil, and animal digestive tracts. In addition to effects on population dynamics and biogeochemistry, virophages may have impacted their environment for millions of years by shaping the genomic landscape of their hosts through gene exchange and recombination. To validate the point, typically, each virophage genome contains a few genes with homologs in eukaryotes, bacteria, giant viruses, or bacteriophages, indicating that virophages mediate horizontal gene transfer. Combined with their ability to integrate into other genomes and their close relationship with eukaryotic polintons, viruses of the family *Lavidaviridae* emerge as an influential group of mobile genetic elements in protozoa and microalgae.

Further reading

Fischer MG (2021) The Virophage Family Lavidaviridae *Curr. Issues Mol. Biol.* 40:1-24

Krupovic, M., Kuhn, J.H., and Fischer, M.G. (2016) A classification system for virophages and satellite viruses. *Arch. Virol.* 161, 233–247. https://doi.org/10.1007/s00705-015-2622-9.

Bekliz, M., Colson, P., and La Scola, B. (2016) The expanding family of virophages. *Viruses* 8:E317.

Zhou, J, Zhang, W, Yan, S., Xiao, J, Zhang, Y, Li, B, Pan, Y, and Wang, Y. (2013). Diversity of virophages in metagenomic data sets. *J. Virol.* 87, 4225–4236. https://doi.org/10.1128/JVI.03398-12.

Krupovic, M, and Koonin, EV. (2015) Polintons: a hotbed of eukaryotic virus, transposon and plasmid evolution. *Nat. Rev. Microbiol.* 13, 105–115. https://doi.org/10.1038/nrmicro3389.

Mougari S *et al.* (2019) Virophages of Giant Viruses: An Update at Eleven. *Viruses* 11, 733; doi: 10.3390/v11080733

Paez-Espino D *et al.* (2019) Diversity, evolution, and classification of virophages uncovered through global metagenomics. *Microbiome* 7:157.

Krupovic M, Kuhn JH, Fischer MG.(2016) A classification system for virophages and satellite viruses. *Arch. Virol.* 161(1):233–47.

Sharma V, Colson P, Pontarotti P, Raoult D. (2016) Mimivirus inaugurated in the 21st century the beginning of reclassification of viruses. *Curr. Opin. Microbiol.* 31, 16–24.

Fischer, MG (2012) Sputnik and Mavirus: More than just satellite viruses. *Nat. Rev. Microbiol.* 10, 88.

30

SUBVIRAL AGENTS

A. DEFECTIVE INTERFERING (DI) PARTICLES

Many viruses containing both RNA and DNA produce defective-particles as a result errors in their nucleic acid synthesis. These particles are produced because a part of the parental genome is deleted thus making them defective for replication. The deleted genome may comprise about 10% the parental genome. DI particles contain normal virion proteins, but have shorter genome. They are replicative defective. They require a helper virus co-infecting the same cell. However, their genome retains the sequences that are needed for recognition by viral polymerases and for packaging of the genome into a virus particle. DI viruses are unable to reproduce themselves without the assistance of the infectious parental virus. Further, the DI genome replicates only in a cell that has been infected with infectious virus of the type from which the DI genome was generated because the latter's genome complements the functions of the deleted genes of defective virus. Replicative defective viruses are produced during the infection with various kinds of RNA viruses such as rhabdoviruses, togaviruses, orthomyxoviruses, and paramy-xoviruses and some of the DNA viruses (herpes viruses, geminiviruses).

A fraction of the DI viruses usually interfere with the multiplication of parent types (standard virus). Hence they are called **defective interfering** (DI) particles. The interference may operate at adsorption level (e.g., paramyxoviruses) or penetration (e.g., retroviruses) or replication (many viruses). Interference usually occurs when the propagation of DI virus is optimal that happens when a high multiplicity of infection when all cells contain an infectious virus genome. DI genomes depress (or interfere with) the yield of infectious progeny, by competing for a limited amount of some product synthesized only by the infectious parent. Interference takes place only when the ratio of DI to infectious genomes reaches a critical level. Upto this point, both infectious and DI genomes replicate to the fullest extent.

Mechanism of interference

Replication of DI particles is dependent upon the infectious virus. Thus the DI genome is parasitic on infectious virus. Interference comes out because more copies of the DI genome can be made in unit time than those of the full length genome. For example, if the DI genome represents one tenth of the standard virus, then, for every full-length genome synthesized, there will be 10 copies made for the DI genome. It is a known and established fact that RNA polymerase is required for replication of both standard virus and DI virus. Initially, there will be sufficient polymerase to allow replication of infectious and DI genomes, as a result there will be no interference. Interference will be apparent only when the number of DI genomes is so great that they sequester most of the polymerase proteins. At this stage, the multiplication of infectious virus, obviously, declines, which in turn results in a concomitant decrease in the dependent DI virus. However, the situation is more complex, because the generation of DI genomes is very dependent on the type of cell infected, so that both cell and virus contribute to a balance situation.

How are defective genomes generated?

The formation of DI genomes of RNA viruses is the consequence of the high variability of these genomes. Electron microscopic examination of genomic and DI RNAs from SS RNA viruses revealed that both were found to be circularized by hydrogen bonding between short complementary sequences at the termini. DI genomes are formed by copy choice mechanism when polymerase (or replicase) having replicated part of the template, skips to another part of the same template or another template. Thus the polymerase begins faithfully but fails to copy the entire genome. Most of the DI viruses (e.g., vesicular stomatitis virus) lack the 3' end of the standard virus genome. There are no DI viruses known that lack the 5' terminus. An important feature for the DI RNA viruses is that their genomes always contain the same termini as those of the parent virus, or consist of inverted repeats of sense and antisense copies of the 5' termini of the normal genome, whereas the remainder of the genome can be substantially deleted without impairing the ability of the DI genome to replicate. This indicates the genome termini are essential and contained the regulatory elements to direct genome synthesis. In other DI viruses, parts of the genome can be duplicated, often several times over, during subsequent replication events, making complex structures that bear little resemblance to the standard genome from which they were derived.

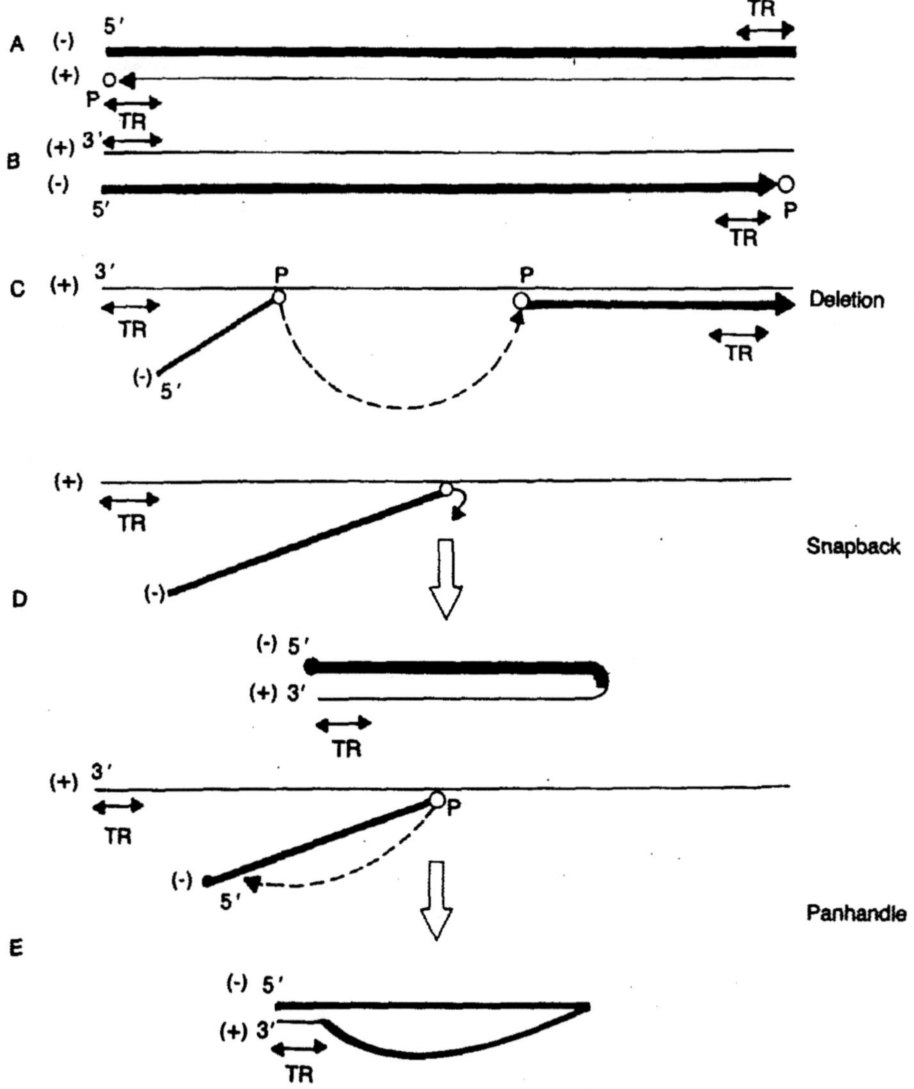

Fig. 30.1. Models of production of D1 particles genomes with VSV virus. (A) Normal transcription of the (–) genome to generate a full-length (+) transcript (B) Transcription of the (+) genome with formation of competent progeny (–) genome. (C) Formation of a deleted D1 genome. (D) Formation of a snapback D1 genome. In the panhandle form (E) the 5' end of the template is transcribed on a length of from 46 to several hundred nucleotides. The 46 nucleotide segment corresponds to the leader, which appears to be especially important in replication. Heavy lines, genomic (–) RNA; the thin lines (+) RNA; TR, initiation of transcription (From Davies *et al.,* Microbiology, 1990).

The skipping of template by RNA-polymerase generates defective genomes of four types (Fig. 30.1) : i. deletions, ii. snapbacks, iii. pan handles and iv. compounds. In deletions the polymerase jumps to site beyond, on the same template, skipping a segment. Snapbacks are formed when the polymerase, having transcribed, part of the (+) strand, switches to use the just made (-) strand as template. The resulting RNA, half of (-), half of (+) polarity, can produce a hairpin on annealing. A pan handle is formed by a similar mechanism, when the polymerase carrying a partial newly made (-) strand switches back to transcribing the extreme 5' end of it. Upon annealing, the strand forms a pan handle. Compound genomes are made by a combination of deletions and snap backs. Specific sequences may be involved in these irregularities. For example, in VSV when a pan handle is formed, the polymerase jumps to a site 46 nucleotides from 5' end of the new (-) strand and then transcribes the remaining stretch which may be identified by its sequence. DI genomes may have some point mutations, not present in regular viruses, and insertions. For example, the presence of cellular tRNA in the DI genomes of sindbis virus.

Significance of DI particles

Interference by DI particles, like interferons and other mechanisms is important in several aspects of viral infection. For instance, in human oral vaccination with attenuated poliomyelitis viruses the three strains must be administered in a precise sequence or at specified concentration ratios to avoid interference of one strain with another. Similarly, the presence of various enteroviruses in the normal intestinal flora may hinder the establis-hment of infection by the vaccinia strains. Viruses already present may also influence the response to naturally infecting virus. For example, dengue virus infections in man is milder in presence of an attenuated strain of yellow fever virus (both flaviviruses).

In animal and human viral infections, persistent infections result when there is an equilibrium among the multiplication of infectious virus, the multiplication of DI virus and cell division. The dynamic cyclic production of infectious and DI viruses eventually give way to a low level, steady state, persistent infection.

DI particles tend to make the infection self-limiting. In a cell culture at high multiplicity of infection, DI particles accumulate with many types viru-ses. If these particles accumulate sufficiently they inhibit the further infec-tion. For example in poliovirus, after high multiplicity serial passage of polio-virus the pure virions begin to become contaminated with slower sedimenting particles containing shorter RNA molecules. These have been designated poliovirus defective interfering particles (poliovirus DI). These poliovirus DIs provide a model for DI particles in other viral systems. It may be more than a laboratory curiosity because large concentrations have been found in poliovirus vaccines and poliovirus DI may be atleast partially responsible for the efficacy of vaccine.

Further Reading

Holland, JJ (1990). Defective viral genomes. In Virology (2nd ed.), pp. 151-165. Fields, B.N. and Knipe, D.M. (ed.), New York, Raven Press.

Holland, JJ (1990). Generation and replication of defective viral genomes. In : Virology (2nd ed.), pp. 77-79. Fields, B.N. and Knipe, D.M. (eds.), New York : Raven Press.

Lazzarani, RA, Kneene, JD and Schubert, M (1981). The origins of defective interfering particles of the negative-strand RNA viruses. *Cell,* 26 : 145.

Vogt, PK and Jackson, AO (eds.) (1999). Satellites and defective viral RNAs. *Current Topics in Microbiology and Immunology,* 239 : 1-180.

B. SATELLITE VIRUSES AND SATELLITE NUCLEIC ACIDS

The term 'satellite virus' was first used by Kassanis in 1962 to describe the small virus that was dependent on tobacco necrosis virus (TNV) for multi-plication in plants. Since then, several other satellites have been described from plant, animal and bacterial viruses. Satellites refer to a virus or nucleic acid that is unable to multiply in cells without the assistance of a specific helper virus. These viruses are referred to as **satellite viruses** and the virus that renders the help for replication is called **helper virus**. Satellite viruses depend on a co-infecting helper virus to provide essential functions. The satellite genomes do not exhibit any significant homology with that of helper virus. Thus, the satellite viruses differ from DI viruses in that they are dependent on taxonomically unrelated helper virus. Satellites are usually differentiated into two categories : **satellite viruses** and **satellite nucleic acids**. *They are different in that satellite viruses encode their own coat protein, whereas satellite nucleic acids rely on the helper virus to provide coat protein to encapsidate their genomes.* The origin of satellites remains obscure, but they are not derived from the helper virus. The presence of satellite virus or satellite nucleic acid may effect the replication of the helper virus and may also decrease or increase the severity of the disease caused by the helper virus.

Satellite viruses or nucleic acids are found in all major viral groups. They have been identified with viruses of DNA or RNA genomes and they may possess linear ds DNA, linear ssDNA and linear dsRNA, linear ssRNA and circular ssRNA genomes. The RNA genomes may be very small ranging from 220 to 1800 nucleotides. Characteristic features of some satellite viruses and nucleic acids along with their host and helper virus are presented in Table 30.1.

Table 30.1. Characteristic features of some satellite viruses and satellite nucleic acids.

S.N.	Name	Host	Satellite virus/ satellite nucleicacid	Genome	Helper virus
1	*Enterobacteriophage P_4*	Bacterium	Satellite nucleic acid	dsDNA	Enterobacteri-ophage P_2
2	Dependovirus	Animal	Satellite virus	ssDNA	Adenovirus or Herpesvirus
3	Tomato leaf curl satellite DNA	Plant	Satellite nucleic acid	ssDNA	Geminiviruses
4	Saccharomyces cerevisiae M virus	Fungus	Satellite nucleic acid	dsRNA	Toti virus
5	Tobacco necrosis virus satellite virus	Plant	Satellite virus	+ssRNA	Tobacco necrosis virus
6	Arabis mosaic virus large satellite RNA	Plant	Satellite nucleic acid	+ssRNA	Arabis mosaic virus
7	Hepatitis delta virus	Animal	Satellite nucleic acid	–ssRNA	Hepatitis B virus
8	Cucumber mosaic virus satellite virus	Plant	Satellite nucleic acid	ssRNA (make no mRNA)	Cucumber mosaic virus

P_4 and P_2 phages as models for satellite and helper viruses

The satellite phage P_4 forms plaques only on bacteria lysogenic for P_2 or related phages. Phage P_4 is ds DNA phage of 7×10^6 daltons, whereas P_2 is 22×10^6 daltons. It is a temperate phage and can infect and lysogenize by insertion even when P_2 is absent. But plaque formation by P_4 depends absolutely on P_2. Furthermore, if any of the 17 genes of the P_2 prophage that are needed for P_2 virion formation or lysis has been inactivated by mutation, P_4 fails to grow. Phage P_4 activates transcription of the late genes of P_2 and uses their protein products to package its own DNA and liberate virions from the infected cell. Replication of P_2 prophage is not induced by P_4. The quantity of protein derived from transcribing the single prophage copy is apparently sufficient. The P_2 genes involved in replication and early regula-tion are not needed for P_4 growth. Apparently P_4 performs these functions with its own genes. Yet the behaviour of P_4 on hosts carrying partial deletions of P_2 prophage indicates that some early gene production of P_2 stimulates transcription of late genes of P_4. Each of the two viruses can thus turn on the later genes of the other. These reciprocal regulatory interactions indicate that P_4 is extremely well adapted for exploiting P_2 and suggest a relationship of long duration between the two.

Hepatitis delta virus (HDV) is a satellite virus on helper hepatitis B virus (HBV). In this relationship, HBV provides HDV with envelope protein. HDV has a covalently closed, circular, negative-sense, ssRNA genome of around 1700 nucleotides. It is about 70% base paired. The RNA has a single open reading frame (ORF) that encodes the core protein. The replication is nuclear and possibly makes use of the helper virus encoded proteins.

Tobacco necrosis virus satellite virus, a plant satellite virus requires tobacco necrosis virus as helper. The single stranded RNA genome of tobacco necrosis virus satellite virus is 1239 nucleotides with one ORF.

In plants, satellite viruses and satellite nucleic acids may attenuate or exacerbate disease caused by the helper virus. Examples of diseases include necrosis and systemic chlorosis, or reduced chlorophyll production leading to leaves that are pale, yellow, or yellow-white. The symptoms induced by satellite RNAs are thought to be a consequence of silencing of host genes. For example, the Y-satellite RNA of cucumber mosaic virus causes systemic chlorosis in tobacco. This syndrome is caused by production of a small RNA from the Y-satellite RNA that has homolosy to a gene needed for chlorophyll biosynthesis. Production of this small RNA leads to degradation of the corresponding mRNA, causing the bright yellow leaves.

Further reading

Francki, RIB (1985). Plant virus satellites. *Ann. Rev. Microbiol.,* **39 :** 151-174.

Bertani, LE and Six, EW (1988). The P2 like phages and their parasite, P4. *Annual Review of Genetics,* **24 :** 465-490.

Berns, KI and Girand, C (1996). Biology of adeno-associated virus. *Curr. Topics in Microbiology and Immunology,* **218 :** 1-23.

Dry, IB *et al.* (1997). A new subviral agent associated with a geminivirus. The first report of a DNA satellite. *Proc. Nat Acad Sci (USA)* **94:** 7078-7093.

Taylor, JM (1999). Replication of human hepatitis delta virus: Influence of studies on subviral plant pathogens. *Adv. Vir. Res.,* **54:** 45-60.

Collmer, CW and Howell, SH (1992). Role of satellite RNA in the expression of symptoms caused by plant viruses. *Ann. Rev. Phytopath.,* **30:** 419-442.

Maramarosch, K (ed.) (1991). Viroids and satellites: Moleculer parasites at frontiers of life. Boca Raton, FL : CRC Press.

Roosnick, MJ *et al.* (1992). Satellite RNAs of plant viruses: structures and biological effects. *Microbiol. Rev.,* **56:** 265-279.

Vogt, PK and Jackson, AO (eds.) (1999). Satellites and defective viral RNAs. *Current topics in Microbiology and Immunology,* **239:** 1-180.

31

VIROIDS

In 1971, two groups of plant pathologists, **T.O. Diener**, and **R.P. Singh** and **Clark** described a new type of infectious agent. They were studying potato tuber spindle disease, which until that time, was thought to be caused by a virus. However, no virions could be detected. Rather, they discovered molecules of RNA in the diseased plant cells. Diener proposed the concept of a **viroid**, an infectious RNA particle smaller than a virus. Since then a number of viroids have been described to infect plants and cause the diseases.

1. GENERAL FEATURES

Investigations over the past thirty years conclusively revealed that these are separate infectious agents of plants different from viruses. They were found to possess the following general features:

1. Viroids are smallest infectious agents known to infect only plants, cause specific maladies and produce characteristic syndrome. No viroid is presently known to infect animals.
2. Viroids are essentially hot climate pathogens. Viroid diseases are reported from tropical climate.
3. Viroids consits of a single circular RNA molecule of low molecular weight with unique structure. Molecular weight ranges from 246 to 390 nucleotides.
4. RNA is circular (however, linear forms are also found), form secondary structure resulting in a DNA like conformation (Fig. 31.1), hence not digested by host nucleases.
5. Genome is nontranslatable.

6. Ninety five percent of the genome is located in the nucleolus of the host nucleus (*Pospiviroidae*). Inside the nucleolus, it forms nucleic acid-protein complexes. Some are associated with chloroplasts (*Avsunviroidae*).
7. Viroid particles are not apparent in infected tissues without the use of special techniques to identify nucleotide sequences in the RNA.
8. Viroids are highly resistant to heat, as well as to UV and ionizing radiation.
9. Unlike statellite nucleic acids, they don't require any helper virus.

Diagnostic procedures of viroids include biological tests, gel electrophoresis, PCR and nucleic acid hybridization tests. Viroids differ from the true viruses in many respects (Table 31.1).

Table 31.1. Differences between viruses and viroids

Viruses	Viroids
Infect plants, animals and bacteria	Infect only plants
Transmitted by all means	Transmitted mechanically, seed, vegetative propagules
Capsids usually present	Absent
Genome DNA or RNA	Circular ssRNA
Encode virus specific polymerases	Totally dependent on host plant enzymes for replication
Plant meristematic tissues are free from viruses	Predominantly meristematic tissue pathogens.
Less stable in callus culture	More stable in callus and cell cultures
Affected by heat, radiation and protein denaturing agents	Not affected

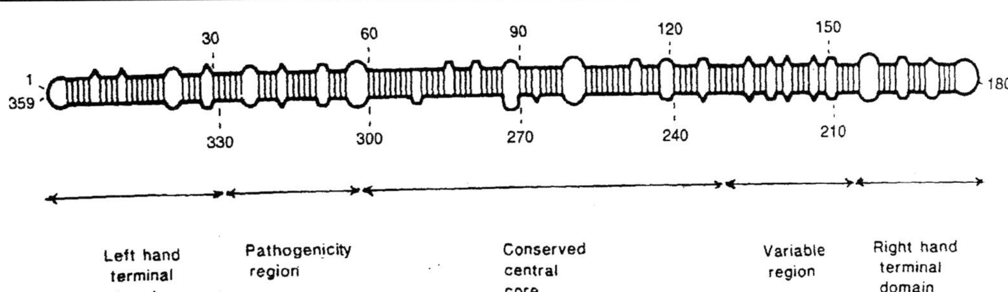

Fig. 31.1. The potato spindle tuber viroid genome

2. REPLICATION

The exact mode of replication of viroids is still an enigma. They replicate via rolling circles. Presently, replication has been explained on the basis of two different schemes : i. DNA dependent replication and ii. RNA dependent replication (Fig. 31.2). The first one is based on the premise that viroids, like retroviruses, produce a novel DNA by reverse transcription which later produces viroid RNA by normal transcription. According to the second concept RNA directed RNA polymerases are present in normal cells to direct RNA synthesis. It is suggested that RNA polymerase II transcribes the viroid RNA. The viroid genomes undergo self-cleavage to form a hammer headed secondary structure at cleavage sites.

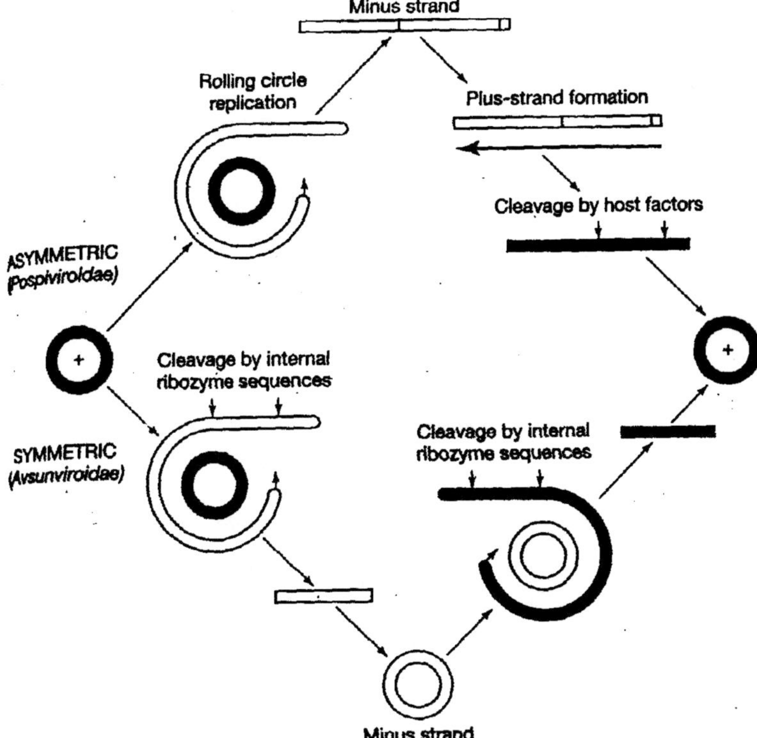

Fig. 31.2. Replication of viroids. The asymmetric rolling circle replication mechanism of *Pospiviroidae* (top), and the symmetric mechanism of *Avsunviroidae* (bottom) (From Dickinson M. Molecular Plant Pathology)

3. CLASSIFICATION

More than thirty viroids have been so far described. These are placed in two families viz. Pospiviroidae with five genera, and

Avsunviroidae with two genera (ICTV, 2005). It may be noted here that the taxa names end with suffix *viroid* instead of virus.

4. VIROID DISEASES

As viroids cause diseases, obviously, they must disrupt host cell metabolism in some way, but because no protein products are produced, it is not clear how viroids and their RNAs cause disease. They may interfere with the cells ability to process mRNA molecules. Without mRNA molecules, proteins cannot be synthesized. If so, cell metabolism would be so disturbed that cell death could result. Although some viroids cause no apparent effect or only mild pathogenic effects in the host, other viroids are known to cause several lethal plant diseases, such as potato spindle tuber disease (Table 31.2 and Fig. 31.3). Interestingly, none of the so called viroids were recognized before 1922 despite centuries of intense cultivation of these crops. Some plant viral pathologists believe that while isolated plants may have contained viroids for an unknown number of years, modern agricultural methods, such as growing large numbers of the same plant in close association and the use of machinery for harvesting, may have allowed viroid diseases to spread, allowing the observers to recognize them. Another view is that viroids may have entered crop plants from unknown wild plants, an idea supported by the observation that the first viroid-infected crop plants appear along the margins of the fields that abut wild plants.

Table 31.2. Some plant diseases caused by viroids

Disease	Viroid
Potato spindle tuber	Potato spindle tuber viroid (PSTVd)
Citrus exocortis	Citrus exocortis viroid (CEVd)
Chrysanthemum stunt	Chrysanthemum stunt viroid (CSVd)
Coconut Cadang-Cadang	Coconut cadang-cadang viroid (CCCVd) (Fig. 31.3)
Tomato bushy top	Tomato busy top viroid (TBTVd)
Avocado sunblotch	Avocadosunblotch viroid (ASBVd)
Carnation stunt	Carnation stunt viroid (CarSVd)
Peach latent mosaic	Peach latent mosaic viroid (PLMVd)

5. ORIGIN OF VIROIDS

A number of hypothesis have been proposed to account for the origin of viroids. One suggests that viroids have accidentally evolved from normal, regulatory nuclear RNAs in their natural cultivated or wild plants into structures of high stability and with the capacity of self replication and intercellular and intracellular mobility. A second hypothesis states that viroids might have evolved from intervening sequences (introns). During mRNA processing the introns became independent, stable and self-replicating. This hypothesis is supported by a sequence homology between most sequenced viroids and the 5' U1 RNA, a smaller nuclear RNA believed to be involved in mRNA splicing. Another hypothesis proposes that viroids evolved from prokaryotic RNA through the infections of higher plants by prokaryotes. This hypothesis is based on the finding that not only eukaryotic RNA polymerases, but also RNA and DNA polymerases from *E. coli* are able to transcribe RNA *in vitro* into complimentary RNA and DNA copies respectively. Still, another hypothesis proposes that viroids may have derived from transposable genetic elements. Keifer *et al.* (1983) found that viroid structures contain sequences that resemble those occurring at the ends of certain movable genetic elements and retroviral provirus. Thus viroids would have evolved from transposable elements or retroviral provirus by deletion of most of the intron regions of the elements as well as intervening host sequences.

Fig. 31.3. Symptoms of cadang-cadang disease of coconut palm caused by the cadang-cadang viroid (Photo courtesy from Hanold and Randles, 1991)

Further reading

Diener, TO (1999). Viroids and the nature of viroid diseases. *Arch. Virol.,* **15 :** 203-220.

Symons, RH (1997). Plant pathogenic RNAs and RNA catalysis. *Nucl. Acids Res.,* **25 :** 2683-2689.

Diener, TO (1979). Viroids and viroid diseases Wiley-Interscience, p. 252.

Maramarosch, K (ed.) (1991). Viroids and satellites : Molecular parasites at the frontier of life. Boca Raton, FL : CRC Press.

Symons, RH (1990). Viroids and related pathogenic RNAs. Seminars in Virology 1, 75-162.

Flores, R, Di Serio, F and Hernandez, C (1997). Viroids : The non-coding genomes. Seminars in Virology, 8 : 65-73.

Reisner, D and Gross, HJ (1985). Viroids. *Ann. Rev. Biochem.,* **54 :** 531-564.

Doherty, EA and Doudna, JA (2000). Ribozyme structures and mechanism. *Ann. Rev. Biochem.,* **69 :** 597-615.

32

PRIONS

Since the beginning of the last century, several cases of a slow but progressive illness in humans were observed independently by Hans Gerhard Creutzfeldt and Alfans Maria Jakob. The disease, now called **Creutzfeldt-Jakob disease** (CJD), is characterized by mental degeneration (dementia) that results in loss of motor function, loss of mental acuity, memory loss and insomnia and eventual death. Since then several such neurological degenerative diseases have been described (Table 32.1).

Table 32.1. Some important diseases caused by prions

Diseases	Host
Cruetzfeldt-Jakob disease (CJD), Kuru	Human
Gerstemann-Straussler-Schenker (GSS) syndrome	Human
Fatal familial insommia (FFI)	Human
Chronic wastage disease (CWD)	Deer
Transmissible mink encephalopathy (TME)	Mink
Scrapie	Goat, sheep
Bovine spongiform encephalopathy (BSE)	Dairy cattle
Feline spongi form encephalopalhy	Cat
Ungulate spongi form encephalopalhy	Nyala, Oryx

Creutzfeldt-Jakob disease (CJD) is a relatively rare presenile dementia that occurs worldwide and has a prevalence of approximately one death per million per annum. Although sporadic, it has a familial pattern in about 10 to 12 per cent of the cases, presenting as an autosomal dominant form of inheritance. Though it is not contagious or a communicable disease it is iatrogenically transmitted to patients and medical personnel and always proved to be fatal. Kuru, a heredofamilial disease is restricted in distribution to a cannabolic human tribe of Easter Highlands of Papua, New Guinea. Kuru is characterized by cerebellar ataxia and a shivering-like tremor that progresses to complete motor incapacity with dysarthria and total loss of speech and death usually in less than one year from onset. The disease incidence has been mainly traced out due to the cannabolic practice of the tribe. Its mortality has declined continuously over the past thirty years because of education and awareness against cannabolic practice. Bovine spongiform encephalopathy (BSE) – commonly called 'madcow disease' – in dairy cattle has been observed to cause slow loss of neuronal function that leads to death. It has been reported to be transmitted to humans through BSE infected beef. Scrapie is a prototypic chronic, fatal ataxia of sheep and goats. Affected animals show progressive ataxia, tremor, wasting and frequently severe pruritus that causes them to rub their hindquarters and flanks against any upright post. This led to the name scrapie.

1. DISCOVERY OF PRIONS

All these neurological disorders, for a long time, were thought to be caused by viruses. However, as the causal agents could not satisfy the Koch's postulates, thus the causative agent for these disorders remained a mystery for a long time and it is only in the last decade that the molecular mechanism underlying these disorders was unraveled. In the course of investigations to unravel the mysterious agent behind these diseases, two hypothesis have been proposed (Mustal and Aguzzi, 2000).

1. The protein only hypothesis, and
2. The virus and virino hypothesis

The first one, protein only hypothesis, was first formulated by Griffith (1967). It states that the infectious agent causing transmissible spongiform encephalopathies consists of the modified form of a normal cellular proteins. It further proposes that the modified form of protein when introduced into a normal host, causes the conversion of normal protein into a modified pathogenic form. The virus and virino hypothesis, that was proposed as an alternative to the 'protein-only hypothesis', states that the infectious agent consists of viral nucleic acids complexed to host derived protein. This theory claims that infectious protein can be recruited by the viral nucleic acids as a sort of coat. The existence of many different strains of prions has been put forward as an argument in favour of this hypothesis.

S.B. Prusiner, working at the University of California, San Francisco, in a series of experiments, demonstrated that scrapie agents were extremely resistant to a class of enzymes called nucleases, which very specifically degrade the nucleic acids. Irradiation with DNA damaging UV rays also revealed no loss in scrapie infectivity. On the other hand, inactivation of the scrapie agent was evident when they were treated with proteases which break the protein molecules. This led to the conclusion that the causal agent of scrapie is proteinaceous in nature having no nucleic acid associated with it. For this reason, Prusiner (1982) introduced the term **prion (proteinaceous infectious particle, PrP)**' for the causal agent of scrapie disease. Since then the 'prion' word became firmly established in literature. Ultimately, in 1996, Prusiner was awarded Nobel prize for medicine for the hypothesis that prions consist of modified form of the normal cellular protein called PrP^c.

S.B. Prusiner (1942 -); Nobel Prize, 1997

2. DISCOVERY OF THE GENE

Absence of any genetic material in the form of either DNA or RNA in the scrapie agent raised questions about the origin of the prion protein. Logically, it was thought that the prion protein might be encoded by a chromosomal gene of the host. In the year 1984, Prusiner determined the sequence of fifteen amino acids from the amino terminal end of the scrapie prion protein. Using the amino acid sequence information, a short polymer of nucleotides was synthesized. This oligonucleotide was used as probe and hybridized with DNA from both normal and scrapie infected brain of hamster to detect the presence of any complementary gene sequence. Surprisingly, there was a positive probe hybridization in both the cases revealing the existence of corresponding gene in the hamster nuclear genome. Following the same process, single prion related genes were discovered in murine and human genomes. Apart from these studies, presence of mRNA which encodes the scrapie prion in both normal and scrapie infected brain of mouse was demonstrated by a group of researchers working at the National Institute of Health, USA in the year 1985 based on hybridization of a synthesized nucleotide probe with the brain mRNA. Subsequently, the chicken prion protein gene was discovered in the year 1992 independantly by a different group of researchers. Finally, the chromosomal location of the prion protein genes was very well established. It has been found that, in humans the prion protein gene is located in the short arm of chromosome 20 whereas in mouse it is on chromosome 2. The prion protein gene is designated as *PrP*.

3. GENE STRUCTURE

The structure of the gene encoding the prion protein has been analyzed in great detail in different organisms employing molecular techniques. The gene encoding the cellular isoform the prion protein (PrP^C or simply PrP) is termed *Prnp* (in the mouse) or *PRNP* (in humans). PrP^C is encoded by a single open reading frame in all known mammalian and avian Prnp genes. The Prnp genes of mouse, sheep and cattle contain 3 exons, while those of hamsters and humans span over 2 exons. Prnp is present in the genome as a single copy and does not show any homology to other known cellular genes.

The expression of the PrP gene has been found to be highly regulated during the development stages. PrP gene expression seems to be controlled by the SP1 transcription factor which binds to G+C rich nonamer-regions upstream of exon 1. Prnp mRNA is constitutively expressed in the adult brains. Maximum concentration of the PrP mRNA is found in the neurons, while axons are the media by which prions are transported to other body parts.

4. THE CELLULAR PRION PROTEIN (PrP^C)

The nascent protein encoded by Prnp gene is about 33-35 kDa in weight with 254 amino acids. Before processing it is anchored to the cell membrane via a glycosyl-phosphatidyl-inositol residue (GPI-anchor). During post-translational processing of PrP^C, an amino-terminal secretory signal peptide of 22 amino acids is cleaved from the 254 amino acid PrP^C precursor protein. Similarly, 23 carboxyl-terminal residues are also processed during addition of the GPI-anchor to a serine residue at position 231. Thus, upon completion of these modifications, mature PrP^{C} contains 209 aminoacids. Further, two oligosaccharides are complexed at N-positions of 181 and 197 residues.

The physiological function of PrP has not yet been clarified. They have not been attributed to any particular physiological function. The same was evident from the experiment with prion knock-out mice. These prion knock-out mice were found to be not only completely resistant to infection when exposed to disease causing prion protein preparations but also they remained apparently healthy, suggesting that the normal prion protein is not essential in mice. Its role in the nervous system remains a mystery. One school of thought is that the function of PrP^{C} might be related to its convertibility into PrP^{SC}. Some structural equivalents of PrP^{SC} may exist in liquid phase in equilibrium with PrP^{C}, in a non-infectious and non-neurotoxic form. Non-genetic propagation of structural information horizontally might be advantageous for a population.

5. PRION PROTEIN SCRAPIE SPECIFIC (PrP^{SC})

The cellular prion protein (PrP^{C}) although present in high concentration, does not cause the disease on its own; neither it is infectious. On the other hand, the pathological isoform PrP^{SC} is invariably associated with transmissible spongiform encephalopathies (TSE). Extensive biochemical characterization failed to identify differences between the primary structure of PrP^{C} and PrP^{SC}. By contrast, physical measurements have demonstrated dramatic conformational differences between the two PrP forms. Fourier transform infrared and circular dichroism spectroscopic studies revealed that protein folding makes the prime difference between the two. The one PrP^{C} undergoes normal folding and, whereas the PrP^{SC} undergoes abnormal folding. In analogy to a well known literary work, the normal PrP^{C} can be compared to the friendly Dr. Jekyll, and the disease causing PrP^{SC} to the dangerous Mr. Hyde, the same entity but in two different manifestations (Anonymous, 1997).

Spectroscopic studies revealed that the folding differences that the α-helical content of PrP^{C} is 40%, with little or no β sheet. In contrast, PrP^{SC} contains 50% β sheet and only 20% α-helix (Fig. 32.1). The PrP^{SC} are extremely stable and resistant to proteolysis, organic solvents, radiation and high temperature (even greater than 100°C). They are sensitive to protein denaturing agents such as phenol and urea. The high β-sheet content of PrP^{SC} is responsible for its amyloidogenic properties, resulting in the characteristic amyloid deposits in the brain, similar to the amyloids found in other neurodegenerative diseases, such as Alzheimer's disease.

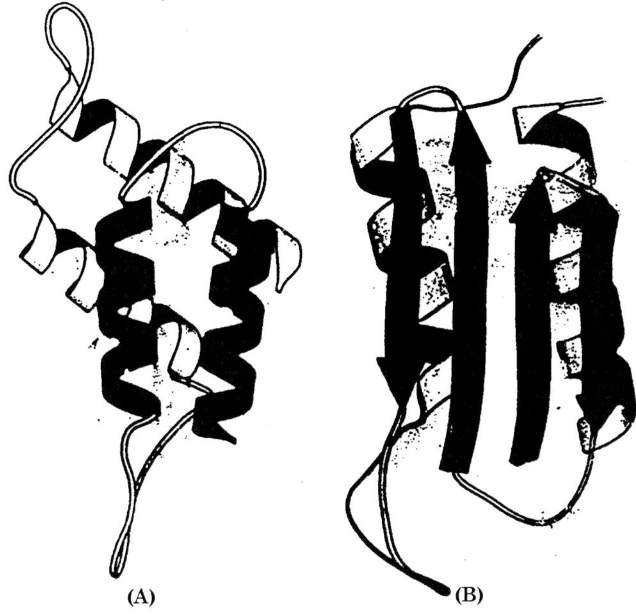

Fig. 32.1. Protein structure models of the two forms of the prion proteins (A) PrP^c (B) PrP^{SC}. Note: beta pleatues in PrP^{SC}.

6. CONVERSION OF PrP^C INTO PrP^{SC}

When PrP^C protein constitute the *milieu* of the physiology of normal individuals, logically the question arises, how do these PrP^C get transformed into PrP^{SC} and further how do these PrP^{SC} are transmitted horizontally as in the case of mad cow disease and kuru disease. The nature of the conversion of PrP^C into PrP^{SC} is still obscure. To date, all attempts to achieve conversion of PrP^C to an infectious form *in vitro* have failed. The conversion may consist of a chemical or of a conformational modification of PrP^C after it has reached the cell surface.

Two different models have been proposed, during the last decade to explain the kinetics of conversion from PrP^C to PrP^{SC}.

(i) The template directed model, and (ii) Nucleation model

i. The template directed model : This model proposed by Prusiner (1991) suggests that the PrP^{SC} form is more stable than PrP^C, but kinetically inaccessible. In this case, PrP^{SC} could promote conversion by catalyzing the rearrangement of a molecule of PrP^C, or of any intermediate form, into the more stable PrP^{SC} conformation. Thus infectivity would rely on the ability of the PrP^{SC} molecule to bind to and catalyze conversion of existing interme-diate molecules. According to this model, the genetically inherited and spontaneous diseases result from mutations that increase the rate at which mutant PrP^{C} spontaneously converts to PrP^{SC} and/or enhance the number of intermediate molecules.

ii. Nucleation model : The basic principle underlying this model is nucleation-dependent polymerization. The conversion between PrP^{C} and PrP^{SC} is reversible in the absence of a preexisting aggregate, with a PrP^{SC} monomer being less stable than PrP^{C}. PrP^{SC} polymers, however, might bind to and stabilize the PrP^{SC} conformation, thereby promoting the conversion of PrP^{C}. The initial nucleation process, thus, represents the barrier to a stable conversion, since formation of low order aggregates is not favoured until a minimum size nucleus is attained.

Still, some propose that there could be a hybrid mechanism. The surface of an PrP^{SC} aggregate, initially formed by a nucleation process, could catalyze the conformational change of PrP^{C} monomers.

Further reading

Prusiner, SB (1996). Molecular biology and pathogenesis of prion diseases. Trends in Biochemical Sciences, **21:** 482-487.

Prusiner, SB, Scott, MR, Dearmond, SJ and Cohen, FE (1998). Prion protein biology. *Cell,* **93:** 337-348.

Tuite, MF (2000). Yeast prions and their prion forming domains. *Cell,* **100 :** 289-292.

Wadsworth, JDF, Jackson, GS, Hill, AF and Collinge, J (1999). Molecular biology of prion propagation. *Curr. Opinion in Genetics and Development,* **9 :** 338-345.

Weissman, C (1999). Molecular genetics of transmissible spongiform encephalopathies., *J. Biol. Chem.,* **274 :** 3-6.

Belay, E (1999). Transmissible spongiform encephalopathies in humans. *Ann. Rev. Microbiol.,* **53 :** 283-314.

Hill, AF, Antoniou, M and Collinge, J (1999). Protease resistant prion protein produced *in vitro* detectable infectivity. *J. Gen. Virol.,* **80 :** 11-14.

PART SIX

TRENDS
in
VIROLOGY

33

VIRAL METAGENOMICS

1. INTRODUCTION

Viruses though submicroscopic constitute the most abundant biological component and a large reservoir of genetic diversity on Earth. Being obligate cellular inhabitants of almost all prokaryotic and eukaryotic organisms, the occupy every habitable niche like their hosts. Their association with host cell ranges from beneficial, non-harming, chronic lysogenic infections to life threatening. Consequently, they exhibit profound effect on the structure and functioning of all ecosystems. Their number over ~10^{31} on planet is approximately 10 times more than prokaryotes. It has been estimated that marine bacterial viruses kill ~20-40% of bacteria per day, move 10^{29} genes per day and exist as prophages with in the genomes of about half the microbes at any given time. It is surprising to know that even the human genome consists of many components of genomes of viruses Viruses control microbial abundance, dynamics and community structure, and microbial genetic diversity and evolution are shaped by virus-mediated gene transfer and host range. In addition, viruses exert a profound effect on food-web interactions and global geochemical cycles.

2. GLOBAL DISTRIBUTION AND ABUNDANCE OF VIRUSES

Direct-count epifluorescence and transmission electron microscopy investigations have shown that viruses are highly abundant in most environments. Bergh et al., (1989) have demonstrated that 1litre of sea water may contain as many as 10^{10} virus-like particles (VLPs), approximately 10 times more than the number of prokaryotes. Terrestrial environments often have 10^9 VLPs per gram. Viruses are the most abundant entities in the biosphere totaling an estimated number of 1.2 X

10^{30}, 2.6 x 10^{30}, 3.5 x 10^{31} and 0.25-2.5 x 10^{31} in the open ocean, in soil and in oceanic and terrestrial sub surfaces respectively. In the human body, 10^{13} human cells are outnumbered by 10-fold bacteria and 100-fold by viruses (Table 33.1).

Table 33.1: Estimated number of virus-like particles on Earth (source Gaemes et al, 2016)

Biome	**Number of Prokaryotic cells**	**Medium virus-to-microbe ratio**	**Virus-like particles biome**	**Percentage of pertotal virus-like particles**
Marine	1.01 x 10^{29}	12.76	1.29 x 10^{30}	2.6828
Freshwater	1.26 x 10^{26}	14	1.76 x 10^{27}	0.0037
Other aquatic	2.44 x 10^{27}	30	7.32 x 10^{28}	0.1524
Sediment	3.80 x 10^{30}	11	4.18 x 10^{31}	87.0131
Soil	2.50 x 10^{29}	19.5	4.88 x 10^{30}	10.1481
Human-associated	2.80 x 10^{23}	0.1	2.80 x 10^{22}	0.0000
Other host-associated	Unknown	25	Unknown	Unknown
Total	4.15 x 10^{30}	12	4.80 x 10^{31}	

3. EXPLORATION OF VIRAL DIVERSITY

Methodologies adapted to explore viral diversity can be broadly categorized into two types: i. Traditional approaches and, ii. Metagenomic approaches.

3.1 Traditional Approaches

Traditionally, virologists have focused on viruses that cause diseases in humans, domestic animals and crops using culture-based and phenotypic characters. Virus discovery during most of the previous century followed a traditional approach of virus isolation and amplification in cell culture or animal and plant models. These virus isolates were then classified based on their morphological and serological properties. The widely used cell culture isolation of viruses failed to identify viruses that were refractory to grow *in vitro*. Moreover, traditional approaches including serological assays and PCR were very specific or focused on a group of viruses. The inherent dependence of these approaches on biological properties or sequences of known viruses resulted in our limited knowledge of the virus world.

Some revolutionary developments in the identification of viruses took place after 1980. Use of sequence dependent (i.e. generic PCR assays and microarray) and sequence independent (i.e. single primer amplification (SISPA) and random priming) approaches for nucleic acid amplification combined with Sanger sequencing or HTS allowed the rapid identification of new viruses. However, most of these tests are designated to be virus specific that make them unsuitable for detection of unexpected and/or completely new viruses as well as novel viral variants. It was largely due to several caveats encountered in the comprehensive identification of all viruses.

Considering these challenges, an ideal sequencing-based virus identification approach should use a selection method to increase the viral sequences or reduce non-viral sequences while remaining unbiased towards the nature and composition of the virus genome. The culture-independent method of metagenomic approach has allowed discovery of unprecedented microbial diversity that remained under estimated by culture-dependent approaches. This vastly useful information gathered through bacterial metagenomics studies, ignited interest of marine biologists to adapt this approach for viral metagenomics. Viral metagenomics is a sequence and culture-independent approach that has proven to be invaluable tool. It can be applied to a wide range of unknown viruses/viral pathogens and revealing novel aspects of biodiversity. Soon after, several groups used this approach or its modified forms, to identify human and animal viruses. Today, metagenomics remains the most widely used approach for new virus identification and virome analysis.

Table 33.2 Virome metrics

Biome	**Number of viromes**	**Number of reads**	**Percentage of reads assembled into ≥ 1-kbp contigs**	**Percentage of reads mapped to reference library**		
				90% identity	**95% identity**	**99% identity**
Marine	192	56,676,517	11.83	38.1	29.4	20.4
Freshwater	48	11,519,523	19.51	15.7	11.4	8.1
Other aquatic	19	3,342,537	14.04	40.2	35.8	27.3
Sediment	21	15,729,082	10.00	54.7	51.7	44.2
Soil	9	2,459,152	15.54	41.9	36.6	29.0
Human-associated	1,158	481,172,486	25.16	30.3	27.2	12.4
Other host-associated	167	34,600,192	3.81	34.6	31.1	25.3
Other	7	396,889	28.22	44.3	36.8	17.4
Total	1,621	605,896,378	16.01	31.8	28.1	14.7

Source : Gaemes *et al.* (2016)

3.2 Metagenomic Approaches

Metagenomic methods can be applied to study viruses in any system. To date, metagenomic approach in the study of viral microbiome (virome) has been applied to a range of biological and environmental samples including human and equine intestinal contents, bat guano, sea water, marine sediment, fresh water, hot springs, soil and plants (Table 33.2). Early results from a large initiative to describe the human microbiome associated with health and disease have recently been published and such findings, together with those of other studies, are likely to lead to the discovery of a wealth of previously unknown viruses. Thus, these studies revealed an extraordinary diversity and prevalence of viruses in different ecosystems, highlighting the key contributions of these microbes to all ecosystems, on earth. Despite the large numbers of viromes that have been examined over the past 20 years, almost every new viromics project yields a large number of sequences that have no significant degree of similarity with those referenced in databases. These sequences collectively are often referred to as **'dark matter'.**

4. METHODS IN VIRAL METAGENOMICS

Viral metagenomic methods have evolved significantly since they were first developed. In early investigations, preliminary sample preparation involved shearing of DNA and cloning. These steps were required in order to obtain sufficient quantity of DNA in view of the low amount of viral DNA in environmental samples (for example, ~10μg/100 l of sea water). As the viral DNA often contains modified nucleotides and because some viral genes (e.g. holins and lysozymes) are toxic to host vector cells, the DNA was randomly sheared to produce small fragments before cloning. Since then the process of sample preparation has been streamlined. The advent of high-throughput sequencing technologies replaced the conventional cloning and facilitated the sequencing speed. Now, several high-throughput sequencing technologies are available that vary by the sequencing principle, sequencing speed, cost and read length and also suit the needs of users. In addition, several characteristics make viruses well studied for high-throughput sequencing and analysis. For example, the hardiness of many viral capsids makes them ideal for concentration and purification. The small size of viral genomes, particularly of phages, can be advantageous for many bioinformatics techniques and full genome assembly.

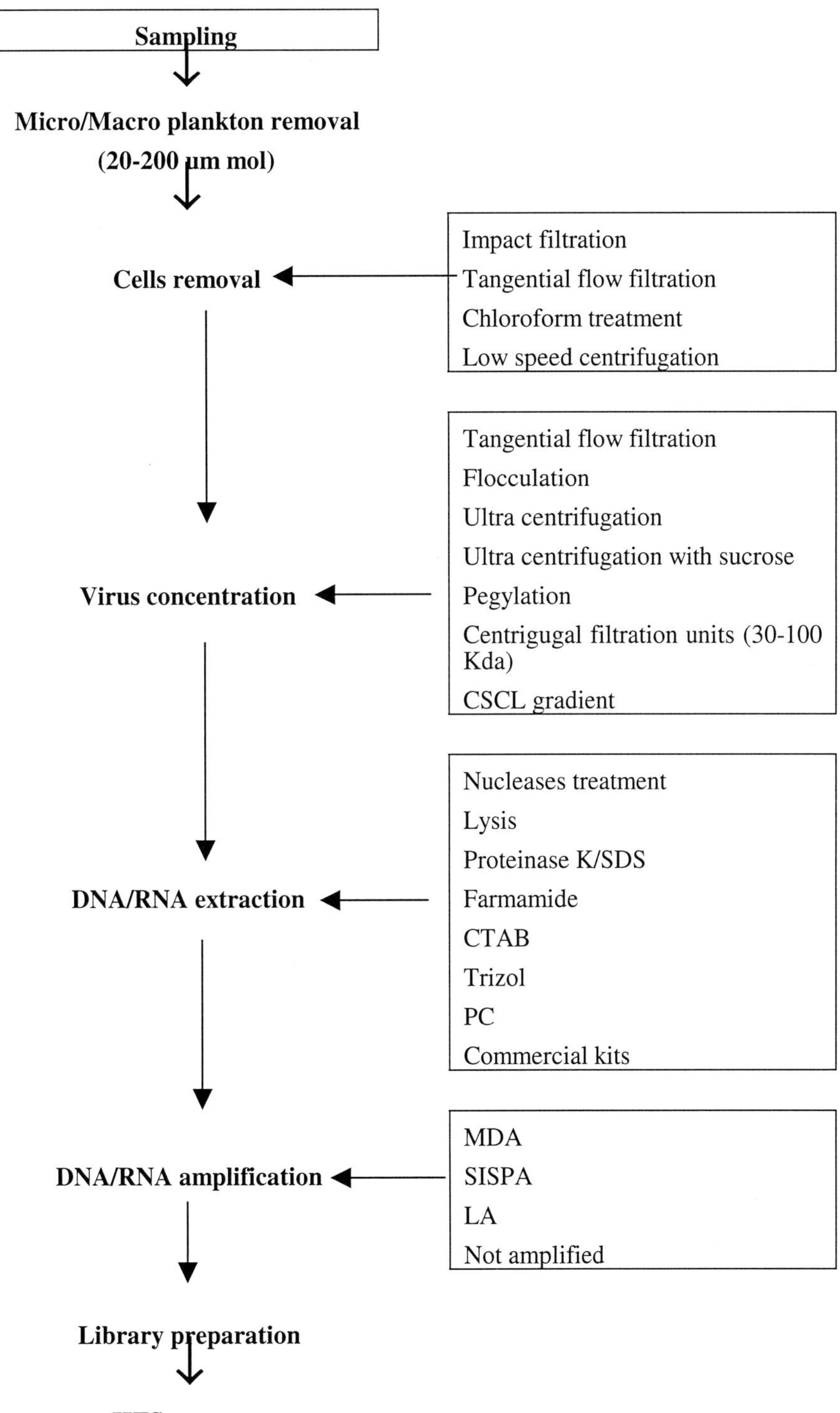

Fig 33.1A Flowchart of viral metagenome generation

An overview of a typical viral metagenomic protocol commonly used in a virus discovery is provided in figure (Fig. 33.1 A, B). Essentially, a metagenomic analysis involves three main steps.

i. Sample preparation.

ii. Library preparation

iii. High-throughput sequencing.

iv. Bioinformatics analysis.

Further, each of these can be divided into different sub-steps depending upon the experimental setup.

4.1 Sample preparation

Sample preparation, in turn, consists of different protocols like sample collection, sample processing, concentration, extraction of nucleic acids, purification etc.

4.2 Library preparation

DNA library preparation is an essential step for high-throughput sequencing by NGS platforms. Over the last decade, many variations in library preparation have been developed. Library preparation steps broadly involve manipulating the sample DNA by fragmentation, end repair and adaptor ligation, size fractionation and amplification. Combinations of recent advances in fragmentation (e.g. acoustic shearing and tagmentation), ligation reactions (adaptor- to- template ratio reference table availability), size fractionation (non-gel sizing) and amplification (linear amplification for deep sequencing and linker amplification protocols) enhance the ability to generate quantitatively representative metagenomic data sets from low-input DNA samples. The recently available Swift Bioscience 1S Plus Kit for preparing sequencing libraries incorporates an adaptase step prior to linker ligase and amplification, which makes it efficient for both ssDNA and dsDNA templates.

The final step of any library preparation procedure is quantification of library before loading the library for sequencing by emulsion PCR (em-PCR). Correct quantification prevents the library DNA from being overloaded, which can lead to mixed signals or under loaded which underutilizes sequencing capacity. Library concentration information also provides the user the opportunity to strategically pool several libraries, when sequencing depth requires less than one run or lane.

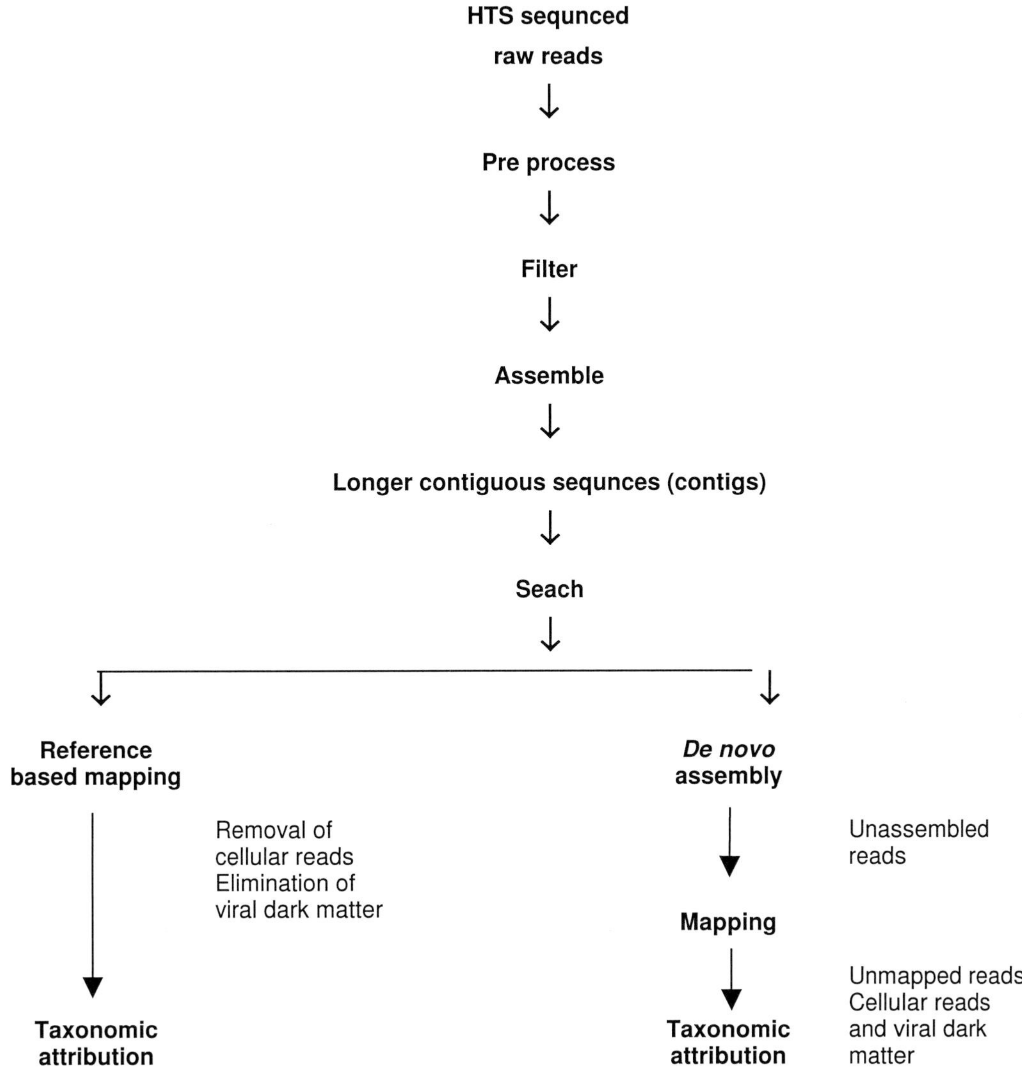

Fig. 33.1B. Flow chart for viral metagenome analysis

4.3 High-throughput sequencing

There are several technologies in widespread use that are capable of reading thousands to billions of DNA sequences per run. The first viral metagenomics projects were carried out using standard cloning techniques together with Sanger's sequencing. The development of pyrosequencing technology reduced most of the problems associated with cloning strategies, although pyrosequencing cost was also initially too expensive for the majority of metagenomic projects. The next generation high-throughput pyrosequencing by 454 Life Science provided greater amounts of sequence data compared to Sanger's sequencing method, enabling detection of even low number of viruses present in the sample with no need of cloning. Even higher throughput technologies such as Solexa Illumina and SOLiD systems are now available which can provide data of as much as 3-6 Giga base pair per run. However, the average read length obtained with these systems is about 50-100bp.

The Illumina/Solexa and Roche 454 next generation sequencing platforms are the most used platforms for viral metagenomes. The Illumina/Solexa method is based on sequencing-by-synthesis chemistry. A typical sequencing run can generate upto 18 giga bases of data with an average read length of 75-100 nucleotides. The sweet potato badnavirus and sweet potato mastrevirus are the examples of viruses discovered using the Illumina/Solexa sequencing platform. The 454 FLX titanium pyrosequencer commercialized by Roche has been the most used for discovery and characterization of novel viruses. The pyrophosphate released by this process is indirectly measured by the production of light. The amount of light released is proportional to the intensity of the light which is captured by charge-coupled device (CCD) camera. 454 FLX platform was used for the identification of uncharacterized mycoviruses, Solenopsis invicta virus 3, Merino walk virus and a new arenavirus among others.

Quality checking and filtering: The bioinformatics pipelines to analyze next generation sequencing data usually start by quality checking. The sequences are trimmed accordingly to their phred quality scores. Phred quality scores are logarithmically related to the base-calling error probabilities. For example, a Phred quality score of 10 corresponds to a base calling accuracy of 90% (10 errors per 100bp) while quality score of 20 equals to a base calling accuracy of 99% (1 error per 1000bp). Filtering protocols depend upon the type of downstream analysis.

High-throughput sequencing technologies produce a number of duplicated reads due to PCR amplification, PCR errors and/or sequencing errors. These duplicated reads pose several challenges such as reduction in computational resources, over estimation of species abundance and under estimation of less abundant species. Identifying and removing these reads, a process called 'de-duplication' is necessary. The Cd-hit-454 tool identifies and distinguishes artificial and natural duplicates in 454 pyrosequencing datasets. CD-HIT-DUP tool identifies duplicates from single or paired Illumina reads. To obtain the dataset that contains reads of interest i.e. virus-related reads, sequences that are not a target of the investigation (e.g. host reads) need to be filtered out. This minimizes the risk of misassembles and also speeds up downstream analysis. For example, the total reads generated from human samples by whole genome amplification (WGA) may contain more than 70% human related reads. Therefore, it has been suggested that all metagenomic sequencing projects also should include sequencing of negative control measures.

In view of contaminant reads and presence of non-target viral reads, the NGS sequences need to be aligned to against reference sequences. Different alignment software's are available to suit to different sequencing platforms. There are hash-based softwares such as SSAH2, MAQ and BFAST as well as suffix/prefix tries based such as BWA-SW, SOAP and Bowtie 2. Hash table-based algorithms require a large amount of operating memory, whereas suffix/prefix tries requiring less computational resources.

4.4 Bioinformatics for viral metagenomics

Bioinformatics analyses of viral metagenomes attempt to answer three questions: how many different types of viruses are there (diversty), what are they (taxonomy), and what are they doing (function). To analyze the viral metagenomic datasets in these lines, several bioinformatics tools and pipelines have been developed (Table 33.3)

4.4.1 Quality control and pre-process

Success of a workflow always depends upon the quality of the input data generated by different sequencing platforms. Thus, the first step is to assess the data quality and exclude technical errors for further analysis. This may consist of several processes depending on the sequencing method and requirements such as sensitivity and speed. The pre-processing may include removing adapter sequences, trimming low quality reads to a set quality score, removal of low quality reads, removing nucleotide repeats, removing short reads, deduplication, matching paired end reads (or removing unmatched reads) and removing reads that contain Ns (unresolved nucleotides). Removing reads may speed up the analysis but may result in not detecting rare viruses.

Table 33.3 : Software for the analysis of viral (meta) genomes.

Name	Application	URL
Kraken	Taxonomic assignment	https://ccb.jhu.edu/software/kraken/
Kaiju	Taxonomic assignment	http://kaiju.binf.ku.dk/
Clark	Taxonomic assignment	http://clark.cs.ucr.edu/
Lambda	Protein homology search	https://seqan.github.io/lambda/
Diamond	Protein homology search	http://ab.inf.uni-tuebingen.de/software/diamond/
NCBI BLAST +	Nucleotide and protein homology search	https://blast.ncbi.nlm.nih.gov/Blast.cgi
VirusHunter	Virus discovery	http://www.ibridgenetwork.org/wustl/virushunter
MetaVir	Taxonomic assignment	http://metavir-meb.univ-bpclermont.fr/
VirSorter	Virus and prophage discovery	https://github.com/simroux/VirSorter
One Codex	Taxonomic assignment	https://www.onecodex.com/
PHYMMBL	Taxonomic assignment	https://ccb.jhu.edu/software/phymmbl/index.shtml
IVA	Viral genome assembly	http://sanger-pathogens.github.io/iva/
Vicuna	Viral genome assembly	http://www.broadinstitute.org/scientific-community/science/projects/viral-genomics/vicuna
PRICE	Viral genome assembly	http://derisilab.ucsf.edu/software/price/
SPAdes	Genome assembly	http://bioinf.spbau.ru/spades
MetaSPAdes	Metagenome assembly	http://bioinf.spbau.ru/spades
MEGAHIT	Metagenome assembly	https://github.com/voutcn/megahit
IDBA-UD	Metagenome assembly	https://github.com/loneknightpy/idba
MetaVelvet	Metagenome assembly	http://metavelvet.dna.bio.keio.ac.jp/
ShoRAH	Viral haplotype reconstruction	https://github.com/ozagordi/shorah
QuRE	Viral haplotype reconstruction	https://sourceforge.net/projects/qure/
PredictHaplo	Viral haplotype reconstruction	http://bmda.cs.unibas.ch/HivHaploTyper/

Table 33.4 : Technical details of classification workflows used in viral metagenomics

Workflow	Software used
GeneSeed-HMM	HMMER, BLAST, EMBOSS, CAP3, Velvet, Newbler, SOAPDenovo, ABySS
Gaint Virus Finder	Perl, Python, BLAST
Kraken	C++, Perl
LMAT	gcc, python, OpenMP, MPI
MEGAN CE	Java, DIAMOND, InterPro2GO, SEED viewer, eggnog viewer, KEGG
MEGAN4	Java
metaViC	Bowtie2, DIAMOND, filter_fastq.pl, GRAM, IDBA-UD, Kronatools, prinseq, QUAST, riboPicker, SPAdes, Trim Galore
Metavir	MUSCLE, CAP3, Gblocks, PhyML, Scriptree
MetLab	Python (2.7) + libraries, GNUMPFR, Prinseq-Lite, Bowtie2, SAMTOOLS, SPAdes, Krona Tools, Kraken, FragGeneScan, HMMsearch, vFamParse
proViDE	Perl, Python
SLIM	Python, BLAST, QUASR, MEGAN, BWA, SPAdes, MUMmer
SMART	C++, Ruby, Flash, Sickle, Google SparseHash, GNUparallel
Taxonomer	Cython, KAnalyze
VIROME	Adobe Flex, MySQL, Blast, tRNA scan SE, MetaGene Annotator, CD-Hit 454
ViromeScan	Bash, R, Perl, Java, Bowrie2, Bmtagger, Picard
VirSorter	Perl, HMMER3, MCL, MetaGeneAnnotator, MUSCLE, BLAST
VirusFinder	Perl, BLAST+, BLAT, Bowtie2, BWA, iCORN, CREST, GATK, SAMtools,SVDetect, Trinity
VirusHunter	Perl, MySQL, Blast, CD-HIT, RepeatMasker
VirusSeeker	Perl, SLURM, BLAST, MegaBLAST, BWA-MEM, cutadapt, ea-utils, PRINSEQ, CD-HIT, Tantan, RepeatMasker, Newbler, Phrap
VirusSeq	Perl, MOSAIK
VirVarSeq	BWA, Q-cpileup, R, Fortran, Perl

4.4.2 Assembling viral genomes

NGS technologies produce billions of short reads from metagenome. More commonly, the short reads are assembled into longer contiguous sequences (contigs). Nowadays, there are hundreds of different assembles, some of them specially developed for metagenomic data (Table 33.4). There are two main assembly algorithms: Overlap Layout Consensus algorithm (OLC) and De Bruijn Graph (DBG). OLC is an intuitive algorithm where reads with an overlap under some conditions can be joined to form larger sequence. The process is run iteratively to construct a whole genome or genomes. OLC programs are more suitable for long reads avoiding assembly errors due to the presence of repeated regions. However, the amount of memory used by these assemblies make them impossible to be applied to the very large numbers of reads obtained with Illumina technology. A better view of the viral community is obtained when viral reads are first assembled into contigs that sometimes could represent complete viral genomes, but still a great number of contigs are not assigned to any known viral family. DBG assemblers split reads in to small pieces called K-mers and hash of these K-mers is build followed by the construction of the genomes through K-mers relationship returning the most probable genome.

Once NGS reads are assembled into partial virus genome contigs these are usually classified into families, genera, species or other functional adhoc operational units (OTU) using BLAST based methodologies. Unfortunately, most currently published metagenome studies do not take the next step and try to classify the sequences phylogenetically so as to determine whether they fall within the bounds of known species, genera or families. It is important to note that a detectable sequence similarity doesn't necessarily imply biological similarity.

4.4.3 Inference of viral diversity

Genetic diversity of virus populations can be studied at different genomic scales: i. position-wise, by identifying single nucleotide variants (SNV), ii. at a local scale, by identifying patterns of SNVs that co-occur at a distance smaller than the average read length, and iii. at a global scale by phasing mutations over distances longer than the read length. Posada-Cespedes et al., (2017) described computational methods for studying viral genetic diversity in accordance with the classification scheme introduced above and focusing on most recent developments.

4.4.4 Genotype abundance, community diversity and structure

To estimate the number of different genotypes (richness) and their relative abundance (evenness) in a metagenomic sample, simple read counts may introduce biases because longer genomes have a higher chance to be sequenced. Another problem is that large parts of metagenomic sequences are classified as unknown, most probably because of short comings in the similarity-based classification methods. Viral metagenome diversity and community diversity is estimated by the number of different genotypes in the sample, defined as richness (alpha diversity) and their relative abundances and distribution, defined as evenness (gamma diversity) among the metagenomic samples. The analysis assumes that more abundant organisms will have longer and higher coverage contigs, whereas less abundant organisms will have many small and low coverage contigs in the sample (alpha diversity). The gamma diversity uses the same assumption but estimates diversity among different metagenomic samples.

5. VIRAL DARK MATTER

Studies made on environmental viral metagenomics consistently report that more than 50% of the retrieved sequences are unknown or unassigned. Environmental viral metagenomic studies have produced more than 810Mbp of data for which there are no similarities in GeneBank. These sequences remain in an ever-growing pool of unknowns. Further analysis of this unknown pool is complicated by the lack of a centralized database for submitting and/or retrieving viral metagenomic data. Moreover, many of the 'known' viral sequences in environmental viromes have low amino acid similarities (<50%) to known viral proteins and thus represent undescribed species. Thus, a pool of sequences unassignable to any database sequences is referred to as dark matter. Formally, dark matter is defined negatively any nucleotide sequence that cannot be taxonomically assigned by alignment to any reference nucleotide or amino acid. Viral dark matter is the subset of dark matter that originates from viruses. It should also be remembered that with total metagenomic data, sequences that do not align to any known organism could be from any organism and need not be viral. From studies that use finely tuned experimental techniques to verify viral particles, 40-90% of sequences are unalignable. There is a big question as to whether this dark matter represents the artifacts of experiments procedure or the largest reservoir of untapped genetic information in the biosphere.

6. VIRAL METAGENOMICS VIS-A-VIS VIRUS TAXONOMY

Viral metagenomic data are changing our perception of virus diversity and are, therefore, challenging the way in which we recognize and classify viruses. The number and diversity of viral sequences that are identified in metagenomic data far exceeds that of experimentally characterized virus isolates. A recent comprehensive metagenomic analysis of thousands of diverse samples has led to the discovery of approximately 125,000 new viral genomes and a 16-fold increase in the number of identified viral genes. Furthermore, numerous viruses are hidden in publicly available microbial genomic datasets. A recently developed tool, VirSorter, identified 12,498 new viral genome sequences in ~15,000 bacterial and archaeal genomes, which increased the number of known prokaryotic viruses ~10-fold and identified viruses that infect 13 prokaryotic phyla. These advances are a striking testimony to the fundamental change in virus discovery: the overwhelming majority of new viral genomes now come from metagenomic data and have never been directly linked to biological agents. Virologists, especially viral taxonomists, have no choice but to work within this new reality.

Historically, the description and classification of a new virus by the International Committee on Taxonomy of Viruses (ICTV) have required substantial information on host range, replication cycle, and the structure and properties of virus particles, which were then used to define groups of viruses. However, the sequenced viruses from metagenomic studies lack information about their hosts and other biological properties that would guide their assignment into species and genera in the family as per ICTV system. The question of whether viruses that are identified by metagenomics can, and should be incorporated into the official ICTV taxonomy scheme on the basis of sequence data alone is a matter of debate. Sequence assemblies that are derived from environmental samples often contain complete, verified genome sequences of new viruses, but do not directly provide information on biological properties. This perceived limitation has raised the concern that virus classification based on sequence information alone would result in taxonomy of sequences rather than of viruses. In a recent workshop, a panel of experts discussed the proposal that, with appropriate quality control, viruses that are known only from metagenomic data can, and should be, incorporated into the official classification scheme of the ICTV. It implies a taxonomy that is based on metagenomic sequence data alone represents a substantial departure from the traditional reliance on phenotypic properties. However, there is a need for development of a robust framework for sequence-based virus for the comprehensive characterization of the global virome. In response to this demand, a workshop of invited experts in the field of virus discovery and environmental surveillance, and members of the ICTV Executive Committee, was held in June 2016 (Boston, Massachusetts, USA), to discuss this possibility and to develop a framework for appropriate approaches to virus classification. The workshop reached a consensus view on classifying viruses solely on the basis of metagenomic sequence data and, consequently, developed a set of proposals. Simmonds et al., (2017) presented some proposals together with an explanation of the rationale for their development and their proposals have been subsequently endorsed by the ICTV Executive Committee. These proposals are diagrammatically summarized in Fig. 33.2.

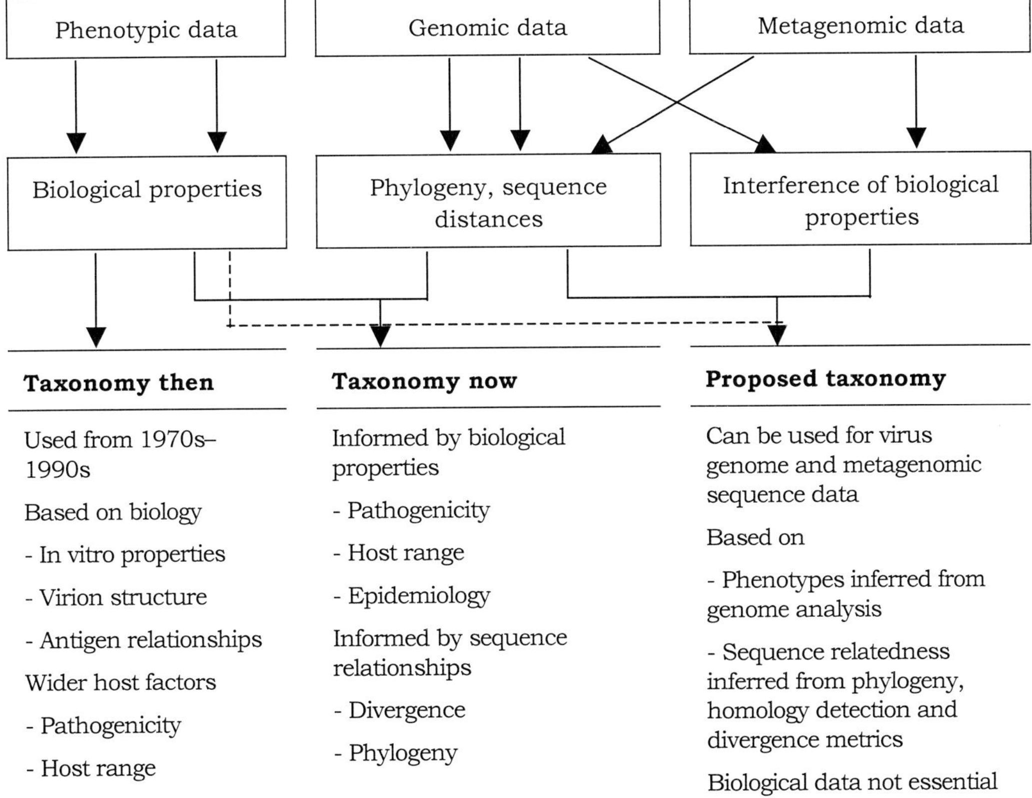

Fig. 33.2 Summary of the proposed classification pipeline**.** The proposed classification pipeline enables both metagenomic sequence data and conventionally derived virus sequences to be classified (Source Simmonds *et al.*, 2017)

Further reading

Angly F *et al.*, (2005) PHACCS, an online tool for estimating the structure and diversity of uncultured viral communities using metagenomic information. BMC Bioinformatics 6:41

Beerenwinkel N, Gunthard HF, Roth V, Metzner KJ (2012) Challenges and opportunities in estimating viral genetic diversity from next-generation sequencing data. Front. Microbiol. 3: 329

Bexfield N, Kellam P (2011) Metagenomics and molecular identification of novel viruses. *Veterinary Journal* 190: 191-198

Breitbart M, Hewson I, Felts B, Mahaffy JM, Nulton J, Salamon P, Rohwer F (2003) Metagenomic analyses of an uncultured viral community from human feces. *J. Bacteriol.* 185:6220-6223

Breitbart M, Salamon P, Andresen B, Mahaffy JM, Segall AM, Mead D, Azam F, Rohwer F (2002) Genomic analysis of uncultured marine viral communities. *Proc Natl Acad* Sci USA 99:14250-14255

Brum JR, Sullivan MB (2015) Rising to the challenge: accelerated pace of discovery transforms marine virology. Nat Rev Microbiol 13: 147-159

Edwards RA, Rohwer F (2005) Viral metagenomics. Nature Reviews Microbiology 3: 504–510

Kapoor A, Lipkin WI (2014) Virus Discovery in the 21st century. In: eLS. John Wiley & Sons, New York, NY.

Krishnamurthy SR, Wang D (2017) Origins and challenges of viral dark matter. Virus Research 239:136-142

Metzker M (2010) Sequencing technologies- the next generation. Nat Rev Genet 11: 31-46

Mokili J, Rohwer F, Dutilh B (2012) Metagenomics and future perspectives in virus discovery. Curr Opin Virol 2: 63-77

Nooij S, Schmitz D, Vennema H, Kroneman A, Koopmans MPG (2018) Overview of virus metagenomics classification methods and their biological applications. Front Microbiol 9: 1-21

Paez-Espino *et al.*, (2016) Uncovering Earth's Virome. Nature 536: 425–430

Posada-Cespedes S, Seifert D, Beerenwinkel N (2017) Recent advances in inferring viral diversity from high-throughput sequencing data. Virus Research 239:17-32

Rebecca R, Bede C, Avraam T, David LR, Mattia P (2016): Challenges in the analysis of viral metagenomes. Virus Evolution 2: 134-157

Simmonds P (2015) Methods for virus classification and the challenge of incorporating metagenomic sequence data. *J Gen Virol* 96: 1193–1206.

Solonenko SA, Sullivan M (2013) Preparation of metagenomic libraries from naturally occurring marine viruses. Methods Enzymol 531: 143-165

Soueidan H, Schmitt L, Candresse T, Nikolski M (2015) Finding and identifying viral needle in the metagenomic haystack: trends and challenges. Front Microbiol 5:1-7

Thurber RV, Haynes M, Breitbart M, Wegley L, Rohwer F (2009) Laboratory Procedures to generate viral metagenomes. Nature Protocols 4:470-483

Zhao G, Wu G, Lim ES, Droit L, Krishnamurthy S, Barouch DH *et al.*, (2017) VirusSeeker, a computational pipeline for virus discovery and virome composition analysis. Virology 503:21-30

34

EMERGING VIRUSES AND VIRAL DISEASES

"It is time to close the book on infectious diseases and pay more attention to chronic ailments such as cancer and heart diseases.... the war against infectious diseases has been won." This was the remark, made in 1967, by the then US Surgeon General William H. Stewart, inspired by looking at the accelerating pace of scientific and technological progress in fields of vaccines and antibiotics (Mani *et el.,* 2012). Macfarlane Burnet (1962), an Australian virologist and Nobel laureate has also expressed the same opinion. However, this over-optimism was short-lived. Several new diseases of infectious origin appeared during last few decades. This is inspite of elaborate protective measures taken to tackle infectious agents, during the last few decades. The global community witnessed several outbreaks of emerging and re-emerging infectious diseases, with high threats to the health security, biodefense, social fabric and economy worldwide (Table 34.1). The occurrence of significant disease outbreaks– such as HIV, SARS (severe acute respiratory syndrome) in 2002, H1N1 swine flu pandemic of Mexico in the 2009 , MERS (Middle East respiratory syndrome) that occurred in Saudi Arabia in 2012, the West African outbreak of Ebola virus (EBOV) in late 2013, the Zika virus (ZIKV) outbreak originating in Brazil in 2015, the 2018 health emergence in Nigeria caused by Lassa virus, and the ongoing Coronavirus disease 2019 (COVID-19) pandemic—have alerted the global community about the impending dangers of emerging infectious diseases. The scientific community, national governments and international agencies have been called for a rapid intervention in preventing and treating emerging infections.

Table 34. 1: Partial list of emerging viral infectious diseases from 1900 to 2020

Year of first description	Name	Deaths	Comments
1918	'Spanish influenza'	In the range of about 50 million to 100 million	1918: other pandemics in 1957-1958 (H2N2), 1968(H3N2) and 2009 (H1N1)
1931	Rift Valley Fever	Overall CFR<1%; ~50% for hemorrhagic fever	Contact with blood or organs of infected animals and mosquito – borne; several outbreaks in 1977, 1997-1998, 200-2016
1937	West Nile Fever	CFR ~5%	Mosquito – borne; worldwide outbreaks(most recent 1999-2010, USA)
1967	Marburg hemorrhagic fever	~470; very high CFR (24-88%,WHO)	Contact with African green monkey; numerous outbreaks in Africa 1969-2018
1969	Lassa fever	~ 5000 deaths annually; CFR 1-2%; Nigerian CFR 25%	Contact with rodents or contaminated food or items; mostly in West Africa (Nigeria 2018)
1969	Acute hemorrhagic conjunctivitis	Rare	First identified in 1969; pandemic in 1981; frequent outbreaks worldwide
1976-2020	Ebola hemorrhagic fever	>15000; CFR 75%	First identified in 1976;first major outbreak in 2013-2016in West Africa and in 2018 in Democratic Republic of Congo; 29 regional epidemics in 2020 in West and Central Africa
1981	HIV/AIDS	~ 37 million	Ongoing pandemic
1996	Avian flu	High CFR (60%)	H5N1 and H7N9 viruses from poultry; several outbreaks from worldwide; last outbreak from China in 2018
1999	Nipah fever	<1000?; very high CFR	Outbreaks in Malaysia, Singapore, Bangladesh and India
2002	SARS	813; CFR~ 10%	Contained- did not turn into pandemic
2009	H1N1; H7N9 'Swine flu'	284000; CFR2.9-9%	Pandemic
2012	MERS	935; CFR 34.4%	Major outbreak in 2012-2019; (camels, humans); detected in 27 countries but mostly in Middle Eastern countries
2014	Chikungunya	Rare	Mosquito - borne
2015	Zika	Unknown	Mosquito - borne
2019-ongoing	COVID – 19 (SARS-CoV-2)	>2.3 million; CFR 2-10%; high in elderly and individuals with comorbidities	Pandemic – animal- to –animal, animal – to – human and human – to – human transmission

1. Emerging and re-emerging diseases – Definitions

The term emerging disease has been in use in scientific publications since the beginning of the 1960s and is used in the modern sense by David Sencer in his 1971 article "Emerging Diseases of Man and Animals" where in the first sentence of the introduction he implicitly defines emerging diseases as "infectious diseases of man and animals currently emerging as public health problems" and as a consequence also includes re-emerging diseases.

World Health Organization (WHO) defined an emerging disease, as a disease which has appeared in a population for the first time, or that might have existed previously, but is rapidly increasing in incidence or geographical range. Emerging infections can be further classified as 'newly emerging', 're-emerging/resurging' or 'deliberately emerging'. Newly emerging infections are those that have not previously been recognized in man. Re-emerging and resurging infections are those that existed in the past but are now rapidly increasing in incidence or in geographical or human host range. Deliberately emerging infections are infections due to microbes (naturally occurring or bioengineered), deliberately introduced by man, usually for vicious use like in bioterrorism.

2. CLASSIFICATION

i. One way to classify emerging infectious diseases is by time and how humans were involved in the emergence:

ii. Newly emerging infectious diseases – diseases that were not previously described in humans, such as HIV/AIDS

iii. Re-emerging infectious diseases – diseases that have spread to new places or which previous treatments no longer control, such as methicillin-resistant *Staphylococcus aureus*

iv. Deliberately emerging infectious diseases – diseases created by humans for bioterrorism

v. Accidentally emerging infectious diseases – diseases created or spread unintentionally by humans, such as vaccine-derived poliovirus

Scientific Advisory Group for Origins of Novel Pathogens (SAGO)

On 16 July 2021, the Director-General of WHO announced the formation of the Scientific Advisory Group for Origins of Novel Pathogens (or SAGO), which is to be a permanent advisory body of the organisation. The Group was formed with a broad objective to examine emerging infectious diseases, including COVID-19. According to the WHO Director-General, "SAGO will play a vital role in the next phase of studies into the origins of SARS-CoV-2, as well as the origins of future new pathogens."

Some emerging and re-emerging viral diseases on global level and India are presented in tables 34.2, 34.3, 34.4 & Fig. 34.1.

Table 34.2: Emerging and re-emerging viral diseases.

Disease	Virus	Family/Genus	Reservoir/spill-over hosts	Transmission
West Nile fever (WNF)	WNV	Flaviviridae *Flavivirus*	Mosquitoes; birds/horses, dogs, rabbits	Mosquitoes
Zika fever	ZIKV	Flaviviridae *Flavivirus*	Mosquitoes; NHPs; domestic animals	Mosquitoes; vertical transmission
Yellow Fever (YF)	YFV	Flaviviridae *Flavivirus*	Mosquitoes; NHPs	Mosquitoes
Dengue fever (DF)	DENV	Flaviviridae *Flavivirus*	Mosquitoes; NHPs	Mosquitoes
Chikungunya fever	CHIKV	Togaviridae *Alphavirus*	Mosquitoes; NHPs	Mosquitoes
Lassa fever (LF)	LASV	Arenaviridae *Mammarenavirus*	Multimammatemouses	Rodent to human
Ebola virus disease (EVD)	EBOV	Filoviridae *Ebolavirus*	Fruit bats/NHPs; antelopes	Human to human
Marburg virus disease (MVD)	MARV	Filoviridae *Marburgvirus*	Bats/NHPs; humans	Human to human
Swine flu	A(H1N1) pdm09	Orthomyxoviridae Influenzavirus A	Pigs Human	Human to human
Severe acute respiratory syndrome (SARS)	SARS-CoV	Coronaviridae *Coronavirus*	Bats/palm civets	Human-to-human
Middle East respiratory sindrome (MERS)	MERS-CoV	Coronaviridae *Coronavirus*	Bats/dromedary camels	Human-to-human
Coronavirus disease 2019 (COVID-19)	SARS-CoV-2	Coronaviridae Coronavirus	Bats; likely malayan pangolins	Human-to-human

Table 34.3: Emerging/re-emerging viral infections and new viruses in India

Family	Virus	Probable/mode of transmission	Outbreak potential	Biosafety-risk group
	Ganjam virus	Tick- Borne	Yes*	2
	Bhanja virus	Tick- Borne	Yes*	2
	SFTS virus	Tick- Borne	Yes	4
	Chobar Gorge virus	Tick- Borne	No	2
		Arthropod-Borne	No	2
	EEV	Arthropod-Borne	Yes*	2
Bunyaviridae	Cat Que virus	Tick- Borne	Yes*	2
	Kaisodi virus	Arthropod-Borne	Yes*	2
	Umbre virus	Arthropod-Borne	No	2
	Oya virus	Tick- Borne	Yes*	2
	ingwavuma virus	Rodent- Borne	No	2
	Chittoor virus			
	Thottapalayam virus			
Nairoviridae	CCHFs virus	Tick- Borne, Human to human	Yes	4
	Yellow fever	Arthropod-Borne	Yes	4
	Zika virus	Arthropod-Borne, Mother to child, sexual route	Yes	2
Flaviviridae	KFD		Yes	4
	JE	Tick- Borne	Yes	2
	Dengue	Arthropod-Borne	Yes	2
	Bagaza virus	Arthropod-Borne	Yes*	2
		Arthropod-Borne		
	Influenza –	Air – Borne	Yes	3
	(H3N2)v alias	Air – Borne	Yes	4
	Influenza Avian	Air – Borne	Yes	2
Paramyxoviridae	(H5N1)	Tick- Borne	Yes*	2
	RSV	Air – Borne	Yes*	2
	Quaranfil virus	Air – Borne	Yes	2
	Parainfluenza 1-4			
	Entero virus – D68			
Paramyxoviridae	Nipah virus	Human to human Direct contact/ Consumption of infected bat/Fruit infected with bat	Yes	4

Picornaviridae	Human rhinovirus A, B and C	Air-Borne	Yes	2
	Hand, foot and mouth disease	Direct contact, faeco- oral route	Yes	2
	Coxsackie- A21 virus	Faeco- oral route	Yes	2
	Coxsackie- A21 virus	Faeco- oral route	Yes	2
	Sapoviruses	Faeco- oral route	Yes	2
	Rota	Faeco- oral route	Yes	2
	Polio and non-polio flacciparalysis	Faeco- oral route	Yes	3
Caliciviridae	Noroviruses	Faeco- oral route	Yes	2
Hepadnaviridae	Hepatitis KIs virus new and vaccine escape mutants of HBV	Blood- Borne	Yes	2
Togaviridae	Rubella virus	Air – Borne	Yes	2
	Chikengunya virus	Arthropod-Borne	Yes	2
Poxviridae	Buffelopox virus (Orthopox virus)	Direct contact	Yes	2
Parvoviridae	Human parvovirus - 4	Parenteral transmission	Yes	2
Arenaviridae	LCMV	Rodent - Borne	Yes*	3
Herpesviridae	CMV	Direct contact	Yes	2
	Chickenpox (varicella) VZV	Air – Borne, Direct contact	Yes	2
Rhabdoviridea	Chandipura virus	Arthropod-Borne	Yes	3
Reoviridae	Kammavanpettai virus (Orbiviruses)	Tick- Borne	No	Unknown

*May cause epidemic; However, no epidemic was reported. Unknown= No clear information on the risk assessment available. SFTS= Severe fever thrombocytopenia syndrome; EEV= Equine encephalosis virus; CCHFs= Crimean – Congo hemorrhagic fevers; KFD= Kyasanur forest disease; JE=Japanese encephalitis; LCMV= Lymphocytic choriomeningitis virus; CMV= Cytomegalovirus; VZV=Varicella–zoster virus; RSV= Respiratory syncytial virus; HBV= Hepatitis –B virus

Table 34.4: Known future threat of emerging viral infections in India

Virus	Probable/mode of transmission	Outbreak potential	Biosafety-risk group
MERS- CoV	Air-borne	Yes	3
Ebola virus	Direct contact	Yes	4
Avian influenza (H7N9)	Direct contact, Air-borne Rare, limited person –to- person spread	Yes	3
Avian influenza (H9N2) human infection China	Direct contact, Air-borne Rare, limited person –to- person spread	Yes	3
Yellow fever virus	Arthropod-borne	Yes	2
Usutu virus like ¬JE (mosquito-borne)	Arthropod-borne	Yes	2
Tilapia novel orthomyxo-like virus causes hepatitis	Indirect transmission by fomites	Unknown	Unknown
Cyclovirus	Faeco-oral	Yes*	Unknown
BannaReo virus encephalitis-(China) like JE	Arthropod-borne	Yes	Unknown
Canine parvovirus causes dog gastroeneritis	Direct or Indirect contact	Yes*	Unknown

*May cause epidemic; However, no epidemic was reported. Unknown= No clear information on the risk assessment available. MERS-CoV= Middle East Respiratory Syndrome Corona virus; JE= Japanese encephalitis.

3. FACTORS CONTRIBUTING TO EMERGENCE OF VIRAL DISEASES

Several factors contribute to the emergence of viral infections and they are discussed hereunder (Fig. 34.2) :

3.1 International travel and trade

Microbes know no geographical boundaries and their dissemination is accelerated by human mobility and the plant and animal materials they carry with them. Movement of people around the globe on various assignments has become rapid, frequent and imperative. At any given time, over a million people, each one a potential carrier of pathogens, are airborne around the planet. Transit times of people and other commercial goods which can be a potential source of pathogens and vectors are very short now, compared to the incubation times of diseases. Carriers of disease can reach theirdestination before the threat they harbour is detectable, reducing health quarantine to a near meaningless measure. Rapid global spread of the recent 2009 H1N1 pandemic influenza, introduction of *Aedes albopictus* (Asian tiger mosquito)into the

United States, Brazil and parts of Africa in shipments of used tyres from Asia and the world-widespread of hanta viruses by rats can be attributed to international travel and commerce.

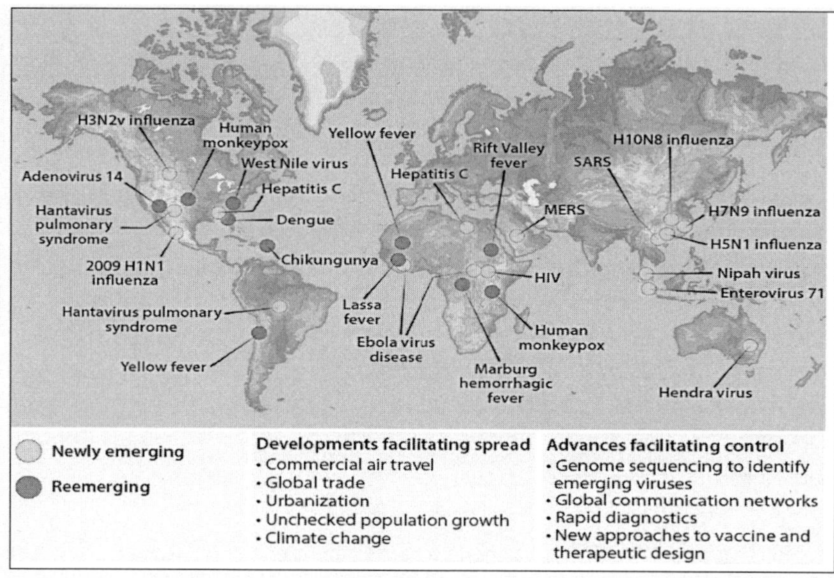

Fig. 34.1 : World map showing the origin and spread of newly emerging and reemerging viral diseases.

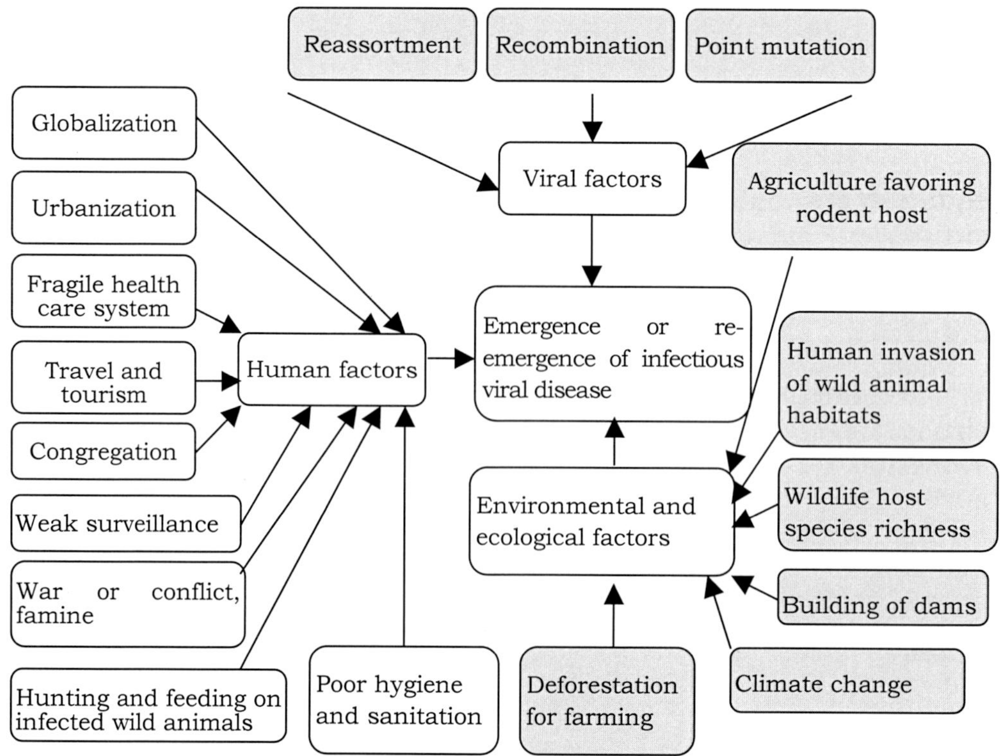

Fig. 34.2 : Factors contributing to the emergence or re-emergence of viral diseases.

3.2 Mass migration of people due to war and natural calamities

Mass migration of people due to war, natural calamities or economic hardships play a key role in disease emergence.Other aspects of modern human behaviour contribute to the emergence of novel infectious diseases, e.g.,water shortage, leading to increased usage of open containers for water storage, creates a breeding ground for mosquito vectors (Dengue), social practices like tattooing in tribal areas in India (HBV, HCV and other blood-borne viruses), rural urbanization, sexual behaviour (HIV), needle sharing among IV drug abusers (HIV, HCV), use of contaminated blood/tissue products (HIV, HBV, HCV,vCJD), contaminated vaccines (polio vaccines contaminated with SV40-linked to increased risk of non-Hodgkin's Lymphoma). Immune deficiency associated with AIDS, and advanced medical care including cancer chemotherapy and organ transplantation, has contributed to an enormous increase in an immunocompromised population world-wide. 'Opportunistic re-emerging infections are frequently seen in such individuals due to reactivation of latent viruses or unrestrained replication of endogenous viruses, leading to enhanced virulence or changes in host or tissue tropism (CMV, EBV, HHV-6, Parvovirus B-19, JC, BK etc.).

3.3 Bioterrorism

Subsequent to the anthrax attacks through the US mail system in 2001, the possible use of viruses as weapons of bioterrorism has been considered. Potential viruses that can be deliberately introduced into susceptible human population as weapons of bioterrorism include smallpox, Ebola, Marburg, Venezuelan equine encephalitis virus, Eastern equine encephalitis virus, Western equine encephalitis virus, Lassa fever virus etc. The potential use of genetically engineered viruses with enhanced virulence is also a possibility.

3.4 Mutations and Recombinations

Microbes constantly evolve and adapt to their environment. Many viruses like SARS-CO2, influenza show a high mutation rate and rapidly evolve into new variants against which the immune system fails to mount a response, and vaccines against existing strains are ineffective. In certain viruses evolution of a new variant may result in new manifestation of disease, change in host or tissue tropism or altered virulence. Emergence of drug-resistant microbes by natural selection confers an added advantage to them. The molecular evolution of Chikungunya virus (CHIKV, A226V mutation in glycoprotein envelope gene) leading to a higher efficiency of replication and dissemination in the vector species *Aedes albopictus*, is reported to have contributed to resurgence of Chikungunya fever (CHIK) in Kerala state in 2007.

Table 34. 5 Summary of mutational rates in viruses

Class	Virus	Genome size (kb)	Average mutation rate
ss(+) RNA	Bacteriophage Qβ	4.22	1.4×10^{-4}
	Tobacco mosaic virus	6.40	8.7×10^{-6}
	Human rhinovirus 14	7.13	6.9×10^{-5}
	Poliovirus 1	7.44	9.0×10^{-5}
	Human norovirus G1	7.65	1.5×10^{-4}
	Tobacco etch virus	9.49	1.2×10^{-5}
	Hepatitis C virus	9.65	3.8×10^{-5}
	Murine hepatitis virus	31.4	3.5×10^{-6}
ss(-) RNA	Vesicular stomatitis virus	11.2	3.7×10^{-5}
	Influenza A virus	13.6	2.5×10^{-5}
	Measles virus	15.9	3.5×10^{-5}
ds RNA	Bacteriophage φ6	13.4	1.6×10^{-6}
Reverse transcribing	Duck hepatitis B virus	3.03	2.0×10^{-5}
	Spleen necrosis virus	7.80	3.7×10^{-5}
	Murine leukemia virus	8.33	3.0×10^{-5}
	Bovine leukemia virus	8.42	1.7×10^{-5}
	Human T cell leukemia virus	8.50	1.6×10^{-5}
	HIV- I (free virions)	9.18	6.3×10^{-5}
	HIV- I (cellular DNA)	9.18	4.4×10^{-6}
	Foamy virus	13.2	2.1×10^{-5}
	Rous sarcoma virus	9.40	1.4×10^{-5}
ss DNA	Bacteriophage φ×174	5.39	1.1×10^{-5}
	Bacteriophage m13	6.41	7.9×10^{-5}
ds DNA	Bacteriophage λ	48.5	5.4×10^{-5}
	Herpus simplex virus	152	5.9×10^{-4}
	Bacteriophage T2	169	9.8×10^{-4}
	Human cytomegalo virus	235	2.0×10^{-5}

The remarkable capacity of some viruses to adapt to new hosts and environments is highly dependent on their ability to generate *de novo* diversity in a short period of time. Rates of spontaneous mutations vary amply among viruses. RNA viruses mutate faster than DNA viruses, single-stranded viruses mutate faster than double stranded virus, and genome size appears to correlate negatively with mutation rate. Higher mutation rates lead to higher genetic diversity. Viral mutations occur at different levels, including polymerase fidelity, sequence context, template secondary structure, cellular microenvironment, replication mechanisms, proofreading, and access to post-replicative repair. Additionally, massive numbers of mutations can be introduced by some virus-encoded diversity-generating elements, as well as by host-encoded cytidine/adenine deaminases. A summary of mutation rates for different viruses is provided in Table 34.5. In general, viral mutation rates roughly range between 10^{-8} and 10^{-4} substitutions per nucleotide per cell infection (s/n/c), with DNA viruses occupying the 10^{-8}–10^{-6} range and RNA viruses the 10^{-6}–10^{-4} range. These differences have several mechanistic bases. First, the polymerases of the vast majority of RNA viruses lack 3' exonuclease proofreading activity and hence are more error-prone than those of DNA viruses.

Reverse transcriptases (RTs) also lack 3' exonuclease activity and, hence, retroviruses(viruses with RNA-containing virions and a cellular DNA stage) and para-retroviruses (viruses with DNA-containingvirions and a cellular RNA stage) mutate and evolve atrates similar to those of non-reverse transcribing RNAviruses (the latter are often called riboviruses).

A general inverse correlation between genome size and mutation rate applies to DNA-based microorganisms including viruses, bacteria and unicellular eukaryotes. According to this rule, the per-genome mutation rate stays relatively constant at a value of approximately 0.003 per round of copy. A similar negative relationship seems to exist in RNA viruses, but their smaller genome size range of variation makes it more difficult to detect such a trend.

An interesting relation between recombination and mutationrates have also been recently found in RNA viruses.These two processes are primarily controlled by the viral polymerase since, in RNA viruses, recombination takes place when the viral polymerase switches between different template genomes present in the same cell. The estimated recombination rates of different riboviruses and retroviruses correlate positively with estimated mutation rates. High mutation rates confer viruses the ability to rapidly produce advantageous mutations, but also inflate the genetic load of the population. In turn, frequent recombination allows beneficial mutations to unlink from deleterious genetic backgrounds, as well different beneficial mutations to be combined into the same genome. As such, recombination is expected to enhance adaptation when a large number of alleles coexist in the same population, a scenario that typically takes place at high mutation rates.

3.5 Human encroachment on wild-life habitats

Humans get infected, quite often, simply because they have put themselves in some virus's way. Deforestation, expansion of human habitat and mining activities have been suggested as risk factors associated with re-emergence of vampire bat rabies in the Amazon basin. The dramatic increase of human cases of Kyasanur Forest Disease (KFD), a tick-borne flavivirus in south west India in 1983 can be attributed to deforestation and agriculture land expansion, which increased exposure of humans to the virus and the vector (ticks). Usage of cleared areas in the forest for cattle grazing favoured the proliferation of the ticks (*Haemaphysalis spinigera*) which prefer to feed on bigger animals, including humans.

As humans encroach further into previously uncultivated environments, new contacts between wild fauna and humans and their livestock increases the risk of cross-species infection. An example of such contact followed the establishment of piggeries close to the tropical forest in northern Malaysia, where, in 1998, the Nipah virus first crossed overfrom fruit bats (flying foxes, *Pteropus* spp.) to pigs and thence to pig farmers. Destruction of natural forest has also encouraged fruit bats to relocate nearer human habitation, like the large colony in the botanic gardens in the heart of Sydney. Indeed, in 1997, Hendra, a related paramyxovirus of Australian fruit bats, fatally infected a veterinarian examining a sick horse.

Rodents continue to be sources of re-emerging infections, as witnessed in the 1990s with hantaviruses in the United States. Rodent-borne hantavirus is prevalent in agricultural systems in South America and East Asia, in arid grasslands in North America and elsewhere. In mid-1993, an unexpected outbreak of acute, some times fatal, respiratory disease occurred in humans in the southwestern United States. This 'hantavirus pulmonary syndrome' was caused by a previously unrecognized virus, maintained primarily within the native deer mouse, and transmitted through excreta. The 1991–1992 El Niño event, with unseasonal heavy summer rains and a proliferation of piñon nuts, hugely amplified local rodent populations which led to the 1993 outbreak. In South America, there have been several outbreaks of hantavirus and arenavirus infections linked to forest clearance and the growth of rodent populations in the new grasslands.

3.6 Changes in agricultural practices

The emergence of Argentine haemorrhagic fever and its expansion to more areas in Argentina, has been linked to agricultural activities like cultivation of corn, that sustain the corn mouse, the main reservoir of the Junin virus,causing the disease. In India, Japanese encephalitis (JE) was introduced where dams and canals were built to bring in water for rice cultivation. A large pig population is conducive for outbreaks of JE, while a large cattle population may support mosquito feeding, but act as dampeners for spread of JE, since viraemia is not present. The worst affected districts in Kerala during the 2007 Chikungunya outbreak, Kottayam and Pathanamthitta, are hilly and heavily forested, with vast rubber plantations. The rainwater that collects in the hemispherical containers fitted to the trunks of rubber trees for latex collection in the region supported prolific breeding of *Aedes albopictus* mosquitoes contributing to resurgence of the disease.

3.7 Wild-Life Trade/Ecotourism

Wild-life constitutes a large and often unknown reservoir of diseases; an estimated 75% of emerging infectious diseases are zoonotic, mainly of viral origin, and likely to be vector-borne. Several factors related to wild-life, like wild-life trade and translocation, live animal and bush meat markets, consumption of exotic foods, ecotourism, access to petting zoos and ownership of exotic pets contribute to emergence of infectious diseases. Outbreaks of Marburg and Ebola have been associated with transport of animals. The rapid economic growth in southern China leading to an increasing demand for animal proteins including those from exotic game food animals such ascivets, and many varieties of these wild game mammals in over crowded cages in wet markets allowed the jumping of the novel SARS coronavirus from animals to human. Currently very few countries have active wild-life disease surveillance and it is the need of the hour to monitor the emergence of novel microbes.

The increasing predilection for meat of exotic species has exacerbated the risk of exposure to infections not previously encountered, and this situation probably triggered the SARS epidemic. Bushmeat, especially in Africa, also poses a serious problem for emerging infectious diseases. Sick animals may be more easily captured. For example, 21 human deaths owing to Ebola virus infection ensued from the butchering of a single chimpanzee. HIV has crossed from chimpanzees to humans on atleast three occasions, and a higher number of zoonotic events from sooty mangabeys are indicated for HIV-2. Whether these cross-species infections arose from butchering the animals or from keeping them as pets is unknown, but a recent survey of primate hunters in Africa showed that they are susceptible, like handlers of primates in captivity, to infection (though not disease) from foamy retroviruses.

3.8 Urbanisation

The world is transitioning to being more urban. In 1950, roughly two-thirds of the world lived in rural settings, and the rest in urban dwellings. By 2050 the UN predicts this will have reversed, with 66% of people living in urbanised settings – and most of this growth will happen in Asia and Africa. Many of the cities with growing numbers of people are already struggling with infrastructure, housing, sanitation, transport and health care facilities. In many cases this means an increasing number of people living in overcrowded and unhygienic environments in which infectious diseases can thrive, without adequate health systems that can deal with these threats. Rapid urbanization boosts certain well-established infectious diseases, such as childhood pneumonia, diarrhea, tuberculosis and dengue, and facilitates dissemination of various 'emerging' diseases—as occurred for SARS in the high-rise housing of Hong Kong. Crowded and dilapidated public housing can potentiate infectious disease transmission through drug abuse and sexually transmitted infections.

3.9 Climate change

Climate change affects every aspect of human existence, from access to water and food to severe weather events and extremes of heat and cold. It can also affect the spread of disease in a number of ways, such as by altering the natural range of disease carrying insects, like mosquitoes.WHO estimates that between 2030 and 2050, climate change will kill an additional quarter of a million people a year through the spread of infectious diseases such as malaria and dengue fever. An increasing risk of flooding, which can be brought about by more frequent extreme weather events, also means that outbreaks of water-borne diseases, like cholera and other diarrhoeal diseases, are also much more likely. And climate change is also radically changing where people live, with climate shock events resulting in significant human displacement, often leading to populations moving into already-crowded cities or sometimes crossing borders into other countries.This can trigger conflict and an increase in the number of people living in refugee camps, where infectious disease epidemics can spiral out of control. Mass congregations like Kumba Mela, political meetings and rallies promote the spread of infectious diseases to new areas fast and on mass scale.

3.10 Increased human-animal contact

The way in which people and animals come into contact today is significantly increasing the risk of outbreaks of zoonotic diseases - those that originate in animals. When pathogens jump the species barrier, from animals to humans, their ability to spread and the severity of the disease they cause is a potentially lethal unknown. The recent jump of corona virus from bats to humans is a live example.

As people venture into new areas, because of population growth or displacement, in search of wild food or encroachment on the environment, they may enter into animal habitats that have never had human contact before. Many recent disease epidemics have originated in this way, and the ongoing coronavirus pandemic is suspected to have originated in bats, just as recent Ebola epidemics have. The national and international trade in wildlife also puts people in contact with animal infections that can then spread from person to person and potentially become a pandemic.

Further reading

Mani RS, Ravi V, Desai A, Madhusudana SN. (2012) Emerging viral infections in India. *Proc Natl Acad Sci India* Sect B *Biol Sci*; 82: 5-21.

Dikid T, Jain SK, Sharma A, Kumar A, Narain JP.(2013) Emerging and re-emerging infections in India: An overview. *Indian J Med Res*; 138: 19-31.

Mourya D T et al (2019) Emerging/re-emerging viral diseases and new viruses on the Indian horizon. *Indian J Med Res* 149, 447-467

Reeta S. Mani , V. Ravi , Anita Desai and SN Madhusudana (2012) Emerging Viral Infections in India. *Proc. Natl. Acad. Sci.* Sect B. *Biol. Sci.* 82(1):5–21

Sanjuan R and Domingo-Calap P (2016) Mechanisms of viral mutation. *Cell. Mol. Life Sci.* 73:4433–4448

Bloom DE and Cadarette D. (2019) Infectious disease threats in the twenty-first century:strengthening the global response. *Front Immunol.* (2019) 10:549. doi: 10.3389/fimmu.2019.00549

Church DL (2004) Major factors affecting the emergence and re-emergence of infectious diseases. Clin Lab Med. 24:559–86. doi: 10.1016/j.cll.2004.05.008

Morens, DM & Fauci, AS (2013) Emerging infectious diseases: threats to human health and global stability. PloSl Pathog. 9, e1003467.

Marston, HD, Folkers, GK, Morens, DM & Fauci, AS (2014) Emerging viral diseases: confronting threats with new technologies. Sci. Transl. Med. 6,253ps210.

Annexure I

ICTV Classification of Viruses

Virus Taxonomy: 2020 Release

EC 52, Online meeting, October 2020

Email ratification March 2021 (MSL #36)

6 realms, 10 kingdoms, 17 phyla, 2 subphyla, 39 classes, 59 orders, 8 suborders, 189 families, 136 subfamilies, 2224 genera, 70 subgenera, 9110 species

Realm I :Adnaviria

(Kingdom-1, phylum-1, class-1, orders-2, families-3)

Kingdom: Zilligvirae	**Realm: Adnaviria**	**1 phylum**
-Phylum: Taleaviricota	Kingdom: Zilligvirae	1 class
-**Class:** Tokiviricetes	Phylum: Taleaviricota	2 orders
-Order: Ligamenvirales	Class: Tokiviricetes	2 families
+**Family:** Lipothrixviridae	Order: Ligamenvirales	4 genera
+**Family**: Rudiviridae	Order: Ligamenvirales	7 genera
-**Order:** Primavirales	Class: Tokiviricetes	1 family
+**Family:** Tristromaviridae	Order: Primavirales	2 genera

Realm II: Duplodnaviria

(Kingdom-1, phyla-2, classes-2, orders-2, families-17)

Kingdom: Heunggongvirae	Realm: Duplodnaviria	2 phyla
1. -Phylum: Peploviricota	Kingdom: Heunggongvirae	1 class
- **Class:**Herviviricetes	Phylum: Peploviricota	1 order
-**Order:**Herpesvirales	Class: Herviviricetes	3 families
+**Family**: Alloherpesviridae	Order: Herpesvirales	4 genera
+**Family:** Herpesviridae	Order: Herpesvirales	3 subfamilies, 1 species
+ **Family:**Malacoherpesviridae	Order: Herpesvirales	2 genera

2. -**Phylum:**Uroviricota	Kingdom: Heunggongvirae	1 class
- **Class:**Caudoviricetes	Phylum: Uroviricota	1 order
-**Order:**Caudovirales	Class: Caudoviricetes	14 families, 1 genus
+**Family**: Ackermannviridae	Order: Caudovirales	2 subfamilies, 7 genera
+**Family:**Autographiviridae	Order: Caudovirales	9 subfamilies, 70 genera
+**Family:**Chaseviridae	Order: Caudovirales	2 subfamilies
+**Family:**Demerecviridae	Order: Caudovirales	3 subfamilies, 4 genera
+**Family:**Drexlerviridae	Order: Caudovirales	4 subfamilies, 9 genera
+**Family:**Guelinviridae	Order: Caudovirales	1 subfamily, 2 genera
+**Family:**Herelleviridae	Order: Caudovirales	5 subfamilies, 7 genera, 1 species
+**Family:**Myoviridae	Order: Caudovirales	8 subfamilies, 153 genera
+**Family:** Podoviridae	Order: Caudovirales	3 subfamilies, 44 genera, 1 species
+**Family:**Rountreeviridae	Order: Caudovirales	2 subfamilies, 2 genera
+**Family:**Salasmaviridae	Order: Caudovirales	3 subfamilies, 4 genera
+**Family**: Schitoviridae	Order: Caudovirales	8 subfamilies, 14 genera
+**Family:**Siphoviridae	Order: Caudovirales	22 subfamilies, 299 genera
+**Family:**Zobellviridae	Order: Caudovirales	1 subfamily, 6 genera
+**Genus:**Lilyvirus	Order: Caudovirales	1 species

Realm III: Monodnaviria

(Kingdoms-4, phyla-5, classes-8, orders-13, Families-17)

I Kingdom: *Loebvirae*	Realm: *Monodnaviria*	1 phylum
1. -Phylum:*Hofneiviricota*	Kingdom: *Loebvirae*	1 class
-Class:*Faserviricetes*	Phylum: *Hofneiviricota*	1 order
-Order:*Tubulavirales*	Class: *Faserviricetes*	3 families
+**Family:***Inoviridae*	Order: *Tubulavirales*	21 genera
+**Family:***Paulinoviridae*	Order: *Tubulavirales*	2 genera
+**Family:***Plectroviridae*	Order: *Tubulavirales*	3 genera
II. Kingdom:*Sangervirae*	Realm: *Monodnaviria*	1 phylum
2. -Phylum:*Phixviricota*	Kingdom: *Sangervirae*	1 class
-Class:*Malgrandaviricetes*	Phylum: *Phixviricota*	1 order
-Order:*Petitvirales*	Class: *Malgrandaviricetes*	1 family
+**Family:***Microviridae*	Order: *Petitvirales*	2 subfamilies
III. Kingdom:*Shotokuvirae*	Realm: *Monodnaviria*	2 phyla
3.-**Phylum:***Cossaviricota*	Kingdom: *Shotokuvirae*	3 classes
-Class:*Mouviricetes*	Phylum: *Cossaviricota*	1 order
-Order:*Polivirales*	Class: *Mouviricetes*	1 family
+**Family:***Bidnaviridae*	Order: *Polivirales*	1 genus
-Class:*Papovaviricetes*	Phylum: *Cossaviricota*	2 orders
-Order:*Sepolyvirales*	Class: *Papovaviricetes*	1 family
+**Family:***Polyomaviridae*	Order: *Sepolyvirales*	6 genera, 5 species
-Order:*Zurhausenvirales*	Class: *Papovaviricetes*	1 family
+**Family:***Papillomaviridae*	Order: *Zurhausenvirales*	2 subfamilies
-Class:*Quintoviricetes*	Phylum: *Cossaviricota*	1 order
-Order:*Piccovirales*	Class: *Quintoviricetes*	1 family
+**Family:***Parvoviridae*	Order: *Piccovirales*	3 subfamilies
4. -Phylum:*Cressdnaviricota*	Kingdom: *Shotokuvirae*	2 classes
-Class:*Arfiviricetes*	Phylum: *Cressdnaviricota*	5 orders
- **Order:***Baphyvirales*	Class: *Arfiviricetes*	1 family

+Family:*Bacilladnaviridae*	Order: *Baphyvirales*	3 genera
-**Order:***Cirlivirales*	Class: *Arfiviricetes*	1 family
+**Family:***Circoviridae*	Order: *Cirlivirales*	2 genera
-**Order:***Cremevirales*	Class: *Arfiviricetes*	1 family
+**Family:***Smacoviridae*	Order: *Cremevirales*	12 genera
-**Order:***Mulpavirales*	Class: *Arfiviricetes*	2 families
+**Family:***Metaxyviridae*	Order: *Mulpavirales*	1 genus
+**Family:***Nanoviridae*	Order: *Mulpavirales*	2 genera
-**Order:***Recrevirales*	Class: *Arfiviricetes*	1 family
+**Family:***Redondoviridae*	Order: *Recrevirales*	1 genus
-**Class:***Repensiviricetes*	Phylum: *Cressdnaviricota*	1 order
-**Order:***Geplafuvirales*	Class: *Repensiviricetes*	2 families
+**Family:***Geminiviridae*	Order: *Geplafuvirales*	14 genera
+**Family:***Genomoviridae*	Order: *Geplafuvirales*	10 genera
IV. Kingdom:*Trapavirae*	Realm: *Monodnaviria*	1 phylum
5. -**Phylum:***Saleviricota*	Kingdom: *Trapavirae*	1 class
-**Class:***Huolimaviricetes*	Phylum: *Saleviricota*	1 order
-**Order:***Haloruvirales*	Class: *Huolimaviricetes*	1 family
+**Family:***Pleolipoviridae*	Order: *Haloruvirales*	3 genera

Realm IV:Riboviria

(Kingdoms-2, phyla-6, classes-21, orders-29, Families-111)

I. -Kingdom:*Orthornavirae*	Realm: *Riboviria*	5 phyla, 2 families, 1 genus
1.-Phylum:*Duplornaviricota*	Kingdom: *Orthornavirae*	3 classes
-Class:*Chrymotiviricetes*	Phylum: *Duplornaviricota*	1 order
-Order:*Ghabrivirales*	Class: *Chrymotiviricetes*	4 families
+Family:*Chrysoviridae*	Order: *Ghabrivirales*	2 genera
+Family:*Megabirnaviridae*	Order: *Ghabrivirales*	1 genus
+Family:*Quadriviridae*	Order: *Ghabrivirales*	1 genus
+Family:*Totiviridae*	Order: *Ghabrivirales*	5 genera
-**Class:***Resentoviricetes*	Phylum: *Duplornaviricota*	1 order
-**Order:***Reovirales*	Class: *Resentoviricetes*	1 family
+**Family:***Reoviridae*	Order: *Reovirales*	2 subfamilies
-**Class:***Vidaverviricetes*	Phylum: *Duplornaviricota*	1 order
-**Order:***Mindivirales*	Class: *Vidaverviricetes*	1 family
+**Family:***Cystoviridae*	Order: *Mindivirales*	1 genus
2.-Phylum:*Kitrinoviricota*	Kingdom: *Orthornavirae*	4 classes
-**Class:***Alsuviricetes*	Phylum: *Kitrinoviricota*	3 orders
-**Order:***Hepelivirales*	Class: *Alsuviricetes*	4 families
+**Family:***Alphatetraviridae*	Order: *Hepelivirales*	2 genera
+**Family:***Benyviridae*	Order: *Hepelivirales*	1 genus
+**Family:***Hepeviridae*	Order: *Hepelivirales*	2 genera
+**Family:***Matonaviridae*	Order: *Hepelivirales*	1 genus
-Order:*Martellivirales*	Class: *Alsuviricetes*	7 families
+**Family:***Bromoviridae*	Order: *Martellivirales*	6 genera
+**Family:***Closteroviridae*	Order: *Martellivirales*	4 genera, 7 species
+**Family:***Endornaviridae*	Order: *Martellivirales*	2 genera
+**Family:***Kitaviridae*	Order: *Martellivirales*	3 genera
+**Family:***Mayoviridae*	Order: *Martellivirales*	2 genera

+Family:*Togaviridae*	Order: *Martellivirales*	1 genus
+Family:*Virgaviridae*	Order: *Martellivirales*	7 genera
-Order:*Tymovirales*	Class: *Alsuviricetes*	5 families
+Family:*Alphaflexiviridae*	Order: *Tymovirales*	6 genera
+Family:*Betaflexiviridae*	Order: *Tymovirales*	2 subfamilies
+Family:*Deltaflexiviridae*	Order: *Tymovirales*	1 genus
+Family:*Gammaflexiviridae*	Order: *Tymovirales*	1 genus
+Family:*Tymoviridae*	Order: *Tymovirales*	3 genera, 2 species
-Class:*Flasuviricetes*	Phylum: *Kitrinoviricota*	1 order
-Order:*Amarillovirales*	Class: *Flasuviricetes*	1 family
+Family:*Flaviviridae*	Order: *Amarillovirales*	4 genera
-Class:*Magsaviricetes*	Phylum: *Kitrinoviricota*	1 order
-Order:*Nodamuvirales*	Class: *Magsaviricetes*	2 families
+Family:*Nodaviridae*	Order: *Nodamuvirales*	2 genera
+Family:*Sinhaliviridae*	Order: *Nodamuvirales*	1 genus
-Class:*Tolucaviricetes*	Phylum: *Kitrinoviricota*	1 order
-Order:*Tolivirales*	Class: *Tolucaviricetes*	2 families
+Family:*Carmotetraviridae*	Order: *Tolivirales*	1 genus
+Family:*Tombusviridae*	Order: *Tolivirales*	3 subfamilies, 1 genus
3. -Phylum:*Lenarviricota*	Kingdom: *Orthornavirae*	4 classes
-Class:*Amabiliviricetes*	Phylum: *Lenarviricota*	1 order
-Order:*Wolframvirales*	Class: *Amabiliviricetes*	1 family
+Family:*Narnaviridae*	Order: *Wolframvirales*	1 genus
-Class:*Howeltoviricetes*	Phylum: *Lenarviricota*	1 order
-Order:*Cryppavirales*	Class: *Howeltoviricetes*	1 family
+Family:*Mitoviridae*	Order: *Cryppavirales*	1 genus
-Class:*Leviviricetes*	Phylum: *Lenarviricota*	2 orders, 9 genera
-Order:*Norzivirales*	Class: *Leviviricetes*	4 families
+Family:*Atkinsviridae*	Order: *Norzivirales*	56 genera
+Family:*Duinviridae*	Order: *Norzivirales*	6 genera
+Family:*Fiersviridae*	Order: *Norzivirales*	185 genera
+Family:*Solspiviridae*	Order: *Norzivirales*	24 genera
-Order:*Timlovirales*	Class: *Leviviricetes*	2 families
+Family:*Blumeviridae*	Order: *Timlovirales*	31 genera
+Family:*Steitzviridae*	Order: *Timlovirales*	117 genera
+Genus:*Chimpavirus*	Class: *Leviviricetes*	1 species
+Genus:*Hohglivirus*	Class: *Leviviricetes*	1 species
+Genus:*Mahrahovirus*	Class: *Leviviricetes*	1 species
+Genus:*Meihzavirus*	Class: *Leviviricetes*	1 species
+Genus:*Nicedsevirus*	Class: *Leviviricetes*	1 species
+Genus:*Sculuvirus*	Class: *Leviviricetes*	1 species
+Genus:*Skrubnovirus*	Class: *Leviviricetes*	1 species
+Genus:*Tetipavirus*	Class: *Leviviricetes*	1 species
+Genus:*Winunavirus*	Class: *Leviviricetes*	1 species
-Class:*Miaviricetes*	Phylum: *Lenarviricota*	1 order
-Order:*Ourlivirales*	Class: *Miaviricetes*	1 family
+Family:*Botourmiaviridae*	Order: *Ourlivirales*	6 genera
4. -Phylum:*Negarnaviricota*	Kingdom: *Orthornavirae*	2 subphyla
- Subphylum:*Haploviricotina*	Phylum: *Negarnaviricota*	4 classes
-Class:*Chunqiuviricetes*	Subphylum: *Haploviricotina*	1 order
-Order:*Muvirales*	Class: *Chunqiuviricetes*	1 family
+ Family:*Qinviridae*	Order: *Muvirales*	1 genus
-Class:*Milneviricetes*	Subphylum: *Haploviricotina*	1 order

- **Order:***Serpentovirales*	Class: *Milneviricetes*	1 family
+ **Family:***Aspiviridae*	Order: *Serpentovirales*	1 genus
-**Class:***Monjiviricetes*	Subphylum: *Haploviricotina*	2 orders
-**Order:***Jingchuvirales*	Class: *Monjiviricetes*	5 families
+**Family:***Aliusviridae*	Order: *Jingchuvirales*	2 genera
+**Family:***Chuviridae*	Order: *Jingchuvirales*	14 genera
+**Family:***Crepuscuviridae*	Order: *Jingchuvirales*	1 genus
+ **Family:***Myriaviridae*	Order: *Jingchuvirales*	1 genus
+ **Family:***Natareviridae*	Order: *Jingchuvirales*	1 genus
-**Order:***Mononegavirales*	Class: *Monjiviricetes*	11 families
+**Family:***Artoviridae*	Order: *Mononegavirales*	2 genera
+**Family:***Bornaviridae*	Order: *Mononegavirales*	3 genera
+**Family:***Filoviridae*	Order: *Mononegavirales*	6 genera
+**Family:***Lispiviridae*	Order: *Mononegavirales*	1 genus
+**Family:***Mymonaviridae*	Order: *Mononegavirales*	9 genera
+**Family:***Nyamiviridae*	Order: *Mononegavirales*	7 genera
+**Family:***Paramyxoviridae*	Order: *Mononegavirales*	4 subfamilies, 3 genera
+**Family:***Pneumoviridae*	Order: *Mononegavirales*	2 genera
+**Family:***Rhabdoviridae*	Order: *Mononegavirales*	3 subfamilies, 7 genera
+**Family:***Sunviridae*	Order: *Mononegavirales*	1 genus
+**Family:***Xinmoviridae*	Order: *Mononegavirales*	1 genus
-**Class:***Yunchangviricetes*	Subphylum: *Haploviricotina*	1 order
-**Order:***Goujianvirales*	Class: *Yunchangviricetes*	1 family
+**Family:***Yueviridae*	Order: *Goujianvirales*	1 genus
-**Subphylum:***Polyploviricotina*	Phylum: *Negarnaviricota*	2 classes
-**Class:***Ellioviricetes*	Subphylum: *Polyploviricotina*	1 order
- **Order:***Bunyavirales*	Class: *Ellioviricetes*	12 families
+**Family:***Arenaviridae*	Order: *Bunyavirales*	4 genera
+**Family:***Cruliviridae*	Order: *Bunyavirales*	1 genus
+**Family:***Fimoviridae*	Order: *Bunyavirales*	1 genus
+**Family:***Hantaviridae*	Order: *Bunyavirales*	4 subfamilies
+**Family:***Leishbuviridae*	Order: *Bunyavirales*	1 genus
+**Family:***Mypoviridae*	Order: *Bunyavirales*	1 genus
+**Family:***Nairoviridae*	Order: *Bunyavirales*	7 genera
+**Family:***Peribunyaviridae*	Order: *Bunyavirales*	4 genera
+**Family:***Phasmaviridae*	Order: *Bunyavirales*	6 genera
+**Family:***Phenuiviridae*	Order: *Bunyavirales*	20 genera
+**Family:***Tospoviridae*	Order: *Bunyavirales*	1 genus
+**Family:***Wupedeviridae*	Order: *Bunyavirales*	1 genus
-**Class:***Insthoviricetes*	Subphylum: *Polyploviricotina*	1 order
-**Order:***Articulavirales*	Class: *Insthoviricetes*	2 families
+**Family:***Amnoonviridae*	Order: *Articulavirales*	1 genus
+**Family:***Orthomyxoviridae*	Order: *Articulavirales*	7 genera
5. **Phylum:***Pisuviricota*	Kingdom: *Orthornavirae*	3 classes
-**Order:***Nidovirales*	Class: *Pisoniviricetes*	8 suborders
-**Suborder:***Abnidovirineae*	Order: *Nidovirales*	1 family
+**Family:***Abyssoviridae*	Suborder: *Abnidovirineae*	1 subfamily
-**Suborder:***Arnidovirineae*	Order: *Nidovirales*	4 families
+**Family:***Arteriviridae*	Suborder: *Arnidovirineae*	6 subfamilies
+**Family:***Cremegaviridae*	Suborder: *Arnidovirineae*	1 subfamily
+**Family:***Gresnaviridae*	Suborder: *Arnidovirineae*	1 subfamily
+**Family:***Olifoviridae*	Suborder: *Arnidovirineae*	1 subfamily
-**Suborder:***Cornidovirineae*	Order: *Nidovirales*	1 family
+**Family:***Coronaviridae*	Suborder: *Cornidovirineae*	2 subfamilies

-**Suborder:***Mesnidovirineae*	Order: *Nidovirales*	2 families
+**Family:***Medioniviridae*	Suborder: *Mesnidovirineae*	2 subfamilies
+ **Family:***Mesoniviridae*	Suborder: *Mesnidovirineae*	1 subfamily
-**Suborder:***Monidovirineae*	Order: *Nidovirales*	1 family
+**Family:***Mononiviridae*	Suborder: *Monidovirineae*	1 subfamily
-**Suborder:***Nanidovirineae*	Order: *Nidovirales*	2 families
+**Family:***Nanghoshaviridae*	Suborder: *Nanidovirineae*	1 subfamily
+**Family:***Nanhypoviridae*	Suborder: *Nanidovirineae*	1 subfamily
-**Suborder:***Ronidovirineae*	Order: *Nidovirales*	2 families
+**Family:***Euroniviridae*	Suborder: *Ronidovirineae*	2 subfamilies
+**Family:***Roniviridae*	Suborder: *Ronidovirineae*	1 subfamily
-**Suborder:***Tornidovirineae*	Order: *Nidovirales*	1 family
+**Family:***Tobaniviridae*	Suborder: *Tornidovirineae*	4 subfamilies
-**Order:***Picornavirales*	Class: *Pisoniviricetes*	8 families
+**Family:***Caliciviridae*	Order: *Picornavirales*	11 genera
+**Family:***Dicistroviridae*	Order: *Picornavirales*	3 genera
+**Family:***Iflaviridae*	Order: *Picornavirales*	1 genus
+**Family:***Marnaviridae*	Order: *Picornavirales*	7 genera
+**Family:***Picornaviridae*	Order: *Picornavirales*	68 genera
+**Family:***Polycipiviridae*	Order: *Picornavirales*	3 genera
+**Family:***Secoviridae*	Order: *Picornavirales*	1 subfamily, 5 genera, 1 species
+**Family:***Solinviviridae*	Order: *Picornavirales*	2 genera
-**Order:***Sobelivirales*	Class: *Pisoniviricetes*	3 families
+**Family:***Alvernaviridae*	Order: *Sobelivirales*	1 genus
+**Family:***Barnaviridae*	Order: *Sobelivirales*	1 genus
+**Family:***Solemoviridae*	Order: *Sobelivirales*	4 genera, 7 species
-**Class:***Stelpaviricetes*	Phylum: *Pisuviricota*	2 orders
-**Order:***Patatavirales*	Class: *Stelpaviricetes*	1 family
+**Family:***Potyviridae*	Order: *Patatavirales*	12 genera, 3 species
-**Order:***Stellavirales*	Class: *Stelpaviricetes*	1 family
+**Family:***Astroviridae*	Order: *Stellavirales*	2 genera
+**Family:***Birnaviridae*	Kingdom: *Orthornavirae*	7 genera
+**Family:***Permutotetraviridae*	Kingdom: *Orthornavirae*	1 genus
+**Genus:*****Botybirnavirus***	**Kingdom:** ***Orthornavirae***	**1 species**
II. Kingdom:*Pararnavirae*	Realm: *Riboviria*	1 phylum
6. -**Phylum:***Artverviricota*	Kingdom: *Pararnavirae*	1 class
- **Class:***Revtraviricetes*	Phylum: *Artverviricota*	2 orders
-**Order:***Blubervirales*	Class: *Revtraviricetes*	1 family
+**Family:***Hepadnaviridae*	Order: *Blubervirales*	5 genera
-**Order:***Ortervirales*	Class: *Revtraviricetes*	5 families
+ **Family:***Belpaoviridae*	Order: *Ortervirales*	1 genus
+ **Family:***Caulimoviridae*	Order: *Ortervirales*	11 genera
+**Family:***Metaviridae*	Order: *Ortervirales*	2 genera
+**Family:***Pseudoviridae*	Order: *Ortervirales*	3 genera, 1 species
+**Family:***Retroviridae*	Order: *Ortervirales*	2 subfamilies
+**Family:***Polymycoviridae*	Realm: *Riboviria*	1 genus
+**Family:***Sarthroviridae*	Realm: *Riboviria*	1 genus
+**Genus:***Albetovirus*	Realm: *Riboviria*	3 species
+**Genus:***Aumaivirus*	Realm: *Riboviria*	1 species
+**Genus:***Papanivirus*	Realm: *Riboviria*	1 species
+**Genus:***Virtovirus*	Realm: *Riboviria*	1 species

Realm V:Ribozyviria 1 family

+**Family:***Kolmioviridae*	Realm: *Ribozyviria*	8 genera

Realm VI:Varidnaviria

(Kingdoms-2, phyla-3, classes-6, orders-12, families-17)

I. Kingdom:*Bamfordvirae*	Realm: *Varidnaviria*	2 phyla
1.- **Phylum:***Nucleocytoviricota*	Kingdom: *Bamfordvirae*	2 classes
-**Class:***Megaviricetes*	Phylum: *Nucleocytoviricota*	3 orders
-**Order:***Algavirales*	Class: *Megaviricetes*	1 family
+**Family:***Phycodnaviridae*	Order: *Algavirales*	6 genera
-**Order:***Imitervirales*	Class: *Megaviricetes*	1 family
+**Family:***Mimiviridae*	Order: *Imitervirales*	2 genera
-**Order:***Pimascovirales*	Class: *Megaviricetes*	3 families
+**Family:***Ascoviridae*	Order: *Pimascovirales*	2 genera
+**Family:***Iridoviridae*	Order: *Pimascovirales*	2 subfamilies
+**Family:***Marseilleviridae*	Order: *Pimascovirales*	1 genus, 2 species
-**Class:***Pokkesviricetes*	Phylum: *Nucleocytoviricota*	2 orders
-**Order:***Asfuvirales*	Class: *Pokkesviricetes*	1 family
+**Family:***Asfarviridae*	Order: *Asfuvirales*	1 genus
-**Order:***Chitovirales*	Class: *Pokkesviricetes*	1 family
+ **Family:***Poxviridae*	Order: *Chitovirales*	2 subfamilies
2. -**Phylum:***Preplasmiviricota*	Kingdom: *Bamfordvirae*	3 classes
-**Class:***Maveriviricetes*	Phylum: *Preplasmiviricota*	1 order
-**Order:***Priklausovirales*	Class: *Maveriviricetes*	1 family
+**Family:***Lavidaviridae*	Order: *Priklausovirales*	2 genera
-**Class:***Polintoviricetes*	Phylum: *Preplasmiviricota*	1 order
-**Order:***Orthopolintovirales*	Class: *Polintoviricetes*	1 family
+**Family:***Adintoviridae*	Order: *Orthopolintovirales*	2 genera
-**Class:***Tectiliviricetes*	Phylum: *Preplasmiviricota*	4 orders, 1 family
-**Order:***Belfryvirales*	Class: *Tectiliviricetes*	1 family
+**Family:***Turriviridae*	Order: *Belfryvirales*	1 genus
- **Order:***Kalamavirales*	Class: *Tectiliviricetes*	1 family
+**Family:***Tectiviridae*	Order: *Kalamavirales*	5 genera
-**Order:***Rowavirales*	Class: *Tectiliviricetes*	1 family
+**Family:***Adenoviridae*	Order: *Rowavirales*	6 genera
-**Order:***Vinavirales*	Class: *Tectiliviricetes*	1 family
+**Family:***Corticoviridae*	Order: *Vinavirales*	1 genus
+**Family:***Autolykiviridae*	Class: *Tectiliviricetes*	2 genera
II- Kingdom:*Helvetiavirae*	Realm: *Varidnaviria*	1 phylum
3. -**Phylum:***Dividoviricota*	Kingdom: *Helvetiavirae*	1 class
-**Class:***Laserviricetes*	Phylum: *Dividoviricota*	1 order
-**Order:***Halopanivirales*	Class: *Laserviricetes*	3 families
+**Family:***Matshushitaviridae*	Order: *Halopanivirales*	1 genus
+**Family:***Simuloviridae*	Order: *Halopanivirales*	1 genus
+**Family:***Sphaerolipoviridae*	Order: *Halopanivirales*	1 genus
-**Class:***Naldaviricetes*		1 order, 1 family
-**Order:***Lefavirales*	Class: *Naldaviricetes*	3 families
+**Family:***Baculoviridae*	Order: *Lefavirales*	4 genera

+ **Family:***Hytrosaviridae*	Order: *Lefavirales*	2 genera
+ **Family:***Nudiviridae*	Order: *Lefavirales*	4 genera
+**Family:***Nimaviridae*	Class: *Naldaviricetes*	1 genus
+**Family:***Alphasatellitidae*		3 subfamilies
+**Family:***Ampullaviridae*		1 genus
+**Family:***Anelloviridae*		31 genera
+**Family:***Avsunviroidae*		3 genera
+**Family:***Bicaudaviridae*		1 genus
+**Family:***Clavaviridae*		1 genus
+**Family:***Finnlakeviridae*		1 genus
+**Family:***Fuselloviridae*		2 genera
+**Family:***Globuloviridae*		1 genus
+**Family:***Guttaviridae*		2 genera
+**Family:***Halspiviridae*		1 genus
+**Family:***Ovaliviridae*		1 genus
+**Family:***Plasmaviridae*		1 genus
+**Family:***Polydnaviridae*		2 genera
+**Family:***Portogloboviridae*		1 genus
+**Family:***Pospiviroidae*		5 genera
+**Family:***Spiraviridae*		1 genus
+**Family:***Thaspiviridae*		1 genus
+**Family:***Tolecusatellitidae*		2 genera
+**Genus:***Dinodnavirus*		1 species
+**Genus:***Rhizidiovirus*		1 species

Annexure II (A)

ICTV Code of Virus Nomenclature

The International Code of Virus Classification and Nomenclature (ICVCN), March 2021

1. Statutory Basis for the International Committee on Taxonomy of Viruses (ICTV)

1.1 The International Committee on Taxonomy of Viruses (ICTV) is a committee of the Virology Division of the International Union of Microbiological Societies. ICTV activities are governed by Statutes agreed with the Virology Division.

1.2 The Statutes define the objectives of the ICTV. These are:

(i) to develop an internationally agreed taxonomy for viruses and other mobile genetic elements (MGEs) that are part of the virosphere (hereafter termed: "viruses");

(ii) to develop internationally agreed names for virus taxa;

(iii) to communicate taxonomic decisions to the international community of virologists;

(iv) to maintain an Index of agreed names of virus taxa.

Comment: The Index is currently maintained online as the Master Species List, accessible at https://ictv.global/msl/.

1.3 The Statutes also state that classification and nomenclature will be subject to Rules set out in an International Code.

Comment: Ratified taxonomic changes will be published in Virology Division News Section in Archives of Virology, and in subsequent ICTV Reports.

2. Principles of Nomenclature

2.1 The essential principles of virus nomenclature are:

(i) to aim for stability; (ii) to avoid or reject the use of names which might cause error or confusion; (iii) to avoid the unnecessary creation of names.

2.2 Nomenclature of virus taxa is independent of other biological nomenclatures. Virus taxon nomenclature is recognized as an exception in the proposed International Code of Bionomenclature (BioCode).

2.3 The primary purpose of naming a taxon is to supply a means of referring to the taxon, rather than to indicate the characters or history of the taxon.

Comment: Taxon names are labels and do not need to be changed in case new data seemingly contradict a historical name.

2.4 The name of a taxon has no official status until it has been approved by the ICTV.

Comment: See section 3.7

Rules of Classification and Nomenclature

I - General Rules

The universal scheme

3.1 Virus classification and taxon nomenclature shall be international and shall be universally applied to all classifiable members of the virosphere.

3.2 The universal virus classification system shall employ the hierarchical levels of realm, subrealm, kingdom, subkingdom, phylum, subphylum, class, subclass, order, suborder, family, subfamily, genus, subgenus and species.

Comment: It is not obligatory to use all levels of the taxonomic hierarchy. The primary classification is of viruses and other MGEs into species that are assigned to genera. When justified, it is customary to assign lower-level taxa within the applicable higher-rank taxa.

An example of the classification of a positive-sense RNA virus is: gill-associated virus (GAV); species *Gill-associated virus*; subgenus *Tipravirus*; genus *Okavirus*; subfamily *Okanivirinae*; family *Roniviridae*; suborder *Ronidovirineae*; order *Nidovirales*; class *Pisoniviricetes*; phylum *Pisuviricota*; kingdom *Orthornavirae*; realm *Riboviria*.

Scope of the classification

3.3 The ICTV is responsible for the classification of members of the virosphere. Members of the virosphere include selfish genetic elements, which are replicons that are subject to selective pressures mostly independent of other replicons and hence have distinct evolutionary histories but depend on cellular hosts for energy and chemical building blocks. The relationship between selfish genetic elements and hosts spans the spectrum from mutualism to aggressive parasitism. Typically, MGEs are selfish genetic elements that move between hosts and/or change their integration sites in host genomes. MGEs are distributed among viruses *sensustricto* and the remaining replicator space of the virosphere (virus-like entities, such as satellite nucleic acids and viroids, and virus-derived elements, such as viriforms).

Comment: The virosphere and its potential subdivisions reflect conceptual sequence spaces with fuzzy boundaries that are not necessarily evolutionarily defined.

1. Viruses *sensu stricto* are defined operationally by the ICTV as a type of MGEs that encode at least one protein that is a major component of the virion encasing the nucleic acid of the respective MGE and therefore the gene encoding the major virion protein itself; or MGEs that are clearly demonstrable to be members of a line of evolutionary descent of such major virion protein-encoding entities. Any monophyletic group of MGEs that originates from a virion protein-encoding ancestor should be classified as a group of viruses. Comment: Long terminal repeat retrotransposons, phages, polintons, satellite viruses, and virophages, as well as capsid-less descendants of RNA viruses, such as agents currently classified in the families *Botourmiaviridae*, *Mitoviridae*, or *Narnaviridae*, are considered to be viruses *sensustricto*in classification and nomenclature.

2. Viroids are defined operationally by the ICTV as a type of MGEs that are uncoated, small, circular, single-stranded RNAs that do not encode proteins and do not depend on viruses for transmission, and that replicate autonomously through an RNA-RNA rolling-circle mechanism mediated by host enzymes and, in some cases, by cis-acting hammerhead ribozymes; or MGEs that are derived from a viroid in the course of evolution. Any monophyletic group of MGEs that originates from a viroid ancestor should be classified as a group of viroids.

3. Satellite nucleic acids are defined operationally by the ICTV as a type of non-viroid MGEs, which are dependent on viruses for replication and transmission; or MGEs that are derived from such entities in the course of evolution. Any monophyletic group of MGEs that originates from a satellite nucleic acid ancestor should be classified as a group of satellite nucleic acids.

4. Viriforms are defined operationally by the ICTV as a type of virus-derived MGEs that have been exapted by their organismal (cellular) hosts to fulfill functions important for the host life cycle; or MGEs that are derived from such entities in the course of evolution. Any monophyletic group of MGEs that originates from a viriform ancestor should be classified as a group of viriforms. Comment: Gene transfer agents (GTAs) and the MGEs previously classified in the family *Polydnaviridae* are considered to be viriforms in classification and nomenclature.

3.4 The ICTV is not responsible for classification and nomenclature of virus taxa below the rank of species. The classification and naming of serotypes, genotypes, strains, variants and isolates of virus species is the responsibility of acknowledged international specialist groups.

Comment: A variety of subspecific grouping may be identified within the members of a single virus species. These may be described as viruses with alternative names (e.g., blackeye cowpea mosaic virus and peanut stripe virus, which are both classified in the species *Bean common mosaic virus*, genus *Potyvirus*, family *Potyviridae*), or as serotypes, genotypes, clades, strains, variants, isolates etc. Naming of such entities is not the responsibility of the ICTV but of international specialty groups. It is the responsibility of ICTV Study Groups to consider how these entities may best be classified into species.

3.5 Artificially created and laboratory hybrid MGEs will not be given taxonomic consideration. Their classification will be the responsibility of acknowledged international specialist groups.

3.6 Taxa will be established only when representative member MGEs are sufficiently well characterized and described in the published literature so as to allow them to be identified unambiguously and the taxon to be distinguished from other similar taxa.

3.7 Names will only be accepted if they are linked to taxa at the hierarchical levels described in Rule 3.2 and which have been approved by the ICTV.

II - Rules about naming Taxa

Status of Names

3.8 Names proposed for taxa are "valid names" if they conform to the Rules set out in the Code and they pertain to established taxa. Valid names are "accepted names" if they are recorded as approved International Names in the most recent ICTV Report or have subsequently become "accepted names" by an ICTV vote of approval for a taxonomic proposal.

Comment: A valid taxon name is one that has been published, one that is associated with descriptive material, and one that conforms to the Rules in the Code. Accepted names will be kept in an "Index" by the ICTV.

3.9 Existing names of taxa shall be retained whenever feasible.

Comment: A stable nomenclature is one of the principal aims of taxonomy and therefore changes to names that have been accepted will only be considered if the accepted name conflicts with the Rules or if a change is necessary to remove ambiguities or confusion.

3.10 The Rule of priority in naming taxa shall not be observed.

Comment: The earlier of candidate names for a taxon may be chosen as a convenience to virologists, but the Rule ensures that it is not possible to invalidate a name in current use by claiming priority for an older name that has been superseded.

3.11 A person's name may be used when devising a name for new taxon. If the person is alive at the time of the proposal, the person's written consent for use of his/her/their name must be provided together with the official taxonomic proposal. Whether the use of a person's name for taxon naming is appropriate will be judged by the responsible ICTV Study Group, the respective ICTV Subcommittee, and the ICTV Executive Committee and approved or disapproved following established taxonomic proposal procedures. Furthermore, a) An individual may not propose his/her/their own name as the basis for any new taxon name; and b) A taxon may not be named wholly or in part after any current member of an ICTV Study Group or Committee.

3.12 Names for taxa shall be easy to use and easy to remember. Euphonious names are preferred.

Comment: In general, short names are desirable and the number of syllables should be kept to a minimum.

3.13 Ligatures, diacritical marks, punctuation marks (excluding hyphens), subscripts, superscripts, oblique bars and non-Latin letters (i.e., those not included in the ISO basic Latin alphabet) may not be used in taxon names. Numbers and hyphens are allowed but hyphens should not be used when attaching numbers or letters to the end of a series of species names and should never be used in names of genera, subfamilies, families or orders.

Comment: The Rule is intended to make text unambiguous and easy to sort electronically; its application should often make names more pronounceable, in agreement with Rule 3.12.

3.14 New names shall not duplicate approved names. New names shall be chosen such that they are not closely similar to names that are in use currently or have been in use in the recent past.

Comment: The name selected for a new taxon should not sound indistinguishable from the name of another taxon at any rank or from any taxon. For example, the existence of the genus Iridovirus means that forms of new name such as "irodovirus" or "iridivirus" are discouraged as they are too easily confused with an approved name. Confusion can also be between species and genus names as both end in "-virus". Thus, for example, a genus typified by the imaginary species Omega virus should not be named Omegavirus because species and genus would then be too readily confused.

3.15 Sigla may be accepted as names of taxa, provided that they are meaningful to virologists in the field, normally as represented by study groups.

Comment: Sigla are names comprising letters and/or letter combinations taken from words in a compound term. The name of the genus Comovirus has the sigla stem "Co-" from cowpea and "-mo-" from mosaic; the name of the family Reoviridae has the sigla stem "R" from "Respiratory, "e" from "enteric" and "o" from "orphan".

Decision making

3.16 In the event of more than one candidate name being proposed, the relevant Subcommittee will make a recommendation to the Executive Committee of the ICTV, which will then decide among the candidates as to which to recommend to ICTV for acceptance.

Comment: When there is more than one candidate name for the same taxon, the choice of name to be approved will usually be based on the recommendations of a particular Study Group working on behalf of the ICTV. The Study Group will be expected to consult widely, so as to ensure the acceptability of names, subject to the Rules in the Code. The policy of the ICTV is that, as far as is possible, decisions on questions of taxonomy and nomenclature should reflect the majority view of the appropriate virologic constituency.

3.17 New names shall be selected such that they, or parts of them, do not convey a meaning for the taxon which would either (1) seem to exclude MGEs that lack the character described by the name but which are members of the taxon being named, or (2) seem to exclude MGEs that are as yet undescribed but which might belong to the taxon being named, or (3) appear to include within the taxon MGEs that are members of different taxa.

3.18 New names shall be chosen with due regard to national and/or local sensitivities. When names are universally used by virologists in published work, these or derivatives shall be the preferred basis for creating names, irrespective of national origin.

3.19 All relevant ICTV Subcommittees and Study Groups will be consulted prior to a decision being taken on any taxonomic proposal submitted to the Executive Committee of the ICTV.

Comment: Proposals concerning a family containing genera whose member viruses infect diverse types of host (e.g., plants and vertebrates, fungi and plants, and so on) must be considered by the Subcommittees responsible for viruses of each host type (i.e., Plant viruses, Vertebrate viruses, and so on). For example, taxonomic proposals concerned with the family Partitiviridae *would be considered by the Fungal Virus Subcommittee and one of its Study Groups but because some genera in the family contain viruses of plants, proposals affecting the family would also be considered by the Plant Virus Subcommittee.*

III - Rules about Species

Definition of an MGE species

3.20 Species shall be created in accordance with the following definition:

"A species is the lowest taxonomic level in the hierarchy approved by the ICTV. A species is a monophyletic group of MGEs whose properties can be distinguished from those of other species by multiple criteria."

Comment: The criteria by which different species within a genus are distinguished shall be established by the appropriate Study Group. These criteria may include, but are not limited to, natural and experimental host range, cell and tissue tropism, pathogenicity, vector specificity, antigenicity, and the degree of relatedness of their genomes or genes. The criteria used should be published in the relevant section of the ICTV Report and reviewed periodically by the appropriate Study Group.

Construction of a name

3.21 A species name shall consist of only two distinct word components separated by a space. The first word component shall begin with a capital letter and be identical in spelling to the name of the genus to which the species belongs. The second word component shall not contain any suffixes specific for taxa of higher ranks. The entire species name (both word components) shall be italicized.

Comment: Latinized binomial species names following the style of Linnaean species names used in other biological taxonomies are permitted (https://ictv.global/ictv/proposals/2018.001G.R.binomialspecies.pdf).

3.22 A species name must provide an appropriately unambiguous identification of the species.

Comment: Species names should be distinctive. They should not be in a form that could be easily confused with the names of other taxa.

IV - Rules about Ranks other than Species

3.23 Every individual MGE is a physical entity and treated as belonging to a number of taxa of hierarchical ranks, some of which may remain undefined.

V - Rules about Viruses

3.24 Other than species, the ranks currently in use in virus taxonomy, from most to least diverse, are realm, subrealm, kingdom, subkingdom, phylum, subphylum, class, subclass, order, suborder, family, subfamily, genus and subgenus. The names for these ranks shall be single words ending with the suffixes "*-viria*", "*-vira*", "*-virae*", "*-virites*", "*-viricota*", "*-viricotina*", "*-viricetes*", "*-viricetidae*",

"*-virales*", "*-virineae*", "*-viridae*", "*-virinae*", "*-virus*" and "*-virus*", respectively.

Comment: No ranks other than those specified in Rule 3.24 are currently approved by ICTV

3.25 The classification of a virus at the species and genus ranks is mandatory. Classification may also encompass any further number of taxa at higher hierarchical ranks

VI - Rules about Virus-like Mobile Genetic Elements

3.26 Rules concerned with the classification of viruses shall also apply to the classification of virus-like MGEs.

1. Satellites

The formal endings for taxon names of satellite nucleic acids are the suffixes "*-satellitia*" for realms, "*-satellita*" for subrealms, "*-satellitae*" for kingdoms, "*-satellitites*" for subkingdoms, "*-satelliticota*" for phyla, "*-satelliticotina*" for subphyla, "*-satelliticetes*" for classes, "*-satelliticetidae*" for subclasses, "*-satellitales*" for orders, "*-satellitineae*" for suborders, "*-satellitidae*" for families, "*-satellitinae*" for subfamilies, and "*-satellite*" for genera and subgenera.

Comment: For example, the species *Ageratum yellow vein betasatellite* is included in the genus *Betasatellite* of the family *Tolecusatellitidae.*

2. Viroids

The formal endings for taxon names of viroids are the suffixes "*-viroidia*" for realms, "*-viroida*" for subrealms, "*-viroidae*" for kingdoms, "*-viroidites*" for subkingdoms, "*-viroidicota*" for phyla, " *-viroidicotina*" for subphyla, "*-viroidicetes*" for classes, "*-viroidicetidae*" for subclasses, "*-viroidales*" for orders, "*-viroidineae*" for suborders, "*-viroidae*" for families, "*-viroidinae*" for subfamilies, and "*-viroid*" for genera and subgenera.

Comment: For example, the species *Potato spindle tuber viroid* is included in the genus *Pospiviroid* of the family *Pospiviroidae.*

3. Viriforms

The formal endings for taxon names of viriforms are the suffixes "*-viriformia*" for realms, "*-viriforma*" for subrealms, "*-viriformae*" for kingdoms,

"*-viriformites*" for subkingdoms, "*-viriformicota*" for phyla, "*-viriformicotina*" for subphyla, "*-viriformicetes*" for classes, "*-viriformicetidae*" for subclasses, "*-viriformales*" for orders, "*-viriformineae*" for suborders, "*-viriformidae*" for families, "*-viriforminae*" for subfamilies, and "*-viriform*" for genera and subgenera.

VII - Rules for Orthography

3.27 In formal taxonomic usage, the accepted names of virus, viroid and satellite realms, subrealms, kingdoms, subkingdoms, phyla, subphyla, classes, subclasses, orders, suborders, families, subfamilies, genera and subgenera are printed in italics and the first letters of the names are capitalized.

Comment: See Rule 3.7 for the definition of an "accepted" name.

3.28 Species names are printed in italics and have the first letter of the first word capitalized. Other words are not capitalized unless they are proper nouns, or parts of proper nouns.

Comment: The species names Tobacco mosaic virus and Murray Valley encephalitis virus are in the correct form and typographical style. Examples of incorrect species name forms are Ustilago maydis virus H (not italicized), Murray valley encephalitis virus (Valley is a proper noun) or tobacco mosaic virus (not capitalized or italicized).

When taxon names are used informally, italics and capital initial letters are not needed. This applies at all taxonomic levels; examples are: (1) "the tobacco mosaic virus polymerase", when describing the properties of the polymerase in members of the species *Tobacco mosaic virus* and (2) "three pestiviruses", to describe viruses that are members of the genus *Pestivirus.*

Annexure II (B)

How to write virus, species, and other taxa names

Names of viruses (the physical things you work with in the lab or that make you sick)are written differently than the names of species and other taxa (logical constructs thathelp us categorize viruses).

A species name* is written in italics with the first word beginning with a capital letter.Other words only begin with a capital if they are proper nouns (including host genusnames but not virus genus names**) or alphabetical identifiers. A species name shouldnot be abbreviated. Examples:

- The genus *Iflavirus* includes the species *Deformed wing virus*.
- Members of the species *West Nile virus* are arboviruses.
- The species *Sandfly fever Naples phlebovirus*has many diverse memberviruses.
- The etiological agents of poliomyelitis (poliovirus types 1, 2 and 3) aremembers of thespecies *Enterovirus C*.
- Anadyr virus, Batai virus, Birao virus, and many more, are members of thespecies *Bunyamwera orthobunyavirus*.
- A new bacteriophage, belonging to the species *Salmonella virus SP6*, has been isolated.
- *Rattus norvegicus polyomavirus 1* is a species in the family *Polyomaviridae*.Some host genus names may also be considered common nouns in English and can bewritten in lower case where they are not the first word of the species name. Example:
- *Indian citrus ringspot virus*

A virus name should never be italicized, even when it includes the name of a hostspecies or genus and should be written in lower case. This ensures that it isdistinguishable from a species name, which otherwise might be identical. The firstletters of words in a virus name, including the first word, should only begin with a capitalwhen these words are proper nouns (including host genus names but not virus genusnames) or start a sentence. Single letters in virus names, including alphanumericalstrain designations, may be capitalized. In most texts, virus names are used much morefrequently than species names and may, therefore, be abbreviated. Examples:

- Isolates of dengue virus 2 were obtained ....

- Detection of West Nile virus in human serum ....
- Salmonella phage SE1 was isolated ....
- Sidaciliaris golden mosaic virus (SCGMV) causes ....
- Aphids transmit potato virus Y (PVY).

A higher taxon name (i.e. above the rank of species) is written as a single word with at axon-specific suffix. Examples:

realm ...*viria*
subrealm...*vira*
kingdom ...*virae*
subkingdom ...*virites*
phylum ...*viricota*
subphylum ...*viricotina*
class ...*viricetes*
subclass ...*viricetidae*
order ...*virales*
suborder ...*virineae*
family ...*viridae*
subfamily ...*virinae*
genus ...*virus*
subgenus ...*virus*

Like a species name, a higher taxon name is written in italics and begins with a capitalletter. This differs from the convention in botany and zoology, in which taxon namesabove the level of genus are not italicized. Taxon names are often preceded by a taxonlevel identifier. Examples:

- ... a new species in the genus *Fabavirus*
- ... members of the subfamily *Comovirinae*
- .... members of the family *Secoviridae*
- The order *Picornavirales*includes viruses infecting hosts of a range ofspecies.

A collective name for a group of viruses belonging to a higher-level taxon is neitheritalicized nor capitalized, even if it was derived from a proper noun. The first letter of acollective name may be capitalized if it begins a sentence.

- ourmiaviruses, ourmiavirus
- Guernseyviruses are distributed worldwide.
- The guernseyviruses are distributed worldwide.

- aparaviruses
- the aparavirus polymerase

Note that if taxa have the same stem (e.g. *Flavivirus* and *Flaviviridae*), this may lead toambiguity because both groups of viruses could be referred to as flaviviruses. Somevirologists use the terms stem + virads, stem + virids, stem + virins, and stem + virus todistinguish members of orders, families, subfamilies and genera, respectively.

Complex example sentences:

- Ebola virus (species *Zaire ebolavirus*; genus *Ebolavirus*; family *Filoviridae*;order *Mononegavirales*) can cause disease in humans and nonhumanprimates.
- Infection of larvae of the silkworm, *Bombyx mori* (family Bombycidae) with thebaculovirus Bombyx mori nucleopolyhedrovirus (BmNPV) (species *Bombyxmori nucleopolyhedrovirus*) is often lethal.
- In the family *Podoviridae*, the subfamily *Autographivirinae*groups together allpodoviruses that contain an RNA polymerase gene in their genome, includingEscherichia phage T7(species *Escherichia virus T7*; genus *Teseptimavirus*)and Klebsiella phage F19 (species *Klebsiella virus F19*; genus *Drulisvirus*).
- Artoviruses form a family in the haploviricotine order *Mononegavirales*.

*The complete rules for naming virus taxa can be found in the ICTV Code: https://ictv.global/code

** A proper noun is a name used for an individual person, place, or organization. Acommon noun denotes a class of objects or a concept. Host genus names are normallyconsidered as proper nouns because they refer to a group of unique entities but some,for example "citrus", have become common nouns because they can also describeintergeneric hybrids. Virus genus names are not considered as proper nouns whenused as part of a species or virus name because they refer to a subset of the genus andnot the genus as a whole.

ABBREVIATIONS

ACMV	African cassava mosaic virus
AIDS	Acquired immunodeficiency virus
APMV	Acanthamoeba polyphagamimivirus
BCTV	Beet curly top virus
BGYMV	Bean golden yellow mosaic virus
BMV	Brome mosaic virus
bp	Base pairs
BSE	Bovine spongiform encephalopathy
CaMV	Cauliflower mosaic virus
CJD	Creutzfeldt Jakob disease
CMV	Cytomegalovirus
CNS	Central nervous system
CP	Coat protein
CroV	Cafeteria roenbergensis virus
DEP	Dilution end point
DI	Defective infective particle
EBV	Epstein barr virus
ELISA	Enzyme linked immunosorbent assay
FA	Fluorescent antibody
FDV	Fiji disease virus
FMDV	Foot and mouth disease virus
HA	Hemagglutin
HBV	Hepatatis B Virus
HCV	Hepatitis C virus

HDV	Hepatitis delta virus
HI	Hemagglutnin inhibition
HIV	Human immunodeficiency virus
HPA	Hybridization protection assay
HPV	Human papilloma virus
HSV	Herpes simplex virus
HTLV	Human T cell leukemia virus
HVCV	Hydra viridis chlorella virus
ICNV	International committee on nomenclature of viruses
ICTV	International committee on taxonomy of viruses
IFN	Interferon
IP	Immuno peroxidase
IUMS	International union of microbiological societies
Kbp	Kilobase pairs
KD	Kilo Daltons
LCR	Ligase chain reaction
LIV	Longevity *in vitro*
LNYU	Lettuce necrotic yellow virus
MIMIVIRE	Mimivirus virophage resistance element
MP	Movement protein
MSV	Maize streak virus
Mu	Millimicron
NAA	Nucleic acid amplification
NASBA	Nucleic acid sequence based amplification
NCLDV	Nucleoeytoplasmic large DNA Viruses
NDV	New castle disease virus
nm	Nanometer
ODYU	Onion dwarf yellow virus
ORF	Open reading frame
PCR	Polymerase chain reaction
PFU	Plaque forming units
Prp^{C}	Prion protein cellular
Prp^{SC}	Prion protein scrapie specific
PstVd	Potato Spindle tumor viroid

PTGS	Post transcriptional gene silencing
PVX	Potato virus X
PVY	Potato virus Y
RAPD	Random amplified polymorphic DNA
RdRp	RNA dependent RNA polymerase
RDV	Rice dwarf virus
RF	Replicative form
RFLP	Restriction fragment length polymorphism
RNV	Rio Negro Virophage
RRSV	Rice ragged stunt virus
RSV	Rous sarcoma virus
RT-PCR	Reverse transcriptase PCR
S	Svedberg units
SBMV	Southern bean mosaic virus
SgRNA	Subgenomic RNA
SIV	Simian immunodeficiency virus
SV-40	Simian virus-40
SYNV	Sonchus yellow net virus
TIBA	Tissue blot immunoassay
TIP	Thermal inhibition point
TMA	Transcription mediated amplification
TMV	Tobacco mosaic virus
TNV	Tobacco necrosis virus
TOMV	Tomato mosaic virus
TSE	Transmissible spongiform encephalopathy
VIGE	Virus mediated gene editing
VIGS	Virus induced gene Silencing
VLP	Virus like particles
VLP	Virus like particles
VNP	Viral nano paticles
VSV	Vesicular stomatis virus
VZV	Varicella zoster virus

GLOSSARY

Abortive infection : Infection without production of infectious progeny.

Ambisense : Coding of proteins on both viral-sense and viral complementary RNAs

Antigenome : Complemantary (+) RNA on which the negative strands is made.

Bacteriophage : Bacterial virus, synonym phage.

Bidirectional replication : Replication in two directions from a common origin.

Capsid : Closed protein shell of virus particle.

Capsomer(e) : Morphological subunit of the capsid, visible in the electron microscope, formed of monomers (usually five or six)

Chaperonin : Nonstructural protein mediating the correct assembly of other proteins.

Chemokine : Chemotactic cytokine that activates and directs leukocyte migration.

Circular (cyclic) permutation : Variable location of DNA termini in a virus population.

Circularization : End-to-end joining of linear DNA molecules to form closed circle.

Clathrin : Protein forming a polyhedral lattice in pits and vesicles involved in endocytosis.

Coat : Protective layer(s) surrounding the viral nucleic acid.

Cohesive ends : Sticky ends; complementary single-stranded ends of a dsDNA molecule.

Complementary : Single-stranded daughter strand synthesized on viral DNA or RNA.

Concatemer (concatamer) : Giant molecule made up of multiple copies of covalently joined similar units.

Core : Inner part of viral particle; Inner tube of contractile phage tails

Cos site : Cohesive end

Creutzfeldt-Jakob disease : A transmissible human spongiform encephalopathy.

Cubic symmetry : A regular polyhedron (tetrahedral, octahedral, icosahedral).

Cytoplasmic polyhedrosis : Insect disease characterized by large cytoplasmic polyhedral inclusion bodies.

Eclipse : Time between disappearance of intracellular infecting viruses, or entry of naked nucleic acid, and appearance of intracellular progeny virions.

Encapsidation : Inclusion of viral DNA or RNA into a capsid.

Endocytosis : Intake via membrane-bound vesicles.

Env : Envelope gene or protein.

Envelope : Lipoprotein membrane surrounding a viral capsid or nucleoprotein.

Exocytosis : Release of contents of secretory vesicles to the exterior.

F-specific : Specific for F (fertility factor-related) sex pili

Helical symmetry : Structure in which the constitutive subunits form a helix.

Icosahedron : Platonic solid with 20 triangular faces, 30 edges, 12 apices, and twofold, threefold, and fivefold axes of symmetry.

Inclusion body : Microscopically visible intracellular aggregate of viruses and/or protein

Induction : Activation of a latent virus infection.

Integration : Insertion of viral DNA into the host genome.

Interferon : Type of (antiviral) regulatory proteins produced by vertebrate cells.

Inverted repeat : Repeat sequence in opposite directions.

Kuru : A transmissible human spongiform encephalopathy.

Latent period : Time between virus infection and appearance of progeny virus.

Lysis : Disruption of host cell.

Lysogeny : Carriage of a prophage by a bacterium.

Lysosome : Intracytoplasmic membranous vesicle containing hydrolytic enzymes.

Lysozyme : Enzyme hydrolyzing certain glycosidic linkages of peptidoglycan.

Lytic : Causing lysis.

Matrix : "Mother substance", e.g., i. Protein located between viral envelope and nucleocapsid (orthomyxo-, paramyxo-, rhabdoviruses. ii. Major protein of inclusion bodies. iii. Ground substance of viroplasm. iv. Template.

Minichromosome : Viral DNA assembled into chromatin with host histones.

Monocistronic : RNA transcript derived from a single gene and coding for a single protein

Negative-sense : (–) sense; nucleic acid complementary to the (+) strand of viral mRNA.

Negri body : Cytoplasmic inclusion body in cells infected with rabies virus.

Nucleoid : Electron-dense central region of certain viruses.

Nucleoplasm : Content of nucleus excluding the nucleolus.

One-step growth : Single cycle of replication in a cell population infected synchronously.

Open reading frame (ORF) : DNA or RNA sequence between an initiation (start) and a termination (stop) codon; corresponds generally to a gene.

Overlapping gene : Part of genome readable in more than one reading frame to give two or more proteins.

Pararetrovirus : Virus practicing transcription of DNA to RNA (e.g., hepadnaviruses, cauliflower mosaic virus).

Peplomer : Glycoprotein projection of viral envelope, also called spike.

Permissive : Able to support a productive virus infection

Phage: Bacteriophage

Phagolysosome : Vesicle for digestion of internalized particles, formed by fusion of phagosomes and lysosomes

Polyprotein : Precursor protein to be cleaved into several identical or different proteins

Positive sense : (+) sense i. RNA strand which functions as mRNA. ii. DNA strand with the same sequence as mRNA.

Prion : Proteinaceous infectious particle

Procapsid : Immature capsid

Prohead : Procapsid of tailed phages

Prophage : Latent phage genome

Provirus : i. dsDNA copy of an ssRNA retroviral genome; ii. Any latent (integrated) virus genome; iii. Immature virus (obsolete).

Receptor : Virus-binding site on cell surface

Replicase : RNA-dependent RNA polymerase; more generally an enzyme involved in replication of viral genomes.

Replicative form : Double-stranded form of ssDNA or RNA during replication

Replicative intermediate : Complex formed by simultaneous synthesis of one or more RNA strands from a single complementary strand

Reverse transcriptase : RNA-dependent DNA polymerase for DNA synthesis on RNA templates

Rise period : Time required for reaching a constant virus titer

Rolling-circle mechanism : Type of DNA or RNA replication in which a sole replication fork move indefinitely around a circular template.

Satellite RNA : Capsid-less RNA (ss or ds, linear or circular) that depends on a helper virus for replication

Satellite virus : Defective virus depending on a helper

Scaffolding protein : Protein around which a capsid is built.

Scrapie : Transmissible spongiform encephalopathy of sheep

Segmented genome : Genome composed of two or more molecules of nucleic acid.

Semi-conservative replication : Type of dsDNA or RNA replication in which both strands are templates, generating progeny viruses with a parent strand and a new strand.

Strand displacement : Type of dsDNA or RNA replication in which one strand only is replicated, the other strand being displaced from the template.

Subgenomic : mRNA segment smaller than genomic RNA; not needed in natural infection, sometimes encapsidated contains genes which are unavailable for translation from genomic RNA

Subviral : i. Smaller than a regular virus particle. ii. Breakdown product from *in vitro* treatment of viruses. iii. Immature virus (infrequent and inappropriate use)

Syncytium : Multinucleate giant cell resulting from cell fusion

Terminal redundancy : Nucleotide duplication at either end of a linear genome

Uncoating : Removal of outer layers of viral particles and exposure of viral nucleic acids

Virion : Complete infectious particles

Viroid : Infectious, capsid-less, naked circular ssRNA

Viroplasm : i. Amorphous inclusion body associated with virus infection. ii. Undefined substance within a virus (e.g., poxvirus).

INDEX